Progress in Molecular and Subcellular Biology 18

Series Editors

Ph. Jeanteur, Y. Kuchino,
W.E.G. Müller (*Managing Editor*)
P.L. Paine

Springer
Berlin
Heidelberg
New York
Barcelona
Budapest
Hong Kong
London
Milan
Paris
Santa Clara
Singapore
Tokyo

Ph. Jeanteur (Ed.)

Cytoplasmic fate of messenger RNA

With 25 Figures

Springer

Prof. Dr. PHILIPPE JEANTEUR
Institut de Génétique Moléculaire
C.N.R.S., B.P. 5051
1919 route de Mende
34033 Montpellier Cedex 01
France

ISBN-13: 978-3-642-64420-7 Springer-Verlag Berlin Heidelberg New York

Library of Congress Cataloging-in-Publication Data. Cytoplasmic fate of messenger RNA/ Ph. Jeanteur, ed. p. cm. – (Progress in molecular and subcellular biology: 18) Includes bibliographical references and index.

ISBN-13: 978-3-642-64420-7 e-ISBN-13: 978-3-642-60471-3

DOI: 10.1007/978-3-642-60471-3

1. Messenger RNA – Metabolism.

2. Cytoplasm. I. Jeanteur, Ph. (Philippe) II. Series. QH506.P76 no. 18 [QP623.5.M47] 574.87'3283–dc20 97-25885

Softcover reprint of the hardcover 1st edition 1997

Cover design: Design & Production GmbH, Heidelberg

Typesetting: Scientific Publishing Services (P) Ltd, Madras

SPIN: 10495126 39/3137/SPS – 5 4 3 2 1 0 – Printed on acid-free paper

Preface

Among all cellular RNA species of the three main types, ribosomal RNA, transfer RNA or messenger RNA, be they from prokaryotic or eukaryotic organisms, the prokaryotic mRNA is unique in that it has no precursor and is synthesized in the same mature form as it is translated into proteins. In fact, ribosomes join the nascent mRNA chain and engage in protein synthesis long before its transcription is complete. Provisions are even made for slowing down the ribosomes at some sites to prevent them from catching up with the RNA-polymerase. Of course, such a situation is only possible in the prokaryotic world where there is no such thing as a nuclear membrane physically secluding the transcription process from the cytoplasm where translation is restricted.

Quite in the opposite extreme, the eukaryotic pre-messenger RNA has to suffer many and sometimes drastic steps of maturation (capping, polyadenylation, splicing, edition) before the decision is made to export it to the cytoplasm. That is where it enters the scope of this book. Once in the cytoplasm, many options are still open to it: its entrance into polysomes may be delayed (as it is in unfertilized eggs) or merely prohibited (ferritin mRNA in iron-starved cells), directed to specific locations within the cytoplasm or be more or less rapidly degraded.

During gametogenesis and early development, translational control is probably the most significant level of gene expression. Most mRNAs are in a translationnally repressed state in oocytes (Osborne and Richter, Amaldi and Pierandrei-Amaldi) and immediately engage in translation after fertilization. The role of polyadenylation in this process as well as that of deadenylation in the subsequent decay of these mRNAs is addressed by Osborne and Richter who emphasize that the complexity of the *cis*-acting RNA sequence elements involved reflects their capacity for regulation as opposed to the relative simplicity of nuclear polyadenylation signals. Although the signals for cytoplasmic polyadenylation are beginning to emerge, those for deadenylation are much less clear. Much also remains to be learned about the biochemistry of cytoplasmic polyadenylation and deadenylation and this issue is addressed by Virtanen and Åström. The general question of how polyadenylation contributes to translational arrest and whether it is a harbinger for degradation remains largely open.

Another set of genes where translational control is a major regulatory level is that for ribosomal proteins (Amaldi and Pierandrei-Amaldi). In addition to being also transcriptionally regulated, most, if not all of these genes resort to translational control to precisely match their production to their stoichiometry in the ribosome. This takes place not only after fertilization but also in cultured cells. mRNAs for these genes contain a 5′ Terminal Oligo Pyrimidine tract and are therefore referred to as TOP mRNAs. Amaldi and Pierandrei-Amaldi address the problem of how this sequence controls their translation by shifting them between free and polysome-bound forms.

A quite elegant example of translation regulation in the developing *drosophila* embryo has very recently been discovered (Dubnau and Struhl 1996; Rivera-Pomar et al. 1996).[1] The homeodomain of bicoid binds to the 3′ UnTranslated Region (UTR) of caudal mRNA and represses its translation, resulting in opposite antero-posterior gradients of these two proteins.

The regulation of iron-controlled proteins provides the best-understood example of a mechanism linking RNA-protein interactions to translational regulation in the case of ferritin and to its degradation in the case of the transferrin receptor. These issues are reviewed in two chapters (Henderson and Kühn; Muckenthaler and Hentze). These regulations are mediated by RNA-protein interactions between Iron Response Elements (IRE) and Iron Response Proteins (IRP). The characteristics of these interactions and the biochemistry of IRP-1 and IRP-2 are the focus of the review by Henderson and Kühn which also emphasizes the link between the RNA-binding and enzymatic properties of these proteins. The chapter by Muckenthaler and Hentze is more concerned with the mechanism by which these RNA-protein complexes interfere with mRNA translation or degradation. Binding of IRPs to the IRE located in the cap proximal part of ferritin mRNA 5′UTR prevents its association with the 43S translation pre-initiation complex. Both chapters suggest that the two IRPs which have different but overlapping specificities might have other, still unknown, mRNA targets (Henderson and Kühn) and more generally that the same type of RNA-protein interaction but involving different partners might operate in situations other than the regulation of iron me-

[1]Dubnau J, Struhl G (1996) Nature 379:694–699. River-Pomar R, Niessing D, Schmidt-Ott U, Gehring WJ, and Jäckle H (1996) Nature 379:746–749.

tabolism, for example, during early development (Muckenthaler and Hentze).

mRNA degradation as a major level of control for at least some sets of genes in somatic cells is becoming more and more widely recognized but probably not yet fully appreciated. Rapid degradation of mRNA is a requisite for the control of some categories of genes (for example, nuclear oncogenes, cytokines, inflammatory proteins) whose expression needs to be transient. The delineation of the RNA sequence elements which target them for regulated degradation is addressed in several of the following chapters. The best characterized among these sequences are the AU-rich elements (ARE) located in the 3′UTR of unstable mRNAs. Jarzembowski and Malter describe the identification and characterization of AU-binding proteins, while the development of cell-free systems for analyzing cytoplasmic mRNA turnover is dealt with by De Maria and Brewer.

The issue of nucleases which contribute to the degradation of mRNA is addressed by two contributions. The chapter by Virtanen and Åström first reviews the biochemistry of polyadenylation and the enzymology of polyA shortening with respect to the specific 3′ polyA exoribonucleases present in various eukaryotic tissues. It also examines the functional significance of polyA in translation and of its regulated shortening in mRNA decay. Starting from the 2-5A-activated RNAse L, which is induced by interferons, Bisbal goes on to speculate that this activatable system might represent the modulation of an otherwise physiological activity required for normal mRNA metabolism.

Finally, Veyrune et al. first discuss the link between ARE, translation, and degradation. They also elaborate on the possibility that directing some mRNAs to specific subcellular locations can contribute to their regulation. For example, it can be speculated that the unusual perinuclear localization of c-myc mRNA can be of functional relevance. This localization appears to depend on the 3′UTR. Whether and how it is connected to the ARE degradation signals remain to be elucidated.

Transcriptional control remains obviously an essential basis for general gene regulation in eukaryotes mostly when it amounts to switching them on, but this is eventually achieved only after a long cascade of events have taken place. The point of this book is to provide evidence that this control is far from accounting for all the facets of gene regulation. Cytoplasmic control is required for the immediate and massive induction of protein synthesis in early egg development. Conversely, when it comes to rapidly switching off specific genes, efficient and targeted mRNA degradation must be called upon. The very existence of these cytoplasmic levels of regulation has only been recognized in recent years. Although they have now reached a status of evidence, it is my expectation that we are still far from fully appreciating the contribution of the RNA itself to the fine and immediate tuning of gene expression.

France, November 1996 PH. JEANTEUR

Contents

List of Contributors

Addresses are given at the beginning of the respective contribution

TOP Genes: A Translationally Controlled Class of Genes Including Those Coding for Ribosomal Proteins

F. AMALDI[1] and P. PIERANDREI-AMALDI[2]

1 Introduction

In the early 1980s, it was recognized that the equimolar accumulation of ribosomal proteins (r-proteins) in eukaryotes is maintained by coordinated regulation at various levels of gene expression, from transcription to protein stability. With the possible exception of yeast, the translational control of mRNA utilization emerged to be the prevalent regulatory mechanism involved. It was also soon realized that the corresponding r-protein genes share a common architecture, mainly in the region surrounding the 5′ end, that was found to be involved in the coregulated expression of gene activity at the transcriptional and translational levels. As described below the most typical common structural feature is the transcription initiation site which is always situated within a 12–25 pyrimidine tract flanked by regions of high G+C content. Consequently the rp-mRNAs (mRNAs for r-proteins) always start at their 5′ end with a tract of several pyrimidines, sequence that has been shown to be implicated in the above-mentioned translational regulation of these mRNAs. A 5′ pyrimidine tract accompanied by a translational control similar to that of rp-mRNA was also observed for some nonribosomal protein genes, generically identified at first as "housekeeping" genes. Today, with the available data, one is able to recognize in the cell the presence of a class of genes that is characterized structurally by the presence of the 5′ pyrimidine sequence and functionally by a coordinated regulation that occurs at the translation level (which is the main focus of this chapter) and, possibly, at the other levels of regulation. This class of genes includes, beside the genes for r-proteins, other genes involved in the production and function of the translation apparatus components. Adopting the term 5′ TOP to designate the 5′ Terminal OligoPyrimidine tract (Meyuhas et al. 1996), we will refer to this class of genes as 5′ TOP genes or, more concisely, TOP genes.

Although the translational regulation of r-protein synthesis and, in some cases, the presence of pyrimidine sequences, have also been described in *Drosophila* (Al-Atia et al. 1985; Schmidt et al. 1985; Patel and Jacobs-Lorena 1992)

[1]Dipartimento di Biologia, Università di Roma "Tor Vergata", via della Ricerca Scientifica, 00133 Rome, Italy
[2]Istituto di Biologia Cellulare, CNR, Via Marx 43, 00137 Roma, Italy

Progress in Molecular and Subcellular Biology, Vol. 18
Ph. Jeanteur (Ed.)

and in *Dictyostelium* (Steel and Jacobson 1987, 1991), we will consider here only vertebrate systems, where the situation appears to be very homogeneous. A number of reviews on translational regulation of r-protein synthesis in eukaryotes have been published over the years (Meyuhas 1984; Jacobs-Lorena and Fried 1987; Mager 1988; Wormington 1988; Amaldi et al. 1989; Amaldi and Pierandrei-Amaldi 1991; Perry and Meyuhas 1991; Woolford and Warner 1991; Kaspar et al. 1993; Pierandrei-Amaldi and Amaldi 1994, Meyuhas et al. 1996).

2 TOP Genes and TOP mRNAs

2.1 How Many TOP Genes?

A proper assignment of a gene to the TOP class, from a structural point of view, requires the sequencing of a genomic fragment encompassing its 5′ end and the precise mapping of the transcription start site obtained by mRNA analysis by primer extension, S1 mapping, or RNase protection techniques. A TOP gene has its main transcription start site located within a 12–25 pyrimidine tract flanked by short C + G-rich sequences and is, in general, preceded by a noncanonical TATA box (for references see: Pierandrei-Amaldi and Amaldi 1994; Meyuhas et al. 1996). It is also observed that, in most cases, the first intron is localized exactly after the initiation ATG codon or very close to it, thus separating a rather short 5′ UTR (5′ untranslated region) from the coding region. In the absence of a sequenced genomic clone, the presence of a 5′ TOP of at least 7–8 nt at the very 5′ end of a cDNA (provided that this is a full-length copy of the mRNA) allows a bona fide assignment of a gene to the TOP class.

Recently, Kato et al. (1994) have analyzed the 5′ end sequence of a large number of full length cDNA clones from two human cDNA banks, and thus identified many mRNAs with the 5′ TOP sequence. On the basis of these data and of many other present in the literature (for references see: Pierandrei-Amaldi and Amaldi 1994; Meyuhas et al. 1996), one can now state that at least the great majority, possibly all, r-protein genes have a TOP gene structure. In fact, no exception has been found up to now in the tens of cases analyzed in various mammals, birds, fish and amphibians. One should mention that only a small subset of these have been studied from the point of view of translational regulation (see below); no clear exception has yet been found to the notion that a good 5′ TOP is accompanied by translational regulation.

Other genes that appear to belong to the TOP class, on the basis of their 5′ end sequence and/or translational behavior, are those coding for translation elongation factors EF-1α, EF-1β and EF-1γ and EF-2 (Thomas and Thomas 1986; Rao and Slobin 1987; Uetsuki et al. 1989; Frydenberg et al. 1991; Loreni et al. 1993; Avni et al. 1994; Jefferies et al. 1994a; Kato et al. 1994). Notice, however, that not all genes coding for translation factors belong to the TOP class; in fact, at least those for initiation factors eIF-4A, eIF-4D and eIF-2α are not translationally controlled (Huang and Hershey 1989). Within the TOP gene class one also finds the ubiquitin-rp fusion protein UbA_{52} (Backer and Board 1991), some nucleolar proteins as nucleolar phosphoprotein B23, some enzymes as nucleoside diphosphate kinase B and ATP synthase c subunit, proteins possibly involved in nucleo-cytoplasmic export as the hnRNP A1 protein and the laminin-binding protein (Kato et al. 1994), and a number of proteins of yet unclear function (Yenofsky et al. 1982, 1983; Chitpatima et al. 1988; Makrides et al. 1988; Rao et al. 1989).

Thus, TOP gene class includes all r-proteins genes and many genes, although not all, coding for other proteins involved in the production and function of the translation apparatus. When more TOP genes have been identified, it will be possible to see the logic behind their being members of the TOP class. Considering 80 different r-proteins and the numerous factors and proteins involved in the production and function of the protein synthesis apparatus, one can evaluate that in a vertebrate cell there are at least 100 or more different TOP genes, which is between 0.2 and 0.5% of total nuclear genes.

2.2 How Much TOP mRNA?

Approximate evaluations of the fraction of mRNA that codes for rp-mRNAs can be obtained by hybridization quantitation, or from the frequency of the corresponding clones in cDNA banks (for instance see: Meyuhas and Perry 1980; Bozzoni et al. 1981; Kato et al. 1994). In general, the value obtained for the steady-state level of the rp-mRNA for a single r-protein in a normal growing cell is about 0.1% of the total cellular mRNA, which makes up 8% for the 80 different r-proteins. This value becomes lower in quiescent cells, and higher (twofold or more) in tumor cells and in transformed cell lines. Of the other TOP mRNAs, the one for elongation factor EF-1α is probably the most represented, corresponding to about 1% of total mRNA, while the mRNAs for elongation factor subunits EF-1β and EF-1γ, and for the nucleolar phosphoprotein B23, appear to be represented approximately as much as r-protein mRNAs (Kato et al. 1994). Thus, extrapolating to all rp-mRNAs and to the other TOP mRNAs, we find that about 15% of the mRNA of a normal growing cell belongs to the TOP class. This value increases to about 30% in transformed cells. It is interesting that these values are in reasonable agreement with the chemical analysis of cap structure. In fact, it has been shown that mRNA starting at the 5′ end with a capped C followed by other pyrimidines amounts approximately to 17% in normal mouse L cells (Schibler et al. 1977) and to more than 30% in the transformed cells of mouse myeloma (Adams and Cory 1975).

3 Phenomenology of the Translational Regulation of TOP mRNAs

3.1 General and Specific Regulation of Protein Synthesis

An efficient regulation of the rate or global protein synthesis is necessary to guarantee that this major consumer of energy is well integrated into the overall metabolic activity of the cell. This global control appears to be attained by phosphorylation and dephosphorylation of a number of protein components of the translation apparatus, mainly of initiation factors (see reviews: Hershey 1991; Merrick 1992; Sonenberg 1994). Among these, eIF-4F and eIF-2B appear to be the limiting components and they are thus in charge of the control of protein synthesis initiation rate. This type of global control results in changes of the ribosome recruitment of mRNA and thus is reflected in the size of polysomes that can be observed by sucrose gradient sedimentation analysis of a cytoplasmic extract. In fast-growing cells one finds few free monomers, most ribosomes being recruited onto heavy polysomes made of mRNAs tightly packed with ribosomes. On the other hand, in resting cells one finds a relatively high amount of free monomers and a small amount of polysomes; in this case, the translated mRNAs are loaded with few ribosomes, thus sedimenting with "light"polysomes. For instance, human lymphocytes isolated from peripheral blood present a polysomal pattern typical of resting cells, with very few polysomes compared to the amount of monomer; at the same time, the translated fraction of mRNA is associated with very light polysomes (Cooper and Braverman 1977, 1980). In vitro growth and mitogen stimulation of the lymphocytes result in an increase of general translation efficiency and in the commitment of most ribosomes on polysomes, while the translated mRNA is shifted from light to heavy polysomes.

This global control of protein synthesis rate affects all mRNAs, although not necessarily to the same extent, due to the differential affinity of the various mRNAs for the translation initiation factors. Besides global control, regulation of single or classes of mRNAs may work to provide temporal regulation of specific gene expression. Examples of specific regulation include ferritin, tubulin, and GCN4 (for references see: Hershey 1991; Thach 1991; Merrick 1992). Translational control of TOP genes discussed here also falls into this type of specific regulation.

3.2 Translational Regulation of TOP mRNAs

Specific translational regulation of rp-mRNA has been described in cultured cell systems during transition of various cell types between growing and non-growing states in response to a number of physiological stimuli including: serum starvation of mouse fibroblasts (Geyer et al. 1982; Kaspar et al. 1990) and *Xenopus* kidney cultured cells (Loreni and Amaldi 1992), dexamethasone treatment of mouse lymphosarcoma cells (Meyuhas et al. 1987) differentiation of mouse myoblasts (Agrawal and Bowman 1987), contact inhibition of human skin fibroblasts (Avni et al. 1994), concanavalin A treatment of bovine lymphocytes (Kaspar et al. 1992). In all these systems, the changes in growth rate are accompanied by changes in rp-mRNA utilization on polysomes. Technically, this is studied analyzing the polysome/mRNP distribution of individual mRNAs by hybridization of the RNA extracted from the fraction of a polysomal gradient with specific probes. In all cases, it has been observed that only a fraction of rp-mRNA is loaded on polysomes, while the remaining part sediments with light mRNPs. The fraction loaded on polysomes increases or decreases according to the growth state of the cells. In fact, the percent of rp-mRNA which sediments with polysomes changes from 60–70 during rapid growth to 20–30 during the downshift. It has been noted (Meyuhas et al. 1996) that all data available in the literature refer to asynchronous populations of cells. Thus it cannot be excluded that the loading changes somewhat during the cell cycle.

Translational control of r-protein synthesis has also been described in various developmental systems. In *Xenopus*, a translational regulation has been observed by studying the events that, after fertilization, lead to the onset of the synthesis of new ribosomes in the embryo. In fact, new rp-mRNA starts to accumulate in the embryo after the "mid-blastula transition" (stage 8) as it occurs for many other mRNAs. For a while, however, the newly transcribed rp-mRNA is poorly translated, being mostly kept in cytoplasmic mRNPs and only 20–30% loaded on polysomes. Only after stage 26, about 20 h later, does the fraction of rp-mRNAs associated with polysomes increase to about 65–80%, and appreciable amounts of new r-proteins start to be synthesized and accumulated (Pierandrei-Amaldi et al. 1982, 1985a; Baum and Wormington 1985; Loreni et al. 1992). A similar uncoupling between the synthesis of rp-mRNA and its utilization has also been observed during *Xenopus* oogenesis (Cardinali et al. 1987). Here the rp-mRNA is already accumulated at its highest level by early stage II; at this stage, its utilization on polysomes is rather low and becomes more and more efficient during the following oocyte growth when the ribosome store is built up. Hormone induction of oocyte maturation results, in spite of a twofold increase in overall protein synthesis, in the cessation of r-protein synthesis due to the dissociation of rp-mRNA from polysomes accompanied by the deadenylation of these transcripts (Hyman and Wormington 1988). In rat, specific translational regulation has been observed for rp-mRNAs during the transition from the rapidly growing state in the fetal liver to the quiescent state in the adult and upon resumption of the hepatocyte proliferation in the regenerating liver (Aloni et al. 1992).

Although in each study only few different rp-mRNAs have been analyzed, different r-proteins have been studied in different systems, allowing the conclusion that most, if not all, rp-mRNA are translationally controlled.

Beside r-protein genes, few other genes that on the basis of their 5′ end structure can be assigned to the TOP class have been shown to be translationally controlled. For instance, during *Xenopus* development and during induction of growth changes in cultured *Xenopus* kidney cells, the translation factor EF-1α mRNA behaves like rp-mRNAs (Loreni et al. 1993). In mammalian cells, both EF-1α and EF-2 mRNAs are translationally controlled in parallel to rp-mRNA (Jefferies et al. 1994b). Other cases analyzed include ubiquitin-rp fused protein UbA_{52} in mouse (Avni et al. 1994) and other proteins of unknown function.

One can conclude that control at the translational level is an important regulation mechanism of the expression of all TOP genes in all higher eukaryotes. A general feature of this control is that the fraction of rp-mRNA associated with polysomes, and therefore translationally active, is regulated according to the protein synthesis activity of the cell. Since protein synthesis activity is generally dependent on cell growth conditions, TOP mRNAs are mostly loaded on polysomes in rapidly growing cells, and mostly stored as inactive mRNPs in resting cells. Also nongrowing cells heavily committed to protein production (for instance for secretion) have TOP mRNAs mostly recruited onto polysomes.

It should be mentioned that there is no clear-cut distinction between TOP genes and other genes. Those described above are endowed with optimal pyrimidine tracts at their 5′ end and are always clearly subjected to translational control. There are cases of mRNAs, however, that contain a poor pyrimidine tract, and whose translational regulation is less evident and/or depends from the cellular context (see below).

4 Some Features of TOP mRNA and of Its Translational Regulation

4.1 TOP mRNAs Are Capped

Considering the peculiar translation properties of TOP mRNAs, it seems relevant to verify if their 5′ ends are regularly capped. As mentioned above, chemical analysis of cap structure of total cellular mRNA has shown that between 15 and 30% consists of cap-C (Adams and Cory 1975; Schibler et al. 1977): this does not provide a proof, but is consistent with, capping of TOP mRNAs. A more direct indication that TOP mRNAs are capped comes from the large number of TOP cDNA clones that Kato et al. (1994) have obtained in a cDNA bank made from total cellular mRNAs with a specific method for capped RNA molecules. Thus, one can conclude that TOP mRNAs are capped; however, this does not necessarily mean that their translation is cap-dependent. An indirect demonstration of cap dependence comes from in vitro translation of extracted rp-mRNA in the presence of cap analogues; these have been shown to inhibit the synthesis of r-proteins as the result of competition for cap-binding initiation factor (Hammond et al. 1991; Shama et al. 1995).

4.2 TOP mRNAs Extracted from Inactive mRNPs Are Translatable

It has been demonstrated for several rp-mRNAs of various systems (Pierandrei-Amaldi and Beccari 1981; Weiss et al. 1981; Baum and Wormington 1985; Hammond et al. 1991), and for other TOP mRNAs such as EF-1α and Q23 (Thomas and Thomas 1986), that the mRNA loaded on polysomes and the one stored in light mRNPs, once extracted, can equally direct the synthesis of the corresponding proteins in cell-free systems, indicating that mRNAs present in the two compartments are not structurally different.

4.3 Bimodal Distribution of TOP mRNA

The distribution pattern of TOP mRNAs along polysome gradients is bimodal, with the recruited fraction of the mRNA fully loaded with ribosomes. This typical distribution has been observed in mammals for several different rp-mRNAs (Agrawal and Bowman 1987; Meyuhas et al. 1987) and in *Xenopus* for the mRNAs coding several r-proteins, nucleolin, and EF-1α (Caizergues-Ferrer et al. 1989; Amaldi and Pierandrei-Amaldi 1991; Loreni et al. 1993). This distribution indicated that each molecule of rp-mRNA is either translationally inactive or fully active with about one ribosome every 100 nucleotides of its translated region. Notice that this pattern of polysomes/mRNPs distribution of TOP mRNAs is different from the one observed for most other cellular mRNAs. In fact, as mentioned above, these are subjected to the global control of protein synthesis that results in shifts of all mRNAs from heavy to light polysomes and vice versa (White et al. 1990; Hershey 1991).

The fact that translational regulation of TOP mRNAs is a specific process disjointed from the global control is demonstrated by the situation in the early stages of *Xenopus* embryogenesis, where a very high general protein synthetic activity, with most mRNAs loaded on heavy polysomes, is accompanied by a specific repression of rp-mRNA translation, with rp-mRNA mostly unloaded from polysomes.

4.4 Quantitative Differences in Polysome Association of Different TOP mRNAs

Although the different TOP mRNAs are similarly translationally controlled, there are differences in the percentage of mRNA associated with polysomes. For instance, between *Xenopus* embryo stages 16 and 32, the L18 mRNA loaded on polysomes increases from 29 to 62%, the S24 mRNA from 44 to 76% (Bagni et al. 1992), and nucleolin mRNA from 62 to 82% (Caizergues-Ferrer et al. 1989; see note on numbering of *Xenopus* r-proteins). This suggests that individual TOP mRNAs might differ in their affinity for some *trans*-acting factors involved in the translational control mechanism. Some individual variability among different animals has also been observed, which might reflect differences in their genetic makeup affecting the amount or activity of these factors (Bagni et al. 1992).

4.5 Relocation of TOP mRNA Between Polysomes and mRNPs Is Reversible and Fast

It has been demonstrated in *Xenopus* cultured cells that, even in the presence of the transcription inhibitor actinomycin D, the levels of S24 mRNA (Loreni and Amaldi 1992), and of EF-1α mRNA (Loreni et al. 1993) remain constant for a few hours after nutritional down- and upshifts, while their loading on polysomes changes. This result indicates that the change from the rapid-growth mRNA distribution to the downshift one, and back to the upshift one, involves a real mobilization of about 50% of TOP mRNA from polysomes to mRNPs and back, without any modification altering its function irreversibly.

The analysis of the kinetics of mRNA distribution changes during nutritional downshifts and upshifts of cultured cells showed that relocation of the several rp-mRNAs (Loreni and Amaldi 1992) and of EF-1α mRNA (Loreni et al. 1993) is very fast, being appreciable after a few minutes of downshift, and only somewhat slower after an upshift. During these nutritional shifts one also observes changes in the amount of TOP mRNAs, which decrease during a downshift and increase during the upshift; however, this is a much slower phenomenon, becoming appreciable only after several hours. These observations indicate that the translational regulation represents a short-term response to the request for translation apparatus production changes. On the other hand, the transcription modulation of the TOP genes affects the TOP mRNA amount in the cell very slowly, providing a long-term response to internal or external signals.

Another conclusion, suggested by the very short time required for TOP mRNA relocation upon nutritional shifts, is that the translational regulation mechanism cannot involve the de novo synthesis of any regulatory factor, which would take much longer, and must rather be based on rapid modification(s) as, for instance, phosphorylation of pre-existing factor(s).

5 Mechanism of Translational Regulation of TOP mRNAs

Many studies have been undertaken, mostly on rp-mRNAs, to identify the elements and factors implicated in the translational regulation of TOP mRNAs. Although a reasonably satisfying model cannot yet be drawn, a number of important points have been established, and are discussed here below.

5.1 Translational Regulation of TOP mRNA Does Not Involve a Feedback Inhibition by the Translation Product

The possibility that the translational regulation of TOP mRNAs involves a feedback inhibition by the translation products themselves, has been ruled out for several r-proteins by various evidence: (1) r-proteins continue to be synthesized in mammalian cells whose rRNA synthesis is inhibited by low doses of actinomycin D (Craig and Perry 1971; Warner 1977); (2) purified individual r-proteins or small groups of them, when microinjected into the cytoplasm of *Xenopus* oocytes or added in an in vitro translation system programmed with a *Xenopus* mRNA, had no inhibitory effect on the synthesis of new r-proteins (Pierandrei-Amaldi et al. 1985b), thus indicating that r-proteins do not feed back the translation of their own mRNA; (3) in *Xenopus* anucleolate mutant embryos (0-nu), where rRNA synthesis cannot occur because of a deletion of the rRNA gene cluster, the rp-mRNAs were recruited on polysomes and translated as in controls (Pierandrei-Amaldi et al. 1985a); (4) in manually enucleated *Xenopus* oocytes, rp-mRNA translation is still normal 24 h after enucleation, in spite of the absence of rRNA synthesis (Pierandrei-Amaldi et al. 1987); (5) overexpression of r-proteins S16 or L32 in transfected mouse myoblasts did not change the translation efficiency of the corresponding mRNAs (Bowman 1987); (6) overexpression of L4, obtained in *Xenopus* oocytes by microinjection of an overdose of the corresponding mRNA, did not affect the translation efficiency of the mRNA itself (Baum et al. 1988). All these data demonstrate that the presence of excess r-proteins, both in vivo and in vitro, has no effect on the translation of their own mRNAs.

5.2 Effect of the Amount of Ribosomes on TOP mRNA Translation

In most experimental systems analyzed, one can observe an inverse proportion between the amount of free ribosomes present in the cell and the percentage of rp-mRNA loaded on polysomes. Thus, during the transition of various cell types between growing and nongrowing states (see above), the percent of TOP mRNA on polysomes changes from 60–70% in growing cells, where the proportion of free ribosomes is very low, to 20–30% in resting cells, where the amount of free ribosomes is high (for instance see Loreni and Amaldi 1992).

A similar inverse relationship is also observed in developmental systems. For instance, during *Xenopus* embryogenesis, the utilization of rp-mRNA is inefficient in the early stages when the maternal store of ribosomes is not yet fully utilized. It is only after stage 26, when the ribosome recruitment on polysomes reaches its normal value of about 70–80%, that the rp-mRNA starts being translated efficiently (Pierandrei-Amaldi et al. 1985a). In the particular situation of the *Xenopus* anucleolate mutant embryos, at late stages when the embryos suffer a critical shortage of ribosomes which all become recruited on polysomes, 100% of the rp-mRNA is associated with polysomes (Pierandrei-

Amaldi et al. 1985a). The reverse situation of an increased ribosome content in the *Xenopus* embryo, induced by microinjection of ribosomes into fertilized eggs, resulted in a repression of rp-mRNA recruitment onto polysomes (Pierandrei-Amaldi et al. 1991).

Although we do not know how the amount of free ribosomes can affect the translation of rp-mRNA, this cannot occur in a simple and direct way. In fact, the unutilized rp-mRNAs sediment as 15–30S mRNPs (Cardinali et al. 1993), slightly faster than the naked mRNAs, thus not bound to ribosome monomers or ribosome subunits.

5.3 The 5′ UTR Is the cis-Acting Element of the Translational Control

As mentioned above, the 5′ UTRs of all vertebrate TOP mRNAs are always short, and start with a sequence of 6–12 pyrimidines (5′ TOP). This fact, together with the observation that, in general, TOP genes have a first intron localized exactly at, or very close to, the initiation ATG codon, suggested that these untranslated sequences might have an important role in the translational regulation. This notion has been proved for few r-protein mRNAs by the use of chimeric genes expressing mRNAs whose reporter translated sequences were preceded by the 5′ UTRs of TOP genes. Thus, by microinjecting *Xenopus* fertilized eggs, it was shown that the 5′ UTR of S24 mRNA conferred the translation behavior typical of TOP mRNAs to the fused reporter sequence (Mariottini and Amaldi 1990). Analogously, by transfecting chimeric genes into mammalian cells, it has been shown that the 5′ UTRs of mammalian S16, L32, and L13a mRNAs are necessary and sufficient to confer the translational regulation typical of TOP mRNA to the attached reporter sequences (Hammond et al. 1991; Levy et al. 1991; Avni et al. 1994). The outcome of the complementary experiment, namely the substitution of the original 5′ UTRs of L32 and S16 mRNAs with those of unrelated mRNAs, is the abolition of the translational control (Hammond et al. 1991; Levy et al. 1991).

In addition, site-directed mutagenesis was used to show that 5′ TOP sequence at the very 5′ end of the mRNA for r-protein S16 is necessary for translational regulation in the mouse system (Levy et al. 1991). Necessary but not sufficient, as shown for the 5′ UTRs of L32 and S16 mRNAs by removal of sequences downstream to the 5′ TOP, a modification that abolishes the translation regulation (Avni et al. 1994).

Information about the detailed structure of the 5′ TOP sequence comes mainly from the comparison of the 5′ ends of the many available TOP mRNAs, and in part also from mutagenesis studies (Levy et al. 1991). From all available data one can summarize the features of a functional 5′ TOP as follows: (1) it is located at the very 5′ end of the mRNA; (2) it is at least 7–8 nt long; (3) it can be "contaminated" by one or two Gs, especially in its second part, but never by an A; (4) it does not have a specific pyrimidine sequence (no consensus can be identified); (5) it contains approximately equal proportions of C and U residues.

As mentioned above, some mRNAs with a 5′ TOP sequence somewhat deviating from this description have been observed: for instance only 4 or 5 nt long, and/or with a C:T ratio different from the usual 1:1, and/or preceded or interrupted by one or more Gs. These non-canonical 5′ TOPs could be responsible for the partial translational regulation observed for some mRNAs. For instance, some mRNAs show polysome association changes which are less drastic than those of typical TOP mRNAs, as it occurs for nucleolin mRNA during *Xenopus* development (Caizergues-Ferrer et al. 1989); other mRNAs show growth-dependent translational regulation in some type of cells but not in others, like the mRNAs for PABP [poly(A)-binding protein] and for elongation factor EF-2 (Meyuhas, pers. comm.).

5.4 Some Proteins Interact with the 5′ UTR of TOP mRNAs

It seems conceivable that some factor(s), specifically interacting with the similar responsive sequences in the 5′ UTR of TOP mRNAs, might be involved directly or indirectly in the translational regulation. In fact, a protein ($p56^{L32}$) of about 56 kDa has been described as binding the 5′ UTR of mouse L32 mRNA, a binding that depends on the presence of an intact 5′ TOP and on a second element 15 nt downstream (Kaspar et al. 1992; Severson et al. 1995). Similarly, in the *Xenopus* system, two proteins have been found to interact specifically in vitro with the 5′ UTR of L4 mRNA: protein A of 57 kDa, probably homologous to the one described in mouse, binds the 5′ TOP sequence, and protein C of 31 kDa binds few nucleotide downstream (Cardinali et al. 1993). At first, two other binding proteins were observed, B and D; however, recent results based on peptide mapping analysis have shown that these are cleavage products of A and C respectively, but still maintain their binding capacity. Although it has been shown that the binding of these proteins to the rp-mRNA 5′ UTR is specific, their involvement in the translational regulation remains to be proved. It has been observed, both in mammals (Kaspar et al. 1992) and in *Xenopus* (Cardinali et al. 1993), that the factor binding activity does not change as a function of growth or developmental stage, suggesting that translational control is modulated by some other factor(s) which interacts with these proteins or possibly with the RNA/protein complex.

5.5 Translational Regulation of TOP mRNA Is Not Due to Limiting Amounts of eIF-4E

The contribution of components of the translation apparatus to rp-mRNA translational control is still an open issue. In fact, a general model of translational regulation is based on the differential affinity of mRNAs for a component of the translation apparatus (Lodish 1974). According to this model, when the amount or the activity of this component is limiting, "low-affinity" mRNAs would be selectively excluded from polysomes. One candidate as determinant of rp-mRNA translational control is the initiation factor 4E (eIF-4E). This has been shown to be a limiting component of the initiation event (Walden et al. 1981; Duncan and McConkey 1982; Hiremath et al. 1985); it is activated by phosphorylation and its overexpression can induce cell transformation (Lazaris-Karatzas et al. 1990). The fact that the abundance and the phosphorylation of eIF-4E appear to be correlated with rp-mRNA polysomal localization has been proposed as an indication of its involvement in translational control (Kaspar et al. 1990). However, this possibility is ruled out by the observation that an overexpression of eIF-4E does not relieve the translational repression of S4 and S16 mRNAs in quiescent mammalian cells (Shama et al. 1995).

It should be noticed that a translational modulation controlled by the amount and/or activity of the initiation factor eIF-4E should be accompanied by shifts of the controlled mRNAs from heavier to lighter polysomes and vice versa. On the contrary, as discussed earlier, TOP mRNAs show a bimodal distribution along polysome gradients with the recruited fraction fully loaded with ribosomes. This observation supports the conclusion that translational regulation of TOP mRNAs depends on some specific mechanism and not on the limitation of general translation factor(s).

5.6 Effect of the Phosphorylation State of Ribosomal Protein S6 on Translational Regulation of TOP mRNA

It has been known for many years that ribosomal protein S6 is rapidly phosphorylated when cells are stimulated to grow or divide. Furthermore, numerous studies in several different systems have documented that there is a correlation between increases in S6 phosphorylation and protein synthesis (for a review see: Ferrari and Thomas 1994). The ribosomes containing the most highly phosphorylated derivatives of S6 appear to have a selective advantage in entering polysomes, suggesting the idea that phosphorylation of S6 may facilitate the initiation process by altering the affinity of the ribosome for mRNA.

Recently, it has been shown that the effect of S6 phosphorylation on ribosome affinity for mRNA is not generalized, but selective for TOP mRNAs. This has been demonstrated by showing that rapamycin, an inhibitor of S6 phosphorylation, selectively suppresses translation of three TOP mRNAs, EF-1α, EF-2 and r-protein S6 mRNAs (Jefferies et al. 1994b; Terada et al. 1994).

A reverse situation, pointing to the same conclusion, is observed upon HSV- 1 (Herpes simplex virus type 1) infection of mammalian cells. While most protein synthesis in the host cell is inhibited after infection, a specific set of proteins, including r-proteins, continue to be produced. Interestingly, one of the early effects of the HSV-1 infection is a stimulation of S6 phosphorylation (Massé et al. 1990; Madjar, pers. comm.).

It is known that $p70^{s6k}$, the kinase specifically responsible for S6 phosphorylation, is activated by insulin and growth factors (for a review see Ferrari and Thomas 1994). Although much of the interconnections still have to be understood, the relevant aspect is that one can now envisage links between the signaling pathways controlling cell growth and proliferation, and the translation efficiency of TOP mRNAs presiding the production of many important elements of the protein synthesis apparatus and of some other housekeeping functions.

5.7 Global vs. TOP mRNA Translational Regulation

A certain degree of independence of the two synthesis regulations, the one controlling the production of proteins making up the protein synthesis apparatus and the other controlling the global protein synthesis rate, is a necessary requirement. We have, in fact, situations in which TOP mRNA translations, and consequently ribosome production, are inefficient in fast proliferating cells with an efficient protein synthesis, as, for instance, during early *Xenopus* development (Pierandrei-Amaldi et al. 1985a). The opposite situation is observed during oogenesis, when a large amount of translation apparatus is produced, in great excess with respect to the need of the oocyte itself, and accumulated for later use during embryogenesis.

The production of the translation apparatus itself cannot be strictly coregulated with global protein synthesis, in order to avoid that a downregulation of this would result in a definitive shutoff of the entire system. It is also important that the component(s) of the translation apparatus implicated in the global control be not under the TOP gene regulation, so as to avoid that a downregulation of translation apparatus production would automatically lead to a downregulation of the global protein synthesis rate. This provides an explanation to the observation that, while most genes coding for protein synthesis apparatus components belong to the TOP gene class, those for some general translation initiation factors, whose amount and activity might regulate the overall protein synthesis rate, do not.

Acknowledgments. Research in the authors' laboratories was supported by EC contracts SCI*-0259-C to F.A. and BIO4-CT95-0045 to P.P.-A., by Progetto Finalizzato Biotecnologie and Progetto Finalizzato Ingegneria Genetica of Consiglio Nazionale delle Ricerche, and by Telethon-Italy (Grant no. E.O2).

Note. In previous papers, the numbering of *Xenopus* ribosomal proteins followed the system introduced in our first study (Pierandrei-Amaldi and Beccari 1981). The large amount of sequence data now accumulated for many species makes it possible to adopt, for a unified nomenclature, the rat system (Wool et al. 1990). Thus the *Xenopus* r-proteins that we previously designated as L1, L14 and S19 are here identified as L4, L18 and S24 respectively.

References

Adams JM, Cory S (1975) Modified nucleosides and bizarre 5′-termini in mouse myeloma mRNA. Nature 255:28–33

Agrawal AG, Bowman LH (1987) Transcriptional and translational regulation of ribosomal protein formation during mouse myoblast differentiation. J Biol Chem 262:4868–4875

Al-Atia GR, Fruscoloni P, Jacobs-Lorena M (1985) Translational regulation of mRNAs for ribosomal proteins during early *Drosophila* development. Biochemistry 24:5798–5803

Aloni R, Peleg D, Meyuhas O (1992) Selective translational control and nonspecific post-transcriptional regulation of ribosomal protein gene expression during development and regeneration of rat liver. Mol Cell Biol 12:2203–2212

Amaldi F, Pierandrei-Amaldi P (1991) Translational regulation of the expression of ribosomal protein genes in *Xenopus laevis*. In: Thach RE (ed) Translationally regulated genes in higher eukaryotes. Karger AG, Basel, pp 93–105

Amaldi F, Bozzoni I, Beccari E, Pierandrei-Amaldi P (1989) Expression of ribosomal protein genes and regulation of ribosome biosynthesis in *Xenopus* development. Trends Biochem Sci 14:175–178

Avni D, Shama S, Loreni F, Meyuhas O (1994) Vertebrate mRNAs with a 5′-terminal pyrimidine tract are candidates for translational repression in quiescent cells: characterization of the translational *cis*-regulatory element. Mol Cell Biol 14:3822–3833

Backer RT, Board PG 1991 The human ubiquitin-52 amino acid fusion protein gene shares several structural features with mammalian ribosomal protein genes. Nucl Acids Res 19:1035–1040

Bagni C, Mariottini P, Terrenato L, Amaldi F (1992) Individual variability in the translational regulation of ribosomal protein synthesis in *Xenopus*. Mol Gen Genet 234:60–64

Baum EZ, Wormington WM (1985) Coordinate expression of ribosomal protein genes during *Xenopus* development. Dev Biol 111:488–489

Baum EZ, Hyman LE, Wormington WM (1988) Post-transcriptional control of ribosomal protein L1 accumulation in *Xenopus* oocytes. Dev Biol 126:141–149

Bowman LH (1987) The synthesis of ribosomal proteins S16 and L32 is not autogenously regulated during myoblast differentiation. Mol Cell Biol 7:4464–4471

Bozzoni I, Beccari E, Luo Z-X, Amaldi F, Pierandrei-Amaldi P, Campioni N (1981) *Xenopus laevis* ribosomal protein genes: isolation of recombinant cDNA clones and study of the genomic organization. Nucl Acids Res 9:1069–1086

Caizergues-Ferrer M, Mariottini P, Curie C, Lapeyre B, Gas N, Amalric F, Amaldi F (1989) Nucleolin from *Xenopus laevis:* cDNA cloning and expression during development. Genes Dev 3:324–333

Cardinali B, Campioni N, Pierandrei-Amaldi P (1987) Ribosomal protein, histone and calmodulin mRNAs are differently regulated at the translational level during oogenesis of *Xenopus laevis*. Exp Cell Res 169:432–441

Cardinali B, Di Cristina M, Pierandrei-Amaldi P (1993) Interaction of proteins with the mRNA for ribosomal protein L1 in *Xenopus:* structural characterization of in vivo complexes and identification of proteins that bind in vitro to its 5′UTR. Nucl Acids Res 21:2301–2308

Chitpatima ST, Makrides S, Bandyopadhyay R, Browerman G (1988) Nucleotide sequence for a major messenger RNA for a 21-kilodalton polypeptide that is under translational control in mouse tumor cells. Nucl Acids Res 16:2350

Cooper HL, Braverman R (1977) Free ribosomes and growth stimulation in human peripheral lymphocytes: activation of free ribosomes as an essential event in growth stimulation. J Cell Physiol 93:213–226

Cooper HL, Braverman R (1980) Protein synthesis in resting and growth-stimulated human peripheral lymphocytes. Evidence for regulation by a non-messenger RNA. Exp Cell Res 127:351–359

Craig N, Perry RP (1971) Persistent cytoplasmic synthesis of ribosomal proteins during the selective inhibition of ribosome RNA synthesis. Nature New Biol 229:75–80

Duncan R, McConkey EH (1982) Rapid alterations in initiation rate and recruitment of inactive RNA are temporally correlated with S6 phosphorylation. Eur J Biochem 123:539–544
Ferrari S, Thomas G (1994) S6 phosphorylation and the $p70^{s6k}/p85^{s6k}$. Crit Rev Biochem Mol Biol 29:385–413
Frydenberg J, Poulse K, Petersen AKB, Lund A, Olesen OF (1991) Isolation and characterization of the gene encoding EF-1aO, an elongation factor 1-a expressed during early development of *Xenopus laevis*. Gene 109:185–192
Geyer PK, Meyuhas O, Perry RP, Johnson LF (1982) Regulation of ribosomal protein mRNA content and translation in growth-stimulated mouse fibroblasts. Mol Cell Biol 2:685–693
Hammond ML, Merrick W, Bowmann LH (1991) Sequences mediating the translation of mouse S16 ribosomal protein mRNA during myoblast differentiation and in vitro, and possible control points for the in vitro translation. Genes Dev 5:1723–1736
Hiremath LS, Webb NR, Rhoads RE (1985) Immunological detection of the messenger RNA cap-binding protein. J Biol Biochem 260:7843–7849
Hershey JWB (1991) Translational control in mammalian cells. Annu Rev Biochem 60:717–755
Huang S, Hershey JB (1989) Translational initiation factor expression and ribosomal protein gene expression are repressed coordinately but by different mechanisms in murine lymphosarcoma cells trated with glucocorticois. Mol Cell Biol 9:3679–3684
Hyman LE, Wormington WM (1988) Translation inactivation of ribosomal protein mRNAs during *Xenopus* oocyte maturation. Genes Dev 2:598–605
Jacobs-Lorena M, Fried HM (1987) Translational regulation of ribosomal protein genes expression in eukaryotes. In: Ilan J (ed) Translational regulation of gene expression. Plenum Press, New York, pp 63–85
Jefferies HB, Thomas G, Thomas G (1994a) Elongation factor-1a mRNA is selectively translated following mitogenic stimulation. J Biol Chem 269:4367–4372
Jefferies HB, Reinhard C, Kozma SC, Thomas G (1994b) Rapamycin selectively represses translation of the 'polypyrimidine tract' mRNA family. Proc Natl Acad Sci USA 91:4441–4445
Kaspar RL, Rychlik W, White MW, Rhoads RE, Morris DR (1990) Simultaneous cytoplasmic redistribution of ribosomal protein L32 mRNA and phosphorylation of eukaryotic initiation factor 4E after mitogenic stimulation of Swiss 3T3 cells. J Biol Chem 265:3619–3622
Kaspar RL, Kakegava T, Cranston H, Morris DR, White MW (1992) A regulatory cis element and a specific binding factor involved in the mitogenic control of murine ribosomal protein L32 translation. J Biol Chem 267:508–514
Kaspar RL, Morris DR, White M (1993) Control of ribosomal protein synthesis in eukaryotic cells. In: Ilan J (ed) Translational regulation of gene expression 2. Plenum Press, New York, pp 335–348
Kato S, Sekine S, Oh S-W, Kim N-S, Umezawa Y, Abe N, Yokoyama-Kobayashi M, Aoki T (1994) Construction of a human full-length cDNA bank. Gene 150:243–250
Lazaris-Karatzas A, Montine KS, Sonenberg N (1990) Malignant trnasformation by a eukaryotic initiation factor subunit that binds to mRNA 5′ cap. Nature 345:544–547
Levy S, Avni D, Hariharan N, Perry RP, Meyuhas O (1991) Oligopyrimidine tract at the 5′ end of mammalian ribosomal protein mRNAs is required for their translational control. Proc Natl Acad Sci USA 88:3319–3323
Lodish HF (1974) Model for the regulation of mRNA translation applied to haemoglobin synthesis. Nature 251:385–388
Loreni F, Amaldi F (1992) Translational regulation of ribosomal protein synthesis in *Xenopus* cultured cells: mRNA relocation between polysomes and RNP during nutritional shifts. Eur J Biochem 205:1027–1032
Loreni F, Francesconi A, Jappelli R, Amaldi F (1992) Analysis of mRNAs under translational control during *Xenopus* embryogenesis: isolation of new ribosomal protein clones. Nucl Acids Res 20:1859–1863
Loreni F, Franscesconi A, Amaldi F (1993) Coordinate translational regulation in the syntheses of elongation factor 1a and ribosomal proteins in *Xenopus laevis*. Nucl Acids Res 21:4721–4725
Mager W (1988) Control of ribosomal protein gene expression. Biochim Biophys Acta 949:1–15

Makrides S, Chitpatima ST, Bandyopadhyay R, Brawerman G (1988) Nucleotide sequence of a messenger RNA for a 40-kilodalton polypeptide that is under translational control in mouse tumor cells. Nucl Acids Res 16:2349

Mariottini P, Amaldi F (1990) The 5′ untranslated region of mRNA for ribosomal protein S19 is involved in its translational regulation during *Xenopus* development. Mol Cell Biol 10:816–822

Massé T, Garcin D, Jacquemont B, Madjar J-J (1990) Herpes simplex virus type-1-induced stimulation of ribosomal protein S6 phosphorylation is inhibited in neomycin-treated human epidermoid carcinoma 2 cells and in *ras*-transformed cells. Eur J Biochem 194:287–291

Merrick WC (1992) Mechanism and regulation of eukaryotic protein synthesis. Microbiol Rev 56:291–315

Meyuhas O (1984) Ribosomal protein gene expression in proliferating and nonproliferating cells. In: Stein GS, Stein JL (ed) Recombinant DNA and cell proliferation. Academic Press, Orlando, pp 243–271

Meyuhas O, Perry RP (1980) Construction and identification of cDNA clones for several mouse ribosomal proteins. Application for the study of r-protein gene expression. Gene 10:113–137

Meyuhas O, Thompson AE, Perry RP (1987) Glucocorticoids selectively inhibit the translation of ribosomal protein mRNAs in P1798 lymphosarcoma cells. Mol Cell Biol 7:2691–2699

Meyuhas O, Baldin V, Bouche G, Amalric F (1990) Glucocorticoids repress ribosome biosynthesis in lymphosarcoma cells by affecting gene expression at the level of transcription, post-transcription and translation. Biochim Biophys Acta 1049:38–44

Meyuhas O, Avni D, Shama S (1996) Translational control of ribosomal protein mRNAs in eukaryotes. In: Hershey JWB, Mathews M, Sonenberg N (eds) Translational control. Cold Spring Harbor Lab Press, Cold Spring Harbor, NY pp 363–388

Patel RC, Jacobs-Lorena M (1992) *cis*-Acting sequences in the 5′ untranslated region of the ribosomal protein A1 mRNA mediate its translational regulation during early embryogenesis of *Drosophila*. J Biol Chem 267:1159–1164

Perry RP, Meyuhas O (1991) Translational control of ribosomal protein production in mammalian cells. In: Thach RE (ed) Translationally regulated genes in higher eukaryotes. Karger AG, Basel, pp 83–92

Pierandrei-Amaldi P, Amaldi F (1994) Aspects of regulation of ribosomal protein synthesis in *Xenopus laevis*. Review. Genetica 94:181–193

Pierandrei-Amaldi P, Beccari E (1981) Messenger RNA for ribosomal proteins in *Xenopus laevis* oocytes. Eur J Biochem 106:603–611

Pierandrei-Amaldi P, Campioni N, Beccari E, Bozzoni I (1982) Expression of ribosomal protein genes in *Xenopus laevis* development. Cell 30:163–171

Pierandrei-Amaldi P, Beccari E, Bozzoni I, Amaldi F (1985a) Ribosomal protein production in normal and anucleolate *Xenopus* embryos: regulation at the posttranscriptional and translational levels. Cell 42:317–323

Pierandrei-Amaldi P, Campioni N, Gallinari P, Beccari E, Bozzoni I, Amaldi F (1985b) Ribosomal protein synthesis is not autogenously regulated at the translational level in *Xenopus laevis*. Dev Biol 167:281–289

Pierandrei-Amaldi P, Amaldi F, Bozzoni I, Fragapane P (1987) Regulation of ribosomal protein genes during *Xenopus* development. In: Firtel RA Davidson EH (eds) Molecular approaches to developmental biology. Alan R Liss Inc, New York, pp 267–278

Pierandrei-Amaldi P, Campioni N, Cardinali B (1991) Experimental changes in the amount of maternally stored ribosomes affect the translation efficiency of ribosomal protein mRNA in *Xenopus* embryo. Cell Mol Biol 37:227–238

Rao TR, Slobin LI (1987) Regulation of the utilization of mRNA for eukaryotic elongation factor Tu in Friend erythroleukemia cells. Mol Cell Biol 7:687–697

Rao TR, Castronovo V, Schmitt MC, Wewer UM, Claysmith AP, Liotta LA, Sobel ME (1989) Evidence for a precursor of the high-affinity metastasis-associated murine laminin receptor. Biochemistry 28:7476–7486

Schibler U, Kelley DE, Perry RP (1977) Comparision of methylated sequences in messenger RNA and the heterogeneous nuclear RNA from mouse L cells. J Mol Biol 115:695–714

Schmidt T, Chen PS, Pellegrini M (1985) The induction of ribosome biosynthesis in a nonmitotic secretory tissue. J Biol Chem 260:7645–7650

Severson WE, Mascolo PL, White MW (1995) Lymphocyte $p56^{L32}$ is a RNA/DNA-binding protein which interacts with conserved elements of the murine L32 ribosomal protein mRNA. Eur J Biochem 229:426–432

Shama S, Avni D, Frederickson RM, Sonenberg N, Meyuhas O (1995) Overexpression of initiation factor eIF-4E does not relieve the translational repression of ribosomal protein mRNAs in quiescent cells. Gene Expr 4:241–252

Sonenberg N (1994) mRNA translation: influence of the 5′ and 3′ untranslated regions. Curr Opin Genet Dev 4:310–5

Steel LF, Jacobson A (1987) Translational control of ribosomal protein synthesis during early *Dictyostelium discoideum* development. Mol Cell Biol 7:965–972

Steel LF, Jacobson A (1991) Sequence elements that affect mRNA translational activity in developing *Dictyostelium* cells. Dev Genet 12:98–103

Terada N, Patel HR, Takase K, Kohno K, Nairn AC, Gelfand EW (1994) Rapamycin selectively inhibits translation of mRNAs encoding elongation factors and ribosomal proteins. Proc Natl Acad Sci USA 91:11477–11481

Thach RE (1991) Translationally regulated genes in higher eukaryotes. Karger AG, Basel

Thomas G, Thomas G (1986) Translational control of mRNA expression during the early mitogenic response in Swiss mouse 3T3 cells: identification of specific proteins. J Cell Biol 103:2137–2144

Uetsuki T, Naito A, Nagata S, Kaziro Y (1989) Isolation and characterization of the human chromosomal gene for polypeptide chain elongation factor-1 alpha. J Biol Chem 264:5791–5798

Walden WE, Godefrog-Colburn T, Thach RE (1981) The role of mRNA competition in regulating translation. J Biol Chem 256:11739–11746

Warner JR (1977) In the absence of ribosomal RNA synthesis, the ribosomal protein of HeLa cells are synthesized normally and degraded rapidly. J Mol Biol 115:315–333

Weiss YC, Vaslet CA, Rosbash M (1981) Ribosomal protein mRNAs increase dramatically during *Xenopus* development. Dev Biol 87:330–339

White MW, Degnin C, Hill J, Morris DR (1990) Specific regulation by endogenous polyamines of translation initiation of S-adenosylmethionin decarboxylase mRNA in swiss 3T3 fibroblasts. Biochem J 268:657–660

Wool IG, Endo Y, Chan YL, Glück A (1990) Structure, function and evolution of mammalian ribosomes. In: Hill WE, Dahlberg A, Garrett RA, Moore PB, Schlessinger D, Warner JR (eds) The ribosome: structure function and evolution. Am Soc Microbiol Press, Washington, pp 203–214

Woolford JLJ, Warner JR (1991) The ribosome and its synthesis. In: Broach J, Jones E, Pringle J (ed) The molecular and cellular biology of the yeast *Saccharomyces:* genome dynamics, protein synthesis, and energetic. Cold Spring Harbor Lab Press, Cold Spring Harbor NY, pp 587–626

Wormington WM (1988) Expression of ribosomal protein genes during *Xenopus* development. In: Browder LW (ed) Developmental biology, vol 5. Plenum Press, New York, pp 227–240

Yenofsky R, Bergman I, Brawerman G (1982) Messenger RNA species partially in a repressed state in mouse sarcoma ascites cells. Proc Natl Acad Sci USA 79:5876–5880

Yenofsky RS, Careghini A, Krowczynska A, Brawerman G (1983) Regulation of mRNA utilization in mouse erythroleukemia cells induced to differentiate by exposure to dimethyl sulfoxide. Mol Cell Biol 3:1197–1203

[illegible]

RNase L: Effector Nuclease of an Activatable RNA Degradation System in Mammals

C. Bisbal

1 Introduction

In eukaryotic cells, the degradation of messenger RNA (mRNA) is an essential determinant in the regulation of gene expression. Indeed, the need to rapidly interrupt cellular responses to various stimuli (growth or developmental) upon the disappearance of the inducing signal necessitates the tight regulation of the degradation process.

The stability of different mRNAs within a given cell can differ by several orders of magnitude, suggesting that the degradation of mRNA is controlled at different levels. The identification of several types of destabilizing elements supports this notion (Jackson 1993; Sach 1993). The nuclease effectors involved in the diverse mechanisms of mRNA degradation, however, remain to be elucidated.

Faced with such a variety of potential target RNA molecules, eukaryotic cells are endowed with a large spectrum of RNases. If the degradation of a specific RNA is to be limited to a particular RNase at any one time, it is evident that the activities of the various RNase must be tightly controlled. The mechanisms implicated in such control might involve: (1) highly specific RNases capable of recognizing specific RNA sequences or structure, (2) compartimentalization of RNases, (3) three-dimensional structure of the RNA such as stem loops, (4) 5′ cap structure or poly(A) on mRNA, (5) association of RNA with proteins, (6) activation or inhibition of specific RNases: RNases can exist in an inactive form that requires an activation process with a subsequent return to the inactive state.

The endoribonuclease L (RNase L) belongs to this latter family of enzymes, and was discovered during studies of the antiviral activity of the interferons (IFN). It is a ubiquitous and latent enzyme. Following activation by an oligonucleotide, 2-5A, it cleaves RNAs at UpNp sequences (Floyd-Smith et al. 1981). In addition, the activity of RNase L can be inhibited by another protein: RLI (RNase L inhibitor; Bisbal et al. 1995).

IGMM CNRS UMR 5535, 1919 route de Mende, 34033 Montpellier Cedex 1, France

Progress in Molecular and Subcellular Biology, Vol. 18
Ph. Jeanteur (Ed.)

2 The Interferons

Interferons (IFNs) are a family of proteins secreted by vertebrate cells in response to various stimuli. Originally identified in 1957 by Isaac and Lindenman as antiviral agents, they have long been known to exert antitumoral and antiproliferative activities (Paucker et al. 1962; Sen and Lengyel 1992).

The IFNs were initially classified according to their cellular origin: leucocyte, fibroblast and immune IFNs. They are now classified according to their sequences and their activities. IFN type I regroups IFN α (leucocyte IFN) and IFN β (fibroblast IFN); IFN type II: IFN γ (immune IFN). Their nucleotide sequences (Nagata et al. 1980; Tanigushi et al. 1980; Tavernier et al. 1981; Gray et al; 1982) and their chromosomal localizations have been elucidated (Owerbach et al. 1981; Trent et al. 1982; Naylor et al. 1983).

Following their induction, IFNs bind to species-specific cell-surface receptors. All type I IFNs bind to the same receptor (Uzé et al. 1990, 1995), while IFN type II binds to a different receptor (Aguet et al. 1988). By binding to their specific receptor, IFNs regulate a great number of genes. This regulation is at the transcriptional level and is mediated by several cellular factors including the Jak/STAT pathway (Williams 1991; Pellegrini and Schindler 1993; Schindler and Darnell 1995). This regulation can be positive or negative. IFNs induce more than 30 genes. Several genes are induced by both type I and II IFNs, but often with differing kinetics and via different biochemical pathways. Other genes are activated exclusively by either type I or II IFNs. The level of this induction is very variable. IFNs can also selectively diminish the level of particular proteins, such as the cellular oncogenes c-myc (Jonak and Knight 1984; Einat et al. 1985), src (Lin et al. 1983) or the immunoglobulin heavy chain in Daudi cells (Meurs and Hovanessian 1988). Some of the proteins whose expression is modified by IFN are presented in Table 1.

3 The 2-5A Pathway

The 2-5A synthetases are a family of interferon-induced proteins capable of polymerizing ATP into short oligomers named 2-5A, which in turn activates RNase L.

The initial and essential finding was made by Ian Kerr's group in 1974 (Kerr et al. 1974a). Protein synthesis in extracts of IFN-treated cells was sensitive to inhibition by double-stranded (ds)RNA and ATP. In these extracts an increase in nuclease activity was observed (Brown et al. 1976; Kerr et al. 1976; Sen et al. 1976). The addition of dsRNA had no effect if the cells were previously infected by a virus (Kerr et al. 1974b).

This observation led to the discovery of the unusual oligonucleotide series named 2-5A : ppp(A2′p5′A)$_n$A, (trimer and tetramer are the most common species in cells; see Fig. 1; Clemens and Williams 1978; Hovanessian and Kerr 1978; Kerr and Brown 1978). This small oligonucleotide, 2-5A, activates a

Table 1. Some proteins induced by INFs

Protein, reference	Inducer	Role
2-5A synthetase Chebath et al. (1987a) Marié and Hovanessian (1992)	IFN α, β > γ	Antiviral activity (2-5A synthesis)
PKR Meurs et al. (1990) Kumar et al. (1994)	IFN α, β > γ	Antiviral activity (phosphorylation of eIF2, IκB)
MxA Staeheli et al. (1986)	IFN α, β > γ, dsRNA	Antiviral activity against influenza virus and VSV
MHC I (heavy chain) Rosa et al. (1986)	IFN α, β, γ	Antigen presentation
MHC II (heavy chain) Rosa (1986)	IFN α, β, γ	Antigen presentation
IRF-1/ISGF-2 Miyamoto et al. (1988)	IFN α, β, γ, dsRNA	Transcription factor
IRF-2 Harada et al. (1989)	IFN α, β, γ, dsRNA	Transcription factor
GAF Decker et al. (1991)	IFN γ	Transcription factor
Stat 1 Fu et al. (1992)	IFN α, β	Transcription factor
Stat 2 Schindler et al. (1992)	IFN α, β	Transcription factor

nuclease that inhibits protein synthesis by cleaving mRNA (Baglioni et al. 1978; Schmidt et al. 1978; Slattery et al. 1979).

In fact, the inhibition of protein synthesis is due to two distinct and independent pathways: the 2-5A pathway and the dsRNA-dependent protein kinase pathway (PKR) (Ratner et al. 1978; Zilberstein et al. 1978). These two pathways are induced in response to interferon and activated by dsRNA (Fig. 2). PKR has now been cloned (Meurs et al. 1990) and its fundamental role in not only antiviral activity of IFN but in cell growth and tumour suppression has been demonstrated (see Jaramillo et al. 1995; Proud 1995 for reviews).

3.1 The 2-5A Synthetases

The 2-5A synthetase genes are induced in all cell types by any kind of IFN. The gene products have latent activity, the only known activator being double-stranded RNA (dsRNA). The first biological function to be attributed to these enzymes was their participation in the pathway that results in resistance to picornavirus.

The importance of this activity was first underlined in mouse L cells by A. Hovanessian and coworkers in 1977. After this initial finding, several forms of this enzyme, with different molecular weights, were detected in different cell

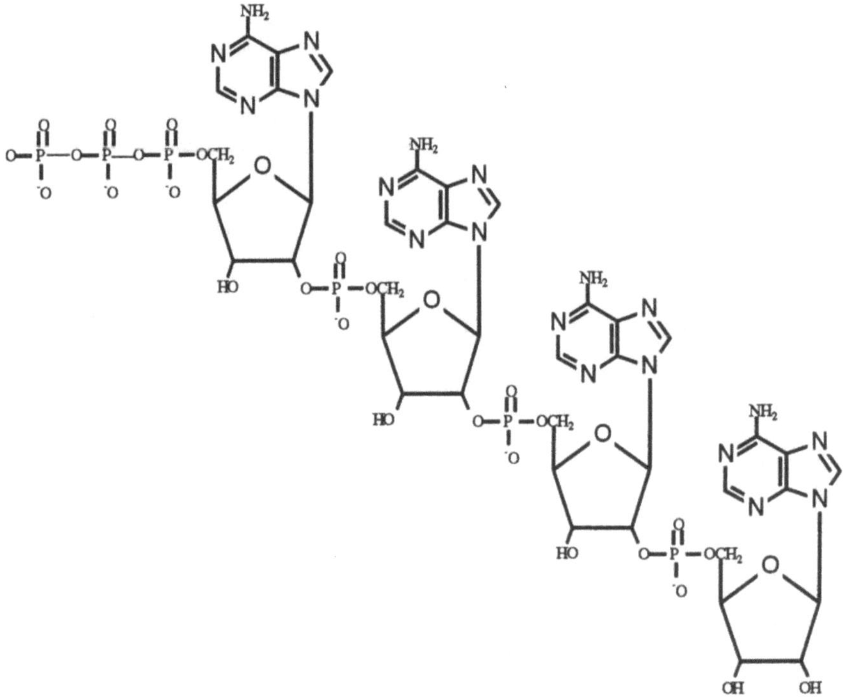

Fig. 1. Structure of the 2-5A tetramer 5′-triphosphate

lines, and the 2-5A synthetases are now considered to be a family of proteins. To date, it is the 40-, 46-, 69- and 100-kDa forms, from human cell lines, that have been best studied. These four forms of 2-5A synthetase are differentially induced by IFN in a given human cell line. In Daudi cells and HeLa cells, for example, the 46-, 69-, and 100-kDa forms are induced, whereas the human amniotic U cells (UAC) express mainly the 100-kDa form (Hovanessian et al. 1987).

The p40 and p46 proteins result from alternative splicing between the 5th and 6th exon of a single 13.5-kb gene localized on chromosome 12 (Merlin et al. 1983; Williams et al. 1986). These two proteins differ only by their C-terminal extremity; p40 has a hydrophobic C-terminal region, while that of p46 is hydrophilic and contains two acidic regions (Benech et al. 1985). These two forms are activated by similar concentrations of dsRNA (100 μg/ml), but they display optimal activity at different pH: pH 5.5 for p46 and pH 6.5 for p40 (Mory et al. 1989).

The human p69 was cloned in 1992 (Marié and Hovanessian 1992). The first 683 amino acids of this p69 synthetase present some homology (50%) with the first 346 amino acids of p40 and p46. p69 is fully active at pH 6.5 at a concentration of 100 μg/ml of dsRNA. p69 exists as a dimer.

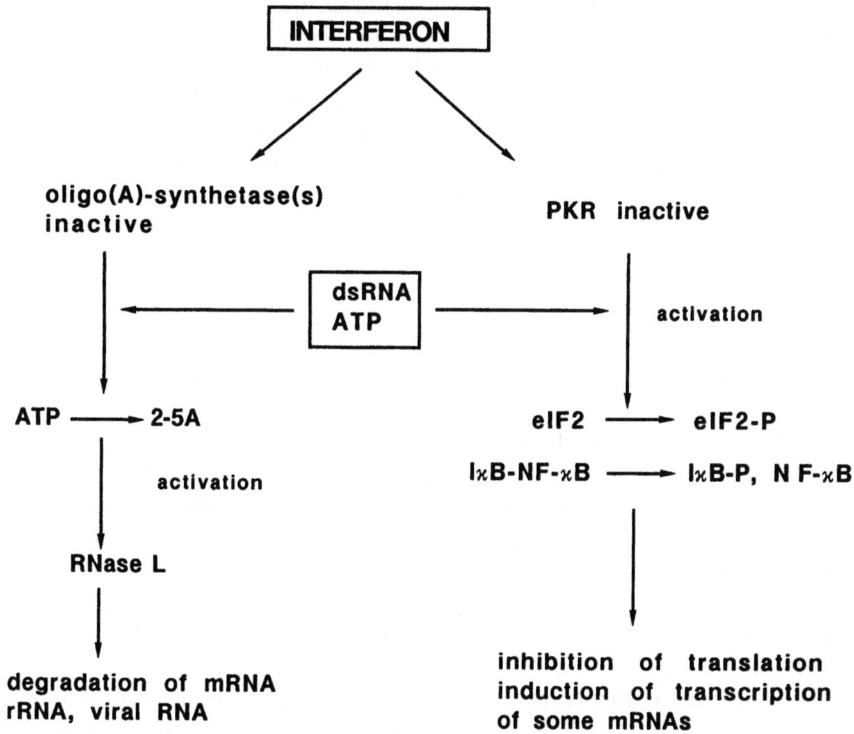

Fig. 2. The two independent pathways induced by IFN and activated by dsRNA

p100 differs somewhat from the three other 2-5A synthetases. Maximal activity is observed at only 5 μg/ml dsRNA and at pH 7.5. In addition, whereas all four forms can polymerize all the different lengths of 2-5A, the principle product of p100 activity is dimeric 2-5A, whereas the three other forms produce mainly tetrameric 2-5A (Hovanessian et al. 1988). As dimeric 2-5A is unable to activate RNase L, the exact role of the p100 enzyme remains unclear.

The subcellular localizations of the various isoforms of 2-5A synthetase are different. p100 is cytoplasmic and associated with the ribosomal fraction; p69 and p46–p40 are also essentially cytoplasmic and associated with the endocellular membranes. They are found associated with the mitochondrial pellet, with polysomes and the endoplasmic reticulum (Chebath et al. 1987a; Hovanessian et al. 1987).

The role of 2-5A synthetase in the antiviral action of IFN against picornavirus has been demonstrated by expression of the cDNAs for the human 40-kDa form in different cell types (Chebath et al. 1897b). The replication of EMCV (encephalomyocarditis virus) or Mengo virus was greatly affected in these cells, whereas the replication of VSV (vesicular stomatitis virus) was not affected. Moreover, 2-5A synthetase is immunoprecipitated complexed with EMCV RNA in infected cells (Gribaudo et al. 1991).

The differences observed in the localization, induction, and conditions of activation of the various isoforms of 2-5A synthetase must allow their independent regulation in specific physiological situations, and suggest that they have different physiological functions. For example, it has been suggested that p100 could play a role in the splicing of pre-mRNA. This enzyme has been detected in the spliceosome complex and in vitro RNA splicing activity was inhibited by treating this complex with antibody to 2-5A synthetase (Sperling et al. 1991).

3.2 2-5A

The only known function of 2-5A is to bind and to activate RNase L.

This oligoadenylate is very unstable, having a half-life of only 5–10 min in vitro (Bayard et al. 1984). It is dephosphorylated by general phosphatases at its 5′ end, leaving the so-called core oligonucleotide that is unable to activate RNase L. Moreover, the core has been reported to inhibit cellular DNA synthesis. This antimitogenic activity does not seem to involve an activation of the RNase L pathway (Eppstein et al. 1982).

2-5A can also be degraded by phosphodiesterases. A 2′, 5′ phosphodiesterase activity has been described in various cell lines (Schmidt et al. 1979; Johnston and Hearl 1987).

The chemical and/or enzymatic synthesis of several analogues of 2-5A with increased metabolic stability has allowed the determination of the regions of the oligomer which interact with RNase L and that are absolutely necessary for the activation of RNase L (Torrence et al. 1994). Only oligomers with more than two adenylates are able to activate RNase L (Silverman 1985). The 5′-di- or tri-phosphate, the 2′,5′-phosphodiester bonds and the 3′- hydroxyl groups of the second adenosine (from the 5′-terminus) are critical for its biological activity. The 5′-terminal adenine base of 2-5A is vital for binding with RNase L, the adenine of the 2′ terminal adenosine is absolutely critical for activation of the nuclease function of the enzyme, but not for its binding.

4 RNase L

RNase L is the only known target of 2-5A and of this cascade of activation.

As cited above, variations in 2-5A or 2-5A synthetases have been observed in various conditions and clearly documented; but, as far as RNase L is concerned, published data are more contradictory. Some studies suggest that IFN treatment or cell growth status increases RNase L activity, whereas others report no alteration in RNase L activity under these circumstances (Sen and Lengyel 1992). These apparent contradictions might be explained by differences in the sensitivity of the methods used to detect RNase L.

4.1 Detection of RNase L

RNase L can be detected (1) by Western blotting with polyclonal or monoclonal antibodies (Salehzada et al. 1991a; Salehzada et al. 1993); (2) by its high affinity for its activator: 2-5A. 2-5A can be radiolabelled by ligating a cytidine 3′, 5′ [5′-^{32}P] biphosphate to its 3′ terminus with T4 RNA ligase, yielding the so-called 2-5ApCp. 2-5ApCp can be used directly in radio-binding assays involving cell extracts or by a covalent labelling procedure (Knight et al. 1980). Alternatively, RNase L may be separated by SDS/PAGE and renatured on nitrocellulose prior to the binding of 2-5ApCp (Salehzada et al. 1991b); (3) by its specific 2-5A-dependent nuclease activity on mRNAs (Nilsen et al. 1981), rRNAs (Silverman et al. 1984) or synthetic poly(uridylic acid) (Silverman 1985).

The two latter techniques all depend on the binding to, and activation of RNase L by exogenous 2-5A. Variations of intracellular 2-5A levels, however, which compete with exogenous 2-5A for the binding to RNase L and its activation could thus interfere with its detection. Changes in intracellular 2-5A concentration have indeed been observed during viral infection, cell growth and cell differentiation in the absence of exogenous IFN treatment (Sen and Lengyel 1992). In addition, inhibitors of RNase L have been documented, for instance during EMCV infection (Cayley et al. 1982). We have cloned and characterized a cellular RNase L inhibitor: RLI (Bisbal et al. 1995). Taken together, these factors could lead to an underestimation of RNase L level by techniques relying upon the estimation of activity in cell extracts as compared to methods allowing the direct detection of RNase L in cell or tissue extracts, such as Western blot with antibodies or 2-5A, that are independent of intracellular 2-5A concentration and of the presence of RNase L inhibitors.

4.2 Subcellular Localization of RNase L

Polyclonal antibodies developed in rabbit against mouse spleen RNase L (Salehzada et al. 1991a) have been used to localize RNase L in intact cells. RNase L has a cytoplasmic distribution, with highest concentrations being observed in the perinuclear region. This perhaps represents a preferential localization in the rough endoplasmic reticulum, especially associated with polysomes. Subcellular fractionation of cell extracts confirms immunofluorescence data and allows RNase L to be localized more accurately to the rough microsomal pellet. About half of the RNase L is only weakly associated with the polysomes, as shown by its removal by high salt washing.

RNase L thus appears to have a subcellular distribution similar to the 69- and 100-kDa forms of 2-5A synthetase. The association of the 100-kDa form of 2-5A synthetase to the rough microsomes is, however, weaker than the association of the 69-kDa form. This co-localization potentially allows for the localized activation of RNase L, thus conferring some degree of specificity of RNase L, which otherwise shows only limited sequence specificity.

4.3 Structure of RNase L

RNase L from Ehrlich ascites tumour cells was first described as a 185-kDa protein when cell extracts with high protein concentration were analyzed by gel filtration (Sephadex G-200; Slattery et al. 1979); but when cell extracts were analyzed with lower protein concentration, the molecular weight of RNase L was found to be 77–90 kDa (Floyd-Smith et al. 1982). Similar results are obtained when the extracts are separated on a glycerol gradient (Floyd-Smith et al. 1982). In all cases, SDS/PAGE analysis gave one band between 70 and 90 kDa.

The presence of the high molecular weight form of RNase L (185-kDa) is independent of the presence of 2-5A, and does not seem to be modified during activation by 2-5A. The removal of 2-5A from the activated enzyme by gel filtration causes a drop in activity to preactivation levels. In addition, enzyme from which the 2-5A was separated by this technique can be reactivated by further addition of 2-5A to the same level as the endonuclease activated for the first time (Slattery et al. 1979).

We have characterized, in different mouse tissues, another protein which binds 2-5A and cleaves poly(U), with a molecular mass of 40 kDa (Bisbal et al. 1989).

Monoclonal antibodies have been produced against human RNase L. Interestingly, one hybridoma clone (mAb3) neutralizes RNase L activity. This antibody specifically recognizes an 80-kDa polypeptide which appears to be associated with, but distinct from, the 2-5A-binding polypeptide previously referred to as RNase L. This 80-kDa protein binds poly(uridylic acid) and has thus been termed RNA binding protein (RNABP). RNABP is immunoprecipitated with the 2-5A-binding polypeptide, but these two polypeptides can be clearly separated by two-dimensional electrophoresis. RNase activity was found to be associated with RNABP. It was thus proposed that the high molecular weight RNase L is a complex of at least these two polypeptides, in which the ribonuclease activity of the polypeptide recognized by mAb3 (RNABP) is regulated by the binding of 2-5A to the other polypeptide (2-5ABP; Salehzada et al. 1993).

At the same time, however, Zhou et al. (1993) cloned an 80-kDa polypeptide (2DR) which both binds 2-5A and cleaves poly(uridilic acid). Purified recombinant 2DR, expressed by a baculovirus system, has been shown to dimerize in the presence of 2-5A, but not in its absence (Dong and Silverman 1995).

This result appears to contradict previous results, cited above. Moreover, 2DR is immunoprecipitated from total cell extracts in association with RNABP by mAb3 (Bisbal et al. 1995). The two 80-kDa polypeptides copurify in a 200-kDa complex by gel filtration of a cell extract. The role of RNABP still remains to be determined. The RNAse activity that we have found to be associated with this protein is perhaps responsible for the 2-5A-independent activity reported by a number of authors for RNase L purified from cell extracts.

A major obstacle in resolving these apparent contradictions is that the subunit structure of the native enzyme has not been clearly described. Studies are still needed to establish whether RNase L occurs in vivo as a monomer, a multimer or bound to other proteins. The results obtained with cell extracts support the concept of a heterodimeric proteins complex. These results, however, do not necessarily exclude these obtained in vitro with the cloned protein purified to homogeneity. Indeed, the fact that purified 2DR dimerizes in vitro does not necessarily reflect its behaviour in situ when other proteins are present.

4.4 RNase L Activity

RNase L is an endoribonuclease that cleaves single-stranded RNA. This nuclease has only a low sequence specificity. Studies of its activity have involved synthetic RNA (polyuridylate, polyadenylate, polycytidylate, polyguanylate) or natural RNA : R5 (bacteriophage T7-specific RNA) or the 3′ terminal region of bacteriophage R17 RNA for which some details of the secondary structure have been established (Floyd-Smith et al. 1981). Activated ribonuclease L can efficiently cleave single-stranded RNA after at least three dinucleotides (UA, UG and UU). The cleavage products have a phosphate group at their 3′ terminus and a hydroxyl residue at their 5′ terminus. For poly(U), a 2-5A-independent cleavage is observed with the purified enzyme. This cleavage could be due to the native RNase L or to either a form of the enzyme that has lost its absolute dependence on 2-5A or by an unknown contaminating nuclease capable of degrading poly(U) but none of the other three homopolyribonucleotides. This 2-5A-independent activity was observed by other groups with purified RNase L or when the cloned enzyme is expressed in a rabbit reticulocyte extract (Salehzada et al. 1993; Bayard et al. 1994; Bisbal et al. 1995).

The specificity of RNase L has also been studied in different cell extracts (mouse Krebs ascites tumour cells, human lymphoblastoid Daudi cells, rabbit reticulocyte lysate) on different natural RNAs (La Crosse sRNA and influenza RNA 7; Wreschner et al. 1981). The enzyme from these different cell types specifically cleaves the RNAs at UpN sequences, yielding 3′-phosphorylated oligonucleotides which terminate in the sequence UpNp. The preference for sites in U-rich regions, however, was not as apparent with the enzymes from the mouse and human cells as it was for the rabbit enzyme, suggesting that although the specificities of these different RNase L molecules are similar, they are not totally identical. This specificity for poly(U) has also been demonstrated for the cloned 2DR (Dong et al. 1994).

For natural RNAs, cleavage was not observed at all UA or UU sequences. It is unknown why this should be so, but it is likely that the specificity is also influenced by the secondary structure of the substrate or by the presence of RNA-binding proteins.

The low specificity of RNase L is a limiting factor in the use of 2-5A as an antiviral agent. Indeed, when 2-5A is introduced into intact cells, it causes antiviral activity but with a nonspecific, global degradation of mRNA and rRNA (Bayard et al. 1984; Bisbal et al. 1987).

In order to increase the specificity of the RNase L, 2-5A has been covalently linked to antisense oligodeoxyribonucleotides (Maitra et al. 1995). RNase L becomes adapted for the selective and specific cleavage of a targeted RNA. PKR (dsRNA-dependent protein kinase) mRNA was ablated from intact HeLa cells after culturing in media with 2 μM 2-5A-antisense. This ablation of PKR activity inhibited induction by double-stranded RNA signalling of transcription factor NF-κB (Maran et al. 1994).

Another limiting factor in the use of RNase L or 2-5A as therapeutic agents is the presence of inhibitors of the system, some of which are induced by viruses. One such inhibitor, RLI (RNase L inhibitor) has now been identified and cloned.

5 RLI

RLI was isolated from an expression library by binding to 2-5ApCp (Bisbal et al. 1995). In vitro translation of this cDNA gives rise to a protein of 68 kDa which associates specifically with cellular RNase L and RNABP, as shown by immunoprecipitation with the mAb3 antibody. RLI mRNA is not regulated by IFN.

This in vitro-translated 68-kDa protein can also form a complex with in vitro-translated 2DR.

The in vitro expression of RLI inhibits the binding of 2-5A by endogenous RNase L or by coexpressed 2DR, and, as a consequence, their 2-5A-dependent activation. This suggests that RLI is a regulatory protein. RLI does not promote 2-5A degradation or stable modification of RNase L, if RLI and RNase L are separated, RNase L regains its capacity to bind 2-5A. The inhibition of RNase L activity by RLI therefore seems to require an association between the two proteins (as tentatively illustrated in Fig. 3). We do not know yet if this association is direct or not.

The overexpression of the RLI cDNA in HeLa cells results in the inhibition of the IFN-activated 2-5A pathway.

As stated above, it is known that RNase L is inhibited by some viruses, and in particular by EMCV. RLI mRNA level is increased during EMCV infection. The transfection of HeLa cells by RLI antisense cDNA inhibits the induction of RLI mRNA by EMCV and reverses the inhibition of RNase L (Martinand et al. 1994). RLI must, therefore, be the inhibitor of the 2-5A/RNase L

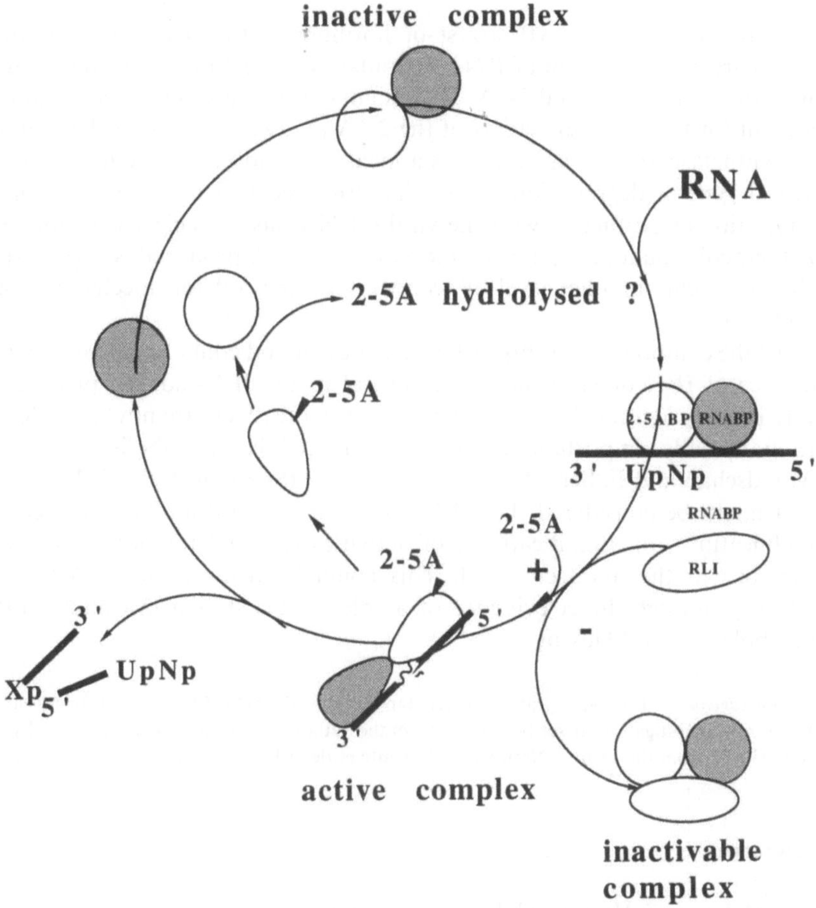

Fig. 3. Hypothetical scheme of RNase L activation by 2-5A and its inhibition by RLI

pathway induced by EMCV. It seems that the induction of RLI mRNA by EMCV is due to the double-stranded part of the viral RNA. RLI is, in fact, induced by poly(I)poly(C) (Martinand et al. 1995). Further investigations are underway to determine whether RLI is implicated in other circumstances in which RNase L is known to be inhibited.

6 Conclusion

Hassel et al. (1993) have confirmed the central role played by RNase L and the 2-5A system in the antiviral and the antiproliferative effects of IFNs, and they have postulated its implication in the control of mRNA stability. 2-5A and RNase L levels are known to vary in various physiological conditions, such as

IFN treatment, cell growth arrest or hormone status. The 2-5A system is a well-characterized system of RNA degradation thought to be regulated mainly by 2-5A synthetase(s) and 2-5A. RLI might constitute an important additional element for the downregulation of the 2-5A pathway. The colocalization of 2-5A synthetase and RNase L allows a local activation of the system to avoid a critical general degradation of cellular mRNAs. During viral infection, the association of synthetase with the viral mRNA has been clearly demonstrated by their coimmunoprecipitation; moreover the 2-5A produced is very unstable. This represents another level of security to switch off this mechanism of degradation.

All the components of this system are present and could be active even in the absence of IFN or viral infection. Several of the IFN-induced proteins that were initially identified in the context of antiviral effects are now considered to be potentially important participants in cellular metabolism like PKR (Mundschau and Faller 1995) or synthetase(s) (Sperling et al. 1991).

It might be considered that IFN-regulated antiviral mechanisms represent modifications, or amplifications, of activities required for normal metabolic functions in the uninfected cell. This could be true of the 2-5A/RNase L pathway, though direct evidence of a role of this system in cellular mRNA metabolism is still lacking.

Acknowledgements. I am very grateful to Dr. Ian Robbins (IGMM, Montpellier) for revising this manuscript. Investigations described herein from the author's laboratory were supported by Grant CRE 930102 from the Institut National de la Santé et de la Recherche Médicale.

References

Aguet M, Dembic Z, Merlin G (1988) Molecular cloning and expression of the human interferon γ receptor. Cell 55:273–280

Baglioni C, Minks MA, Maroney PA (1978) Interferon action may be mediated by activation of a nuclease by pppA2′p5A2′p5′A. Nature 273:784–787

Bayard B, Bisbal C, Silhol M, Cnockaert J, Huez G, Lebleu B (1984) Increased stability and antiviral activity of 2′-O-phosphoglyceryl derivatives of (2′–5′)oligo(adenylate). Eur J Biochem 142:291–298

Bayard B, Bette-Bobillo P, Aliau S (1994) Affinity purification and characterization of (2′–5′)oligo(adanylate)-dependent RNase from mouse spleen. Eur J Biochem 223:403–410

Benech P, Mory Y, Revel M, Chebath J (1985) Structure of two forms of the interferon-induced (2′–5′) oligo A synthetase of human cells based on cDNAs and gene sequences. EMBO J 4:2249–2256

Bisbal C, Silhol M, Lemaitre M, Bayard B, Salehzada T, Lebleu B (1987) 5′-modified agonist and antagonist $(2'\text{-}5')(A)_n$ analogues: synthesis and biological activity. Biochemistry 26:5172–5178

Bisbal C, Salehzada T, Lebleu B, Bayard B (1989) Characterization of two murine (2′–5′)(A)n-dependent endonucleases of different molecular mass. Eur J Biochem 179:595–602

Bisbal C, Martinand C, Silhol M, Lebleu B, Salehzada T (1995) Cloning and characterization of a RNase L inhibitor. J Biol Chem 270:13308–13317

Brown GE, Lebleu B, Kawakita M, Shaila S, Sen GC, Lengyel P (1976) Increased endonuclease activity in an extract from mouse Ehrlich ascites tumor cells which had been treated with a partially purified interferon preparation: dependence on double-stranded RNA. Biochem Biophys Res Commun 69:114–122

Cayley P, Knight M, Kerr IM (1982) Virus-mediated inhibition of the ppp(A2′p)nA system and its prevention by interferon. Biochem Biophys Res Commun 104:376–382

Chebath J, Benech P, Hovanessian AG, Galabru J, Revel M (1987a) Four different forms of interferon-induced 2′,5′ oligo(A)synthetase identified by immunoblotting in human cells. J Biol Chem 262:3852–3857

Chebath J, Benech P, Revel M, Vigneron M (1987b) Constitutive expression of (2′–5′)oligo A synthetase confers resistance to picornavirus infection. Nature 330:587–588

Clemens MJ, Williams BRG (1978) Inhibition of cell-free protein synthesis by pppA2′p5′A2′p5′A a novel oligonucleotide synthesized by interferon-treated L cell extracts. Cell 13:565–572

Decker T, Lew D, Mirkovitch J, Darnell JE jr (1991) Cytoplasmic activation of GAF, an IFN-γ-regulated DNA binding factor. EMBO J 10:927–932

Dong B, Silverman R (1995) 2-5A-dependent RNase molecules dimerize during activation by 2-5A. J Biol Chem 270:4133–4137

Dong B, Xu L, Hassel B, Lee X, Torrence P, Silverman R (1994) Intrinsic molecular activities of the interferon-induced 2-5A-dependent RNase. J Biol Chem 269:14153–14158

Einat M, Resnitzki D, Kimchi A (1985) Close link between reduction of c-myc expression by interferon and G0/G1 arrest. Nature 313:597–600

Eppstein DA, Marsh VY, Schryver BB, Larsem MA, Barnett JW, Verheyden JPH, Prisbe E (1982) Analogs of (A2′p)nA. Correlation of structure of analogs of ppp(A2′p)2A and (A2′p)2A with stability and biological activity. J Biol Chem 257:13390–13397

Floyd-Smith G, Slattery E, Lengyel P (1981) Interferon action: RNA cleavage pattern of a (2′–5′)oligoadenylate-dependent endonuclease. Science 212:1030–1032

Floyd-Smith G, Yoshie O, Lengyel P (1982) Covalent linkage of (2′–5′)pppApApA(^{32}P)pCp to (2′–5′)(A)n-dependent ribonuclease in cell extracts by ultraviolet irradiation. J Biol Chem 257:8584–8587

Fu XY, Schindler C, Improta T, Aebersold R, Darnell JE jr (1992) The proteins of ISGF-3, the interferon α-induced transcriptional activator, define a gene family involved in signal transduction. Proc Natl Acad Sci USA 89:7840–7843

Gray PW, Leung DW, Pennica D, Yelverton E, Najaria P, Simonsen CC, Derynnck D, Sherwood PJ, Wallace DM, Berger SL, Levinson AD, Goeddel DV (1982) Expression of human immune interferon cDNA in *E. coli* and monkey cells. Nature 295:503–508

Gribaudo G, Lembo D, Cavallo G, Landolfo S, Lengyel P (1991) Interferon action: binding of viral RNA to the 40-kilodalton 2′–5′-oligoadenylate synthetase in interferon-treated HeLa cells infected with encephalomyocarditis virus. J Virol 65:1748–1757

Harada H, Fujita T, Miyamoto M, Kimura Y, Maruyama M, Furia A, Miyata T, Taniguchi T (1989) Structurally similar but functionally distinct factors, IRF-1 and IRF-2, bind to the same regulatory elements of IFN and IFN-inducible genes. Cell 58:729–739

Hassel B, Zhou A, Sotomayor C, Maran A, Silverman R (1993) A dominant negative mutant of 2-5A-dependent RNase suppresses antiproliferative and antiviral effects of interferon EMBO J 12:3297–3304

Hovanessian AG, Kerr IM (1978) Synthesis of an oligonucleotide inhibitor of protein synthesis in rabbit reticulocyte lysates analogous to that formed in extracts from interferon-treated cells. Eur J Biochem 84:149–159

Hovanessian AG, Brown RE, Kerr IM (1977) Synthesis of low molecular weight inhibitor of protein synthesis with enzyme from interferon-treated cells. Nature 268:537–540

Hovanessian AG, Laurent AG, Chebath J, Galabru J, Robert N, Svab J (1987) Identification of 69 kd and 100 kd forms of 2-5A synthetase in interferon-treated human cells by specific monoclonal antibodies. EMBO J 6:1273–1280

Hovanessian G, Svab J, Marié I, Robert N, Chamaret S, Laurent AG (1988) Characterization of 69–100-kDa forms of 2-5A synthetase from interferon-treated human cells. J Biol Chem 263:4945–4949

Isaacs A, Lindenmann J (1957) Virus interference. I. The interferon. Proc R Soc Lond Biol Sci 147:258–267
Jackson RJ (1993) Cytoplasmic regulation of mRNA function: the importance of the 3′ untranslated region. Cell 74:9–14
Jaramillo M, Abraham N, Bell JC (1995) The interferon system : a review with emphasis on the role of PKR in growth control. Cancer Invest 13(3):327–338
Johnston MI, Hearl WG (1987) Characterization of a 2′-phosphodiesterase from bovine spleen. J Biol Chem 262:8377–8382
Jonak GJ, Knight E jr (1984) Selective reduction of c-myc mRNA in Daudi cells by human β interferon. Proc Natl Acad Sci USA 81:1747–1750
Kerr IM, Brown RE (1978) pppA2′p5′A2′p5′A an inhibitor of protein synthesis synthesized with an enzyme fraction from interferon-treated cells. Proc Natl Acad Sci USA 75:256–260
Kerr IM, Brown RE, Ball LA (1974a) Increased sensitivity of cell-free protein synthesis to double-stranded RNA after interferon treatment. Nature 250:57–59
Kerr IM, Friedman RE, Brown RE, Ball A, Brown JC (1974b) Inhibition of protein synthesis in cell-free systems from interferon-treated infected cells: furter characterization and effect of formylmethionyl-t-RNA. J Virol 13:9–21
Kerr IM, Brown RE, Clemens MJ, Gilbert CS (1976) Interferon-mediated inhibition of cell-free-protein synthesis in response to double-stranded RNA. Eur J Biochem 69:551–561
Knight M, Cayley PJ, Silverman RH, Wreschner DH, Gilbert CS, Brown RE, Kerr IM (1980) Radioimmune, radiobinding and HPLC analysis of 2-5A and related oligonucleotides from intact cells. Nature 288:189–192
Kumar A, Haque J, Lacoste J, Hiscott J, Williams BRG (1994) Double-stranded RNA-dependent protein kinase activates transcription factor NF-κB by phosphorylating IκB. Proc Natl Acad Sci USA 91:6288-6292
Lin SL, Garber EA, Wang E, Galiguiri LA, Schellekens H, Goldberg AR, Tamm I (1983) Reduced synthesis of pp60 Src and expression of the transformation-related phenotypes in interferon-treated Rous sarcoma virus-transformed rat cells. Mol Cell Biol 3:1656–1664
Maitra RK, Li G, Xiao W, Dong B, Torrence P, Silverman R (1995) Catalytic cleavage of an RNA target by 2-5A antisense and RNase L. J Biol Chem 270:15071–15075
Maran A, Maitra R, Kumar A, Dong B, Xia W, Li G, William BRG, Torrence P, Silverman R (1994) Blockage of NF-κB signaling by selective ablation of an mRNA target by 2-5A antisens chimeras. Science 265:789–792
Marié I, Hovanessian AG (1992) The 69-kDa 2-5A synthetase is composed of two homologous and adjacent functional domains. J Biol Chem 267:9933–9939
Martinand C, Bisbal C, Silhol M, Lebleu B. Salehzada T (1994) Constitutive expression of sense and antisense constructions of RLI in human cells modify the RNase L pathway. J Interferon Res 14(Suppl 1):102
Martinand C, Salehzada T, Silhol M, Lebleu B, Bisbal C (1995) The RNase L inhibitor (RLI) is induced by poly(I)poly(C). J Interferon Res 15(Suppl 1):148
Merlin G, Chebath J, Benech P, Metz, Revel M (1983) Molecular cloning and sequence of partial cDNA for interferon-induced (2′5′)oligo(A)synthetase mRNA from human cells. Proc Natl Acad Sci USA 80:4904–4908
Meurs E, Hovanessian AG (1988) α-Interferon inhibits the expression of heavy μ messager RNA in Daudi cells. EMBO J 7:1689–1696
Meurs E, Chong K, Galabru J, Thomas N, Kerr I, Williams BRG, Hovanessian AG (1990) Molecular cloning and characterization of the human double-stranded RNA-activated protein kinase induced by interferon. Cell 62:379–390
Miyamoto M, Fujita T, Kimura Y, Maruyama M, Harada H, Sudo Y, Miyata T, Taniguchi T (1988) Regulated expression of a gene encoding a nuclear factor, IRF-1, that specifically binds to IFN-β gene regulatory elements. Cell 54:903–913
Mory Y, Vaks B, Chebath J (1989) Production of two human 2′–5′-oligoadenylate synthetase enzymes in *Escherichia coli.* J Interferon Res 9:295–304

Mundschau L, Faller D (1995) Platelet-derived growth factor signal transduction through the interferon-inducible kinase PKR. J Biol Chem 270:3100–3106

Nagata S, Mantei N, Weissmann C (1980) The structure of one of the eight or more chromosomal genes for human interferon α. Nature 28:401–408

Naylor SL, Sakagushi AY, Shows TB, Law ML, Goeddel DV, Gray PW (1983) Human immune interferon gene is located on chromosome 12. J Exp Med 157:1020–1027

Nilsen TW, Maroney PA, Baglioni C (1981) Double-stranded RNA causes synthesis of 2′–5′-oligo(A) and degradation of messenger RNA in interferon-treated cells. J Biol Chem 256:7806–7811

Owerbach D, Rutter WJ, Shows TB, Gray P, Goeddel DV, Lawn RM (1981) Leucocyte and fibroblast interferon genes are located on human chromosome 9. Proc Natl Acad Sci USA 78:3123–3127

Paucker K, Cantell K, Henle W (1962) Quantitative studies on viral interference in suspended L cells. III. Effect of interfering viruses and interferon on the growth rate of cells. Virology 17:324–334

Pellegrini S, Schindler C (1993) Early events in signalling by interferons. Trends Biochem Sci 18:338–342

Proud CG (1995) PKR: a new name and new roles TIBS 20:241–246

Ratner L, Wiegand RC, Farrell PJ, Sen GC, Cabrer B, Lengyel P (1978) Interferon, double stranded RNA and RNA degradation. Fractionation of the endonuclease TNT system into two macromolecular components: role of a small molecule in nuclease activation. Biochem Biophys Res Commun 81:947–954

Rosa F, Cochet M, Fellous M (1986) Interferon and major histocompatibility complex genes: a model to analyse eucaryotic gene regulation ? Interferon 7:47–87, Harcourt Brace Jovanovich, Academic Press, London

Sach B (1993) Messenger RNA degradation in eucaryotes. Cell 74:413–421

Salehzada T, Silhol M, Lebleu B, Bisbal C (1991a) Polyclonal antibodies against RNase L. Subcellular localization of this enzyme in mouse cells. J Biol Chem 266:5808–5813

Salehzada T, Silhol M, Lebleu B, Bisbal C (1991b) Regeneration of enzyme activity after Western blot: activation of RNase L by 2-5A on filter. Importance for its detection. Anal Biochem 196:410–414

Salehzada T, Silhol M, Steff AM, Lebleu B, Bisbal C (1993) 2′,5′-oligoadenylate-dependent RNase L is a dimer of regulatory and catalytic subunits. J Biol Chem 268:7733–7740

Schindler C, Darnell JE (1995) Transcriptional responses to polypeptide ligands: the Jak-STAT pathway. Annu Rev Biochem 64:621–651

Schindler C, Fu XY, Improta T, Aebersold R, Darnell JE Jr (1992) Proteins of transcription factor ISGF-3 : one gene encodes the 91- and 84-kDa ISGF-3 proteins that are activated by interferon α. Proc Natl Acad Sci USA 89:7836–7839

Schmidt A, Zilberstein A, Shulman L, Federman P, Berissi H, Revel M (1978) Interferon action: isolation of nuclease F, a translation inhibitor activated by interferon-induced (2′–5′)oligoisoadenylate. Febs Lett 95:257–264

Schmidt A, Chernajovsky Y, Shulman L, Federman P, Berissi H, Revel M (1979) An interferon-induced phosphodiesterase degrading (2′,5′) oligoisoadenylate and C-C-A terminus of tRNA. Proc Natl Acad Sci USA 76:4788–4792

Sen GC, Lengyel P (1992) The interferon system, a bird's eye view of its biochemistry. J Biol Chem 267:5017–5020

Sen GC, Lebleu B, Brown GE, Kawakita M, Slattery E, Lengyel P (1976) Interferon, double-stranded RNA and mRNA degradation. Nature 264:370–373

Silverman RH (1985) Functional analysis of 2-5A-dependent RNase and 2-5A using 2′5′ oligoadenylate-cellulose. Anal Biochem 144:450–460

Silverman R, Skehel J, James T, Wreschner D, Kerr IM (1984) rRNA cleavage as an index of ppp(A2′p)nA activity in interferon-treated encephalomyocarditis virus-infected cells. J Virol 46:1051–1055

Slattery E, Ghosh N, Samanta H, Lengyel P (1979) Interferon, double-stranded RNA and RNA degradation: activation of an endonuclease by (2′–5′)A_n. Proc Natl Acad Sci USA 76: 4778–4782

Sperling J, Chebath J, Arad-Dann H, Offen D, Spann P, Lehrer R, Goldblatt D, Jolles B, Sperling R (1991) Possible involvement of (2′–5′)oligoadenylate synthetase activity in pre-mRNA splicing. Proc Natl Acad Sci USA 88:10377–10381

Staeheli P, Haller O, Boll W, Lindenmann J, Weissman C (1986) Mx protein: constitutive expression in 3T3 cells transformed with cloned Mx cDNA confers selective resistance to influenza virus. Cell 44:147–158

Taniguchi T, Mantei N, Swarztein M, Nagata S, Muramatsu M, Weissmann C (1980) Human leukocyte and fibroblast interferons are structurally related. Nature 285:547–549

Tavernier J, Derynck R, Fiers W (1981) Evidence for a unique human fibroblast interferon (IFN-β1) chromosomal gene, devoid of intervening sequences. Nucl Acids Res 9:461–471

Torrence PF, Xiao W, Li G, Khamnei S (1994) Development of 2′,5′-oligonucleotides as potential therapeutic agents. Curr Med Chem 1:176–191

Trent YM, Olson S, Lawr RM (1982) Chromosomal localization of human leucocyte, fibroblaste and immune interferon genes by means of in situ hybridization. Proc Natl Acad Sci USA 79:7809–7813

Uzé G, Lutfalla G, Gresser I (1990) Genetic transfer of functional human interferon α receptor into mouse cells: cloning and expression of its cDNA. Cell 60:225–234

Uzé G, Lutfalla G, Mogensen K (1995) α and β interferons and their receptor and their friends and relations. J Interferon Res 15:3–26

Williams BRG (1991) Transcriptional regulation of interferon-stimulated genes. Eur J Biochem 200:1–11

Williams BRG, Saunders ME, Williard HF (1986) Interferon-regulated human 2′–5′ synthetase gene maps to chromosome 12. Somat Cell Mol Genet 12:403–408

Wreschner DH, McCauley JW, Skehel JJ, Kerr IM (1981) Interferon action-sequence specificity of the ppp(A2′p)nA-dependent ribonuclease. Nature 289:414–417

Zhou A, Hassel B, Silverman R (1993) Expression cloning of 2-5A-dependent RNase: a uniquely regulated mediator of interferon action. Cell 72:753–765

Zilberstein A, Kimchi A, Schmidt A, Revel M (1978) Isolation of two interferon-induced translational inhibitors: a protein kinase and an oligo-isoadenylate synthetase. Proc Natl Acad Sci USA 75:4734–4738

3′ Untranslated Regions of *c-myc* and *c-fos* mRNAs: Multifunctional Elements Regulating mRNA Translation, Degradation and Subcellular Localization

J.L. Veyrune[1], J. Hesketh[2] and J.M. Blanchard[1]

1 *c-fos* and *c-myc* mRNA Degradation and Translation

1.1 3′UTR and mRNA Decay

c-fos and *c-myc* proto-oncogenes code for very unstable mRNAs (Dani et al. 1984, 1985; Greenberg and Ziff 1984; Kruijer et al. 1984; Muller et al. 1984; Treisman 1985; Bonnieu et al. 1989; Rahmsdorf et al. 1987; Piechaczyk and Blanchard 1994). Many experiments have suggested that the information required for destabilization may lie within multiple discrete regions of these mRNAs, including both coding (Kabnick and Housman 1988; Shyu et al. 1989, 1991; Laird et al. 1991; Wisdom and Lee 1990, 1991; Bernstein et al. 1992; Herrick and Ross 1994) and 3′ noncoding (Treisman 1985; Fort et al. 1987; Rahmsdorf et al. 1987; Wilson and Treisman 1988; Shyu et al. 1989; Jones and Cole 1987; Bonnieu et al. 1988, 1990; Brewer and Ross 1988; Brewer 1991; Aghib et al. 1990; Herrick and Ross 1994) portions of the molecule (Fig. 1).

The role of the 3′ untranslated region (3′UTR) has been extensively studied, and *c-fos* 3′UTR has proven to be an excellent paradigm for such work. Indeed, *c-fos* 3′UTR destabilizes to a large extent the otherwise stable *β*-globin mRNA when appended to it in a 3′ position, and A/U-rich elements (ARE) present within this region have been identified to be important for mRNA decay (Fort et al. 1987; Shyu et al. 1989, 1991; Wilson and Treisman 1988; Rahmsdorf et al. 1987; Bonnieu et al. 1989). Such sequences have been initially identified as shared by a set of mRNAs induced during the acute phase of inflammation (Caput et al. 1986). The pentamer AUUUA was found frequently in multiple reiterations in the 3′UTR of a wide variety of unstable mRNAs coding for cytokines (GM-CSF, interleukins, IFNs, TNF, etc.) or proto-oncogenic transcription factors (*c-jun, c-fos, c-myc,* etc.). Interestingly, ARE have also been shown to participate in the fast removal of poly(A) (Fort et al. 1987; Wilson and Treisman 1988; Shyu et al. 1991; Chen et al. 1994, 1995; Chen and Shyu 1994; Schiavi et al. 1994).

[1]Institut de Génétique Moléculaire de Montpellier, CNRS, UMR 5535, BP 5051, 1919 route de Mende, 34033 Montpellier Cedex 1, France
[2]Rowett Research Institute, Greenburn Road, Bucksburn, Aberdeen AB2 9SB, Scotland, UK

Progress in Molecular and Subcellular Biology, Vol. 18
Ph. Jeanteur (Ed.)

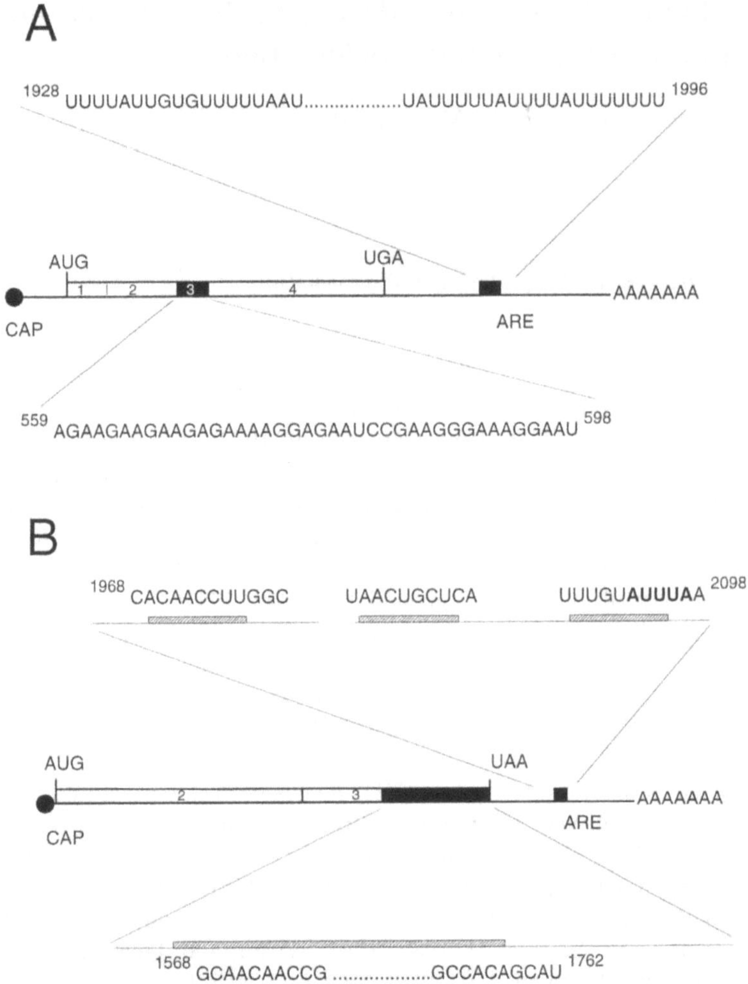

Fig. 1. Schematic representation of *c-fos* and *c-myc* mRNAs. Regions involved in both *c-fos* (A) and *c-myc* (B) mRNAs instability are represented by *black boxes*. Two regions of *c-fos* mRNA known to be involved in messenger destabilization. The ARE localized in the 3′ untranslated region (3′UTR) and the coding sequence included in exon 3 (*black boxes*). Corresponding sequences are represented. Positions are relative to human *c-fos* mRNA (**A**) and *c-myc* mRNA (**B**) cap sites. *Shaded boxes* represent conserved sequences: three for the 3′UTR destabilizing element (Vriz and Mechali 1989) (note the AUUUA motif in the third conserved region) and one for the coding region. For this last region, exon 3 from 11 species (*Serinus canaria, Carassius auratus, Felis silvestris, Pan troglodytes, Asterias vulgaris, Homo sapiens, Brachydanio rerio, Gallus gallus, Mus musculus, Salmo gairdneri* and *Xenopus laevis*) were compared. These sequences can be found in Genbank with the following mnemonics respectively: SEICMYC, CRACMYC, CATMYCCA3, CHPCMYC, ASRCMYC, HSMYCC, ZEFZCMYC, CHKMYC, MUSCMYC2, SMOMYC and XELMYC

Thus, ARE appear to be involved in multiple functions and this is reflected at the level of their structure. Previous results have suggested that *c-fos* ARE can be divided into two structurally and functionally distinct domains (Chen et al. 1994). Domain I corresponds to the 5′ part of the ARE, contains three AUUUA motifs and functions alone as a destabilizer. Single A to U mutations in the AUUUA motifs did not prevent poly(A) decay while degradation of the body of the mRNA was slowed down fivefold. The 3′ part of the ARE contains domain II, is U-rich and, contrary to domain I, does not work as a destabilizer by itself but can increase the destabilization efficiency conferred by this latter domain. Remarkably, domain II is able to work as a degradation enhancer when placed downstream of heterologous ARE (Chen and Shyu 1994). The presence of AUUUA motifs is not always sufficient and sometimes their mutation does not alleviate a rapid turnover. Mutation of undividual AUUUA motifs in either *c-myc* (Bonnieu et al. 1990) or *β*-interferon (Peppel et al. 1991) did not result in a significant stabilization of the corresponding mRNAs. Also, the pentamer does not mediate a rapid turnover when inserted alone within the 3′UTR of either rabbit *β*-globin (Akashi et al. 1994) or bovine growth hormone mRNAs (Lagnado et al. 1994). At least three copies are required to observe a destabilizing effect. GM-CSF mRNA stability has been shown to be markedly increased by treatment of a T cell line with phorbol ester, and largely dependent on the ARE present within the 3′UTR (Shaw and Kamen 1986; Akashi et al. 1994). Interestingly, normal T cells activation by a combination of anti CD3 and anti CD28 antibodies resulted in the stabilization of many ARE-containing cytokine mRNA, whereas neither *c-myc* nor *c-fos* mRNAs stability was affected (Lindsten et al. 1989). Similarly, the high level of expression of GM-CSF observed in a monocytic tumor cell line results from the constitutive stabilization of its mRNA, although no mutation could be detected in its sequence (Schuler and Cole 1988). The GM-CSF 3′UTR fused to a transfected reporter gene conferred a similar stabilizing effect, whereas *c-myc* or *c-fos* 3′UTR failed to do so. The observation of different fates for *c-fos, c-myc* and GM-CSF mRNA in the same cellular context has led to the suggestion that ARE can be divided into functionally different subsets. A classification can be proposed in function of both the number of AUUUA motifs in the 3′UTR and of the type of extracellular stimuli the cognate gene responds to (Chen and Shyu 1994). Thus, ARE present in several mRNAs coding for transcription factors contain relatively few copies of AUUUA motifs within, or followed by, a U-rich region, whereas those contained in mRNAs encoding cytokines frequently have multiple AUUUA elements in tandem. This complexity is also reflected by the proteins which can interact with these regions, some of them showing constitutive binding (Malter 1989; Brewer 1991; Vakalopoulou et al. 1991; Bickel et al. 1992; Myer et al. 1992; You et al. 1992; Hamilton et al. 1993; Zhang et al. 1993) and others inducible binding (Bohjanen et al. 1991, 1992). However, their precise role in mRNA decay is not yet known.

1.2 3′UTR and mRNA Translation

The cytoplasmic life span of many unstable mRNAs such as those encoding *c-fos* and *c-myc* is dramatically increased when translation is blocked with inhibitors of protein synthesis (Muller et al. 1984; Treisman 1985; Fort et al. 1987; Wilson and Treisman 1988), leading to the so-called superinduction phenomenon. Interestingly, the ARE present in the 3′ untranslated region seem also to preclude translation in *Xenopus* oocytes (Kruys et al. 1987, 1988, 1989) where, contrary to what is observed in somatic cells, numerous unstable mRNA are not degraded. This observation suggests a link may exist between mRNA degradation and translation. It has previously been reported that the 3′UTR of human interferon β mRNA (hu-IFNβ) has an inhibitory effect on its translation in *Xenopus* oocytes without affecting its stability in these cells (Kruys et al. 1987, 1988). The translational blockade is imposed by ARE present in hu-IFNβ and can be substituted by those in the 3′UTR of GM-CSF and *c-fos* mRNA (Kruys et al. 1989). Indeed, an inverse correlation was observed between the amount of interferon produced after microinjection, and the number of AUUUA in the 3′UTR. When *c-fos* RNAs bearing 3′ deletions of increasing lengths were synthesized in vitro and translated in the rabbit reticulocyte lysate, their translational efficiency was increased upon the destruction of the ARE present in the 3′UTR. The latter element was shown to inhibit translation in vitro in a poly(A)- and degradation-independent manner (Blanchard and Veyrune, unpubl. observ.). The translational inhibitory effect of ARE has also been assayed in vivo by transfecting cells with constructs where the natural *c-fos* promoter has been replaced by the rat β-actin promoter. This allows the analysis of FOS protein synthesis in exponentially growing cells, in the absence of the endogenous gene product because the cognate gene is silent under these conditions. *c-fos* mRNA recruitment into large polysomes has been analyzed in pools of stably transfected mouse Ltk$^-$ cells expressing comparable amounts of *c-fos* RNAs transcribed from either a wild-type mouse *c-fos* gene or a mutant in which 3′ ARE has been deleted (Fort et al. 1987). Wild-type mRNA was systematically found in polysomes smaller than the deletion mutant, and when compared for their FOS protein content, cells expressing the 3′UTR deletion mutant expressed five to ten times more protein (our unpubl. observ.). These data show that ARE present in *c-fos* 3′UTR exert a strong negative effect on translation efficiency and extend thus to mammalian somatic cells the previous observation carried out in *xenopus* eggs (Kruys et al. 1987, 1988, 1989).

These data are compatible with a role of ARE in translational control and are reminiscent of previous reports of cells expressing measurable levels of *c-fos* mRNA without any detectable amount of FOS protein (Mitchell et al. 1985). In monocytic or myelomonocytic cells, a rapid induction of *c-fos* mRNA by phorbol esters is followed by a progressive decrease to a plateau level which can last for several hours, during which no FOS protein can be detected. Moreover, in transgenic mice expressing *c-fos* constructs, phenotypic effects were observed only after the removal of the normal *c-fos* 3′UTR, an effect which was not explained solely by an increase of mRNA stability (Ruther et al. 1987).

1.3 Is There a Link Between Translation and 3′UTR-Directed Degradation?

The mechanism(s) by which protein synthesis inhibitors stabilize certain mRNA remains obscure. Several possibilities (which are not mutually exclusive) involving either highly unstable proteins or translation per se have been proposed. Moreover, these different hypotheses are complicated by the fact that some protein synthesis inhibitors can also act as nuclear signalling agonists (Edwards and Mahadevan 1992). Moreover, there is an intrinsic difficulty in distinguishing between *cis* and *trans* effects of inhibitors, which resides in their bulk effect on protein synthesis. Different approaches have been designed to tackle this problem (Fig. 2).

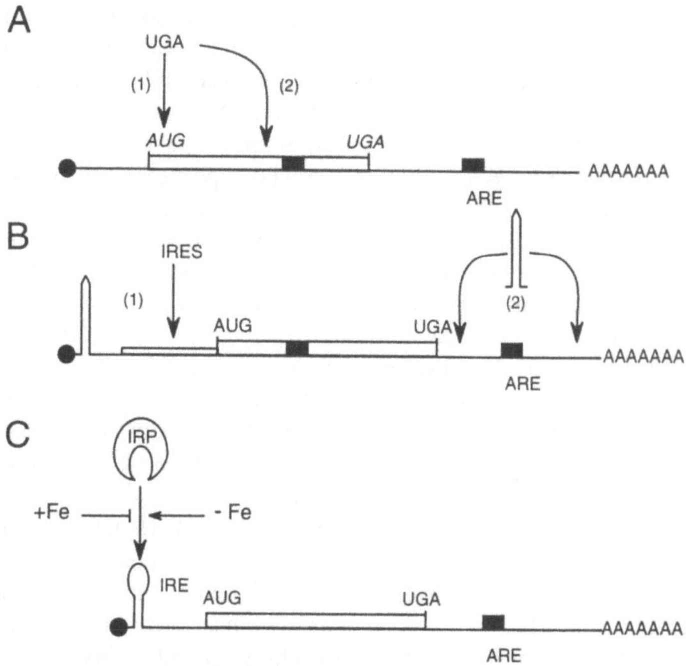

Fig. 2. Different approaches used to impede translation. **A** Mutation of the first codon AUG (1) (Wisdom and Lee 1991), or introduction of a stop codon upstream of the coding region instability determinant (2) (Herrick and Ross 1994). **B** Introduction of a stable secondary structure upstream of the initiation codon (1) (Aharon and Schneider 1993) or the ARE present in the 3′UTR (2) (Curatola et al. 1995). In the former case, an internal ribosome entry site (IRES) positioned downstream of the hairpin structure releases the block. **C** Introduction of the iron response element (IRE) of ferritin mRNA close to the 5′ cap (Koeller et al. 1991; Veyrune et al. 1995; Winstall et al. 1995)

Mutation of *c-myc* translation initiation codon into a stop codon stabilized *c-myc* mRNA. However, this effect was shown to be mainly due to the coding carboxy-terminal instability determinant (Wisdom and Lee 1991; Herrick and Ross 1994). Interestingly, superinduction of *c-myc* mRNA by cycloheximide is not observed for the mutated, translation-incompetent mRNA. A similar approach has been used to analyze the role of translation on GM-CSF mRNA degradation (Savant and Cleveland 1992). In this case, a large divalent cation-independent complex is found associated only with an unstable mRNA. Translocation of ribosomes through the ARE abolished its destabilizing effect, showing the necessity for this element to be present in the 3′UTR of the mRNA. Another approach was to introduce a stable secondary structure upstream of the initiation codon, in order to impede translation (Kozak 1986). Chimeric mRNA expressing the hepatitis B virus S-antigen protein as a marker and containing 62 nucleotides of the GM-CSF 3′UTR have been assayed for both their translatability and their decay rate as a function of the structure of their 5′UTR (Aharon and Schneider 1993). All constructs contained the adenovirus late 5′UTR, also known as the tripartite leader. Introduction of complementary sequences which form a stable hairpin structure ($\Delta G = -35$ to -40 kcal) within the unstructured tripartite leader give rise to translation-impaired mutant mRNAs with a greatly reduced ability to bind ribosomes (Dolph et al. 1990). Decay rates were determined by standard actinomycin D chase experiments after transient transfection into Cos-1 and HeLa cells. Substituting the mutant 5′UTR for that of the wild type decreased translation by 15–20-fold, whereas no variation in mRNA steady state was observed. However, introduction of the GM-CSF ARE resulted in a drastic inability to accumulate mRNA when translation was facilitated. Moreover, rescue of translation by the introduction of an internal ribosome entry site downstream of the hairpin in translation-impaired mutants restored a rapid turnover (Aharon and Schneider 1993). Interestingly, inhibition of ARE-induced cotranslational destabilization by secondary structures is observed at any position 5′ to the ARE, even several hundred nucleotides downstream from the translation termination codon (Curatola et al. 1995). In contrast, such an effect is not observed when the insertion site is 3′ relative to the ARE.

Similarly, the ability of various *c-fos* regions to act as instability sequences and their functional dependence upon translation have been addressed. First, a stable stem-loop structure has been included in the 5′ UTR of a *c-fos* mRNA lacking the ARE in its 3′UTR, and it has been shown that several coding portions of the mRNA act as translation-dependent instability determinants in serum-starved NIH3T3 cells restimulated with serum (Schiavi et al. 1994). More recently GM-CSF and *c-fos* ARE have been compared for their destabilizing abilities under the same experimental conditions when appended 3′ to the β-globin coding region. Only a minor effect was observed on both stability and deadenylation in quiescent, restimulated NIH 3T3 cells (Chen et al. 1995).

The ferritin mRNA iron response element (IRE), which when grafted 5′ to a reporter sequence has previously been shown to give rise to translational control dependent on intracellular iron concentration (Hentze et al. 1987, 1988; Aziz and Munro 1987; Casey et al. 1988; Mattia et al. 1989), has also been used to address this problem. When iron is abundant (cells treated with an iron source such as hemin or ferric ammonium citrate), the IRE promotes mRNA translation, whereas when iron is scarce (cells treated with a chelator such as desferrioxamine), the RNA is found in free mRNPs. This effect is specific for the mRNA containing the IRE and global protein synthesis is not affected by iron availability. Degradation of the mRNA encoding the transferrin receptor and containing the ARE present in *c-fos* 3′UTR was shown to be independent of iron concentration in stably transfected B6 mouse fibroblasts (Koeller et al. 1991). However, these data are in contradiction with two more recent reports based on a similar approach, using a chimeric *β*-globin/*c-fos* mRNA containing the human ferritin IRE (Veyrune et al. 1995; Winstall et al. 1995). Vectors containing a *β*-globin/*c-fos* gene have been either transiently transfected into mouse NIH3T3 cells (Winstall et al. 1995) or stably transfected into mouse Ltk^- cells (Veyrune et al. 1995). In the former case, RNA is analyzed by an RNAse protection assay, whereas in the latter case, RNA is studied by Northern blotting. In both cases, cells are treated for 16 h with ferric ammonium citrate, hemine or desferrioxamine and further treated with actinomycin D for various amounts of time, to analyze the stability of the chimeric mRNAs. Desferrioxamine treatment reduced dramatically the amount of reporter mRNA present in polysomes compared with that of cells treated with iron, where the same RNA is significantly shifted toward polysomes. *β*-globin/*c-fos* mRNA half-life is increased severalfold in cells treated with desferrioxamine, while that of an mRNA, expressed from vectors devoid of IRE or containing a mutated IRE, remains unchanged. Removal of the ARE-containing fragment from the *β*-globin/*c-fos* construct results in a stable mRNA (Fort et al. 1987; Bonnieu et al. 1989) insensitive to intracellular iron availability. This discrepancy could be partly explained by the existence of secondary structures involving sequences located in both the 3′UTR and the coding region (E. Thiel, Pers. comm.; J.L. Veyrune and J.M. Blanchard, unpubl. observ.), which could counteract the effects of the IRE and/or the ARE.

In the latter two series of experiments, the entire *β*-globin coding sequence has been inserted downstream from the IRE, which would suggest that the coding portion of this mRNA is permissive for the regulation by the IRE element.

The question of the requirement of ARE-dependent mRNA degradation for translation is therefore still a matter of debate. In most of the experiments the distinction between initiation and elongation has not been thoroughly analyzed. As pointed out in a recent review (Sachs 1993), it might well be that one of the reasons for the observed discrepancies resides in variations in the inhibition of translation by the various 5′ inhibitory elements. Should the degradative enzymes be associated with either initiation or elongation translation factors, the overall decay kinetics will probably be different. Such a precedent of a ribonuclease associated with a protein required for translation has been found in yeast (Sachs and Deardorff 1992).

Whereas the reason why the coding region instability sequences are active only in starved cells restimulated with serum remains unclear, elements present within both coding and 3′UTR are involved in poly(A) shortening. This common denominator is important with regard to a potential link between degradation and translation, because poly(A) has also been involved in the control of both degradation and translation. Along these lines, a recent report showing that poly(A) addition promoted 5′ cap ribose methylation during progesterone-induced oocyte maturation in *Xenopus laevis* could shed some light on the translational requirement for inducing mRNA decay (Kuge and Richter 1995). Should this observation be generalized to somatic cells, it would be interesting to look for the reverse reaction, i.e. whether deadenylation could give rise to cap ribose demethylation and thus explain the translational inhibitory role of ARE-containing 3′UTR, which remains mysterious.

2 Localization of mRNAs in the Cytoplasm and Their Association with the Cytoskeleton

Until recently, it was thought that proteins were synthesised on polyribosomes that were either attached to the endoplasmic reticulum or 'free' in the cytoplasm (Blobel and Dobberstein 1975); membrane and secreted proteins were envisaged as made on the membrane-bound polysomes attached to the rough endoplasmic reticulum and all other proteins on free polysomes. Such a model implies that, other than direction of polyribosome complexes to the endoplasmic reticulum by the signal sequence in the nascent polypetide chain, there is no targeting or localization of specific mRNAs to different subcellular sites. However, in the past 10 years evidence has been produced which shows that specific mRNAs for non-membrane proteins localized to particular parts of the cytoplasm, particularly in oocytes and early embryos of *Xenopus* and *Drosophila* (St 1995), but also in a variety of somatic mammalian cells (Singer 1993; Hesketh 1994, 1996). These observations, together with mounting evidence for an association of mRNAs and polyribosomes with the cytoskeleton, suggest that some mRNAs are synthesized in specific locations within cells. Such spatial organization of the protein synthetic apparatus may provide a mechanism by which proteins can be synthesized close to where they are needed by the cell: this could be of particular importance in the case of proteins which are rapidly degraded, such as the products of the *c-myc* and *c-fos* genes (Carillo et al. 1994; Watt and Molloy 1993). In the second part of this review we discuss in general terms the evidence for mRNA localization and cytoskeleton-polysome interactions and then describe in more detail the results obtained concerning localization of *c-myc* and *c-fos* mRNAs and the role of their 3′UTRs in mRNA targeting/localization.

2.1 Localization of mRNAs

There is increasing evidence that in a wide variety of cell types certain specific mRNAs accumulate at particular cytoplasmic sites (see Table 1). Such mRNA localization appears to be a general phenomenon which occurs not only in germ cells and early embryos but also in a wide variety of somatic cell types: for example, it is evident in cells with a complex and polarized architecture such as

Table 1. Examples of localized mRNAs

mRNA	Organism and/or cell type	mRNA localization	References[a]
bicoid	*Drosophila* oocytes	Anterior pole	[1–7]
Cyclin B	*Drosophila* oocytes	Posterior pole	[8–11]
fs(1)K10	*Drosophila* oocytes	Anterior pole	[12, 13]
nanos	*Drosophila* oocytes	Posterior pole	[14–17]
oskar	*Drosophila* oocytes	Posterior pole	[14, 18–21]
orb	*Drosophila* oocytes	Posterior pole	[22, 23]
BC1	Neuron	Dendrites and cell body	[24–26]
GAP43	Neuron	Dendrites and cell body	[27, 28]
MAP2	Neuron	Dendrites and cell body	[27, 29–31]
Oxytocin	Neuron	Axon and cell body	[32]
Vasopressin	Neuron	Axon and cell body	[33]
Myelin basic protein	Oligodendrocyte	From Cell body to processes	[34–36]
An1, 2 and 3	*Xenopus* oocyte	Animal pole	[37–39]
TGF-β5	*Xenopus* oocyte	Vegetal pole	[40]
Vg1	*Xenopus* oocyte	Vegetal pole	[41-44]
Xcat-2	*Xenopus* oocyte	Vegetal pole	[45]
Xlsirts	*Xenopus* oocyte	Vegetal pole	[42, 46]
Xwnt-11	*Xenopus* oocyte	Vegetal pole	[47]
a-actin	Muscle cells	Perinuclear	[48]
β-actin	Muscle cells	Perinuclear and peripheral	[49-52]
g-actin	Muscle cells	Perinuclear	[49]
Vimentin	Muscle cells	Costameres	[53]
c-myc	Fibroblast	Perinuclear	[54-56]

[a] 1, Ferrandon et al. (1994); 2, Macdonald and Struhl (1988); 3, Luk et al. (1994); 4, MacDonald (1990); 5, Macdonald et al. (1993); 6, Pokrywka and Stephenson (1991); 7, Stephenson and Pokrywka (1992); 8, Raff et al. (1990); 9, Dalby and Glover (1992); 10, Dalby and Glover (1993); 11, Hovland et al. (1995); 12, Cheung et al. (1992); 13, Pokrywka and Stephenson (1995); 14, Ephrussi et al. (1991); 15, Gavis and Lehmann (1992); 16, Gavis and Lehmann (1994); 17, Mosquera et al. (1993); 18, Webster et al. (1994); 19, Kim et al. (1991); 20, Kim et al. (1993); 21, Kim et al. (1995); 22, Lantz et al. (1992); 23, Lantz and Schedl (1994); 24, Tiedge et al. (1991); 25, Tiedge et al. (1992); 26, Tiedge et al. (1993); 27, Bruckenstein et al. (1990); 28, Kleiman et al. (1990); 29, Garner et al. (1988); 30, Litman et al. (1993); 31, Litman et al. (1994); 32, Trembleau et al. (1990); 33, Trembleau et al. (1994); 34, Ainger et al. (1993); 35, Amur et al. (1993); 36, Ueno et al. (1994); 37, Linnen et al. (1993); 38, Gururajan et al. (1994); 39, Gururajan et al. (1991); 40, Kondaiah et al. (1990); 41, Forristall et al. (1995); 42, Kloc and Etkin (1994); 43, Yisraeli and Melton (1988); 44, Yisraeli et al. (1990); 45, Elinson et al. (1993); 46, Kloc and Etkin (1995); 47, Ku and Melton (1993); 48, Kislauskis et al. (1993); 49, Hill et al. (1994); 50, Kislauskis et al. (1994); 51, Latham et al. (1994); 52, Sharpless et al. (1993); 53, Cripe et al. (1993); 54, Hesketh et al. (1991a); 55, Hesketh et al. (1994); 56, Veyrune et al. (1996).

oligodendroglia and neurons where myelin basic protein, tau and MAP-2 in mRNAs have been found to be compartmentalized within the cells (Ainger et al. 1993; Litman et al. 1993; Garner et al. 1988). In addition, in skeletal muscle vimentin mRNA is localized to the cytoplasm close to the costameres (Morris and Fulton 1994). Actin isoform mRNAs also show specific localization to different parts of the cytoplasm and in the highly polarized intestinal epithelium cells β-actin mRNA is localized in the apical region of the cytoplasm (Cheng and Bjerknes 1989). However specific mRNA localization is not restricted to such highly polarized cells and in spreading fibroblasts β-actin mRNA is found in the cell periphery in the cytoplasm underlying the lamellipodia (Sundell and Singer 1990). Under normal growth conditions, β-actin mRNA is found to be peripheral, whilst the γ-actin mRNA is largely in the perinuclear cytoplasm (Hill and Gunning 1993; Latham et al. 1994); but in cells deprived of serum, the β-actin mRNA in the peripheral cytoplasm disappeared and was found only in the perinuclear cytoplasm. Re-addition of serum, or addition of platelet-derived growth factor or forskolin, caused a rapid (within 10 min) reappearance of β-actin mRNA in the peripheral cytoplasm: this was unaffected by inhibition of RNA synthesis and thus appeared to represent relocalization of existing mRNA. It appears therefore that β-actin mRNA localization can respond to physiological stimuli such as growth factors and is regulated through a receptor-mediated phosphorylation pathway, but as yet the precise signal pathway remains to be elucidated.

There are no available data concerning the subcellular distribution of *c-myc* mRNA in the cytoplasm under conditions of normal expression. However, in cells transfected with the normal second and third exons of the *c-myc* gene in situ hybridization has shown that the *c-myc* transcripts are found to be localized in the perinuclear cytoplasm (Hesketh et al. 1994) but not in the peripheral cytoplasm (see Fig. 3). In contrast, after transfection of cells with multiple copies of the *c-myc* gene under control of a heat-shock promoter, heat treatment was found to induce *c-myc* gene transcription and the mRNA was transported to the cytoplasm, but no localization of the mRNA was reported (Pallavicini et al. 1994). The apparent lack of *c-myc* mRNA localization in this latter study may be due to the small cytoplasmic volume evident in these cells (the Ltk$^-$ fibroblasts used in the other study appear to have a more spread out morphology, as seen in Fig. 3 and Hesketh et al. 1994), or to abnormalities in cytoskeletal organization and mRNA processing induced by exposure to heat (Glass et al. 1985; Yost and Lindquist 1988).

The increasing number of studies which show examples of mRNA localization now provide sound evidence to support the hypothesis that in a variety of cells and tissues there is subcellular localization and transport of specific mRNAs and of ribosomes. In several cases, localization of the mRNA corresponds closely to the protein localization: thus, vimentin mRNA is found close to the costameres (Morris and Fulton 1994), structures into which vimentin protein is incorporated, MAP-2 mRNA and protein are both in neuronal dendrites (Garner et al. 1988), and in spreading fibroblasts β-actin mRNA is

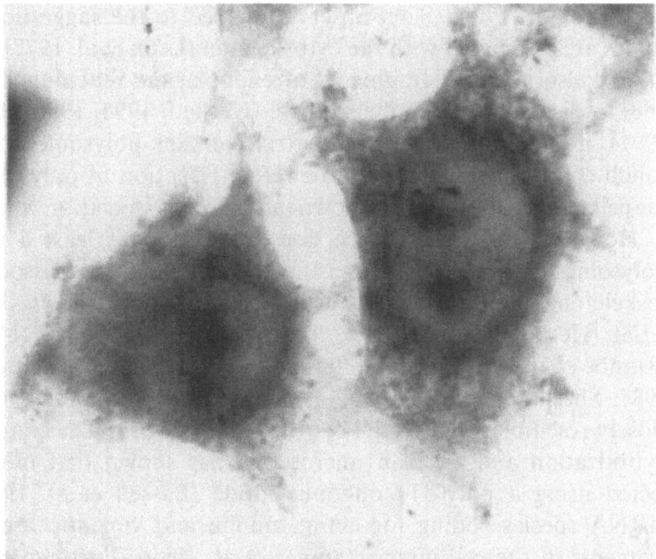

Fig. 3. In situ hybridization showing the localization of *c-myc* mRNA. Ltk$^-$ fibroblasts were transfected with exons 2 and 3 of the murine *c-myc* gene and transcripts detected using digoxigenin-labelled riboprobes and anti-digoxigenin linked to alkaline phosphatase (Wisdom and Lee 1991). Note the distinct perinuclear distribution of the *c-myc* mRNA and the lack of staining in the peripheral cytoplasm

found underlying the lamellipodia (Sundell and Singer 1990), close to where a high requirement for actin synthesis under these conditions is known to be. These observations suggest that such mRNA sorting/targeting provides a mechanism for the synthesis of certain proteins close to their site of function, i.e. for a targeting of the protein synthetic apparatus. Furthermore, it appears that mRNA localization can respond to growth factor stimulation (Latham et al. 1994) and thus it potentially provides a mechanism for regulating the site of mRNA translation and protein synthesis within cells.

2.2 Association of mRNAs and Polysomes with the Cytoskeleton: Cytoskeletal-Bound Polysomes

Treatment of cultured eukaryotic cells with low concentrations of non-ionic detergents releases soluble cellular components and plasma membrane components, but leaves an insoluble cell matrix containing cytoskeletal filaments (Osborn and Weber 1977). The cell matrix also contains polysomes and mRNAs, as shown by electron microscopy, immunohistochemistry, in situ hybridization and sucrose gradient analysis (Lenk et al. 1977; Van Venrooij et al. 1981; Hesketh et al. 1991b; Taneja et al. 1992). Indeed, it has been estimated that approximately 75% of cellular mRNAs are retained in the cell matrix

(Taneja et al. 1992). Such observations led to the suggestion that some polysomes are associated with the cytoskeleton (Lenk et al. 1977). However, the cell matrix also contains fragments of endoplasmic reticulum (Dang et al. 1983) and, as discussed in detail elsewhere (Hesketh 1994, 1996; Hesketh and Pryme 1991), it is clear that the cell matrix contains polysomes associated with the rough endoplasmic reticulum: therefore retention of polysomes or mRNAs in the cell matrix does not on its own indicate an association with the cytoskeleton.

However, further evidence does show that at least a proportion of the polysomes and mRNAs present in the cell matrix are associated with the cytoskeleton. Immunohistochemical techniques have shown that ribosomal material (Hesketh et al. 1991b; Toh et al. 1980), 5′ mRNA binding protein (Zumbe et al. 1982) and both initiation and elongation factors (Gavrilova et al. 1987; Shestakova et al. 1993a,b) are localized within cells in a pattern which closely resembles that of the cytoskeletal network. The combination of in situ hybridzation and electron microscopy has shown that mRNAs in toto, detected using a poly(dT) oligonucleotide (Bassell et al. 1994a), and specific mRNA species coding for actin, tubulin and vimentin are clustered around filaments in the cell matrix (Singer et al. 1989). Furthermore, ultrastructural data both from high resolution structural studies using high-voltage and deep-etch electron microscopy and from in situ hybridization not only show the presence of polysome-like structures or mRNAs close to or surrounding filamentous structures, but also suggest that they are present particularly at the intersections or cross-over points where filaments interact (Wolosewick and Porter 1979; Bassell et al. 1994a).

Crucial evidence for an association of polysomes with the cytoskeleton has also come from biochemical analysis following treatments which cause breakdown of the cytoskeleton, particularly the microfilaments. Thus, pretreatment of cells with cytochalasins not only causes a depolymerization of actin and a collapse of the microfilament network, but also a loss of polysomes from the cell matrix and an increase in the proportion of polysomes recovered in the soluble fraction (Ramaekers et al. 1983; Bird and Sells 1986; Ornelles et al. 1986; Taneja et al. 1992). Similarly, actin depolymerization induced by DNase I (Adams et al. 1983) or by treatment of the cell matrix with 130 mM KCl has been found to lead to a redistribution of polysomes such that there is loss from the cell matrix and increased recovery in the soluble fraction (Vedeler et al. 1991; Hesketh and Pryme 1988). Treatment of the cell matrix with 100–200 mM KCl causes cytoskeletal disorganization (Lenk et al. 1977), actin depolymerization (Hesketh and Pryme 1988), progressive loss of RNA (Bird and Sells 1986) and release of polysomes (Hesketh and Pryme 1988). The material released by salt is enriched in actin and vimentin and both the actin and the polysome contents of the fraction are reduced by pretreatment with cytochalasin or phalloidin (Vedeler et al. 1991). On the basis of such data, it appears that those polysomes released by salt treatment of the cell matrix are derived from a population of cytoskeletal-bound, probably microfilament-associated, polysomes.

The effects of microfilament-disruptive agents on polysome and mRNA distribution suggest that microfilaments play an important role in mRNA/polysome-cytoskeleton interactions. This is supported by data from light and electron microscopy: in lens cells, electron microscopy shows polysomes associated with microfilaments (Ramaekers et al. 1983) and in fibroblasts 72% of polyadenylated mRNA is found in juxtaposition to actin filaments (Bassell et al. 1994a); co-localization studies suggest that the distribution of ribosomes (Toh et al. 1980; Hesketh et al. 1991b), mRNA-binding proteins (Zumbe et al. 1982) and initiation factors (Wolosewick and Porter 1979) partly resembles that of actin filaments. An important role for microfilaments in the spatial organization of polysomes and mRNAs is consistent with the increasing evidence that elongation factor 1 is an actin-binding protein (Condeelis 1995).

Although most evidence indicates an association of polysomes and mRNAs with microfilaments, there is also data to suggest that microtubules and intermediate filaments may be involved. In the case of intermediate filaments, the lack of specific agents which disrupt intermediate filaments has limited the information available on intermediate filament-mRNA interactions. However, cytoskeletal fractions from oocytes of *Xenopus* or *Drosophila* have been shown to be enriched in both cytokeratin and specific mRNAs such as Vg1 and Xcat (Pondel and King 1988; Forristall et al. 1995). In general, polysomes are found in areas of the cytoplasm which are free of intermediate filaments (Lenk et al. 1977; Hesketh and Pryme 1991), although mRNA cap-binding protein and the elongation factor eEF-2 have been found to be co-localized with intermediate filaments (Zumbe et al. 1982; Shestakova et al. 1993a) and in situ hybridzation using a non-specific poly(dT) oligonucleotide shows some 29% of mRNAs are found close to intermediate filaments (Bassell et al. 1994a). At present, the extent and importance of mRNA-intermediate filament interactions are not clear.

On the other hand, there is increasing evidence that the microtubules are an important component in the interaction between the cytoskeleton and specific mRNAs. Microtubule preparations have been found to contain ribosomes and polysomes (Hamill et al. 1994) and inhibitor studies show that mRNA localization in oocytes is sensitive to colchicine (Yisraeli et al. 1990; Pokrywka and Stephenson 1994). A cytoskeletal fraction from *Drosophila* oocytes is enriched in bicoid mRNA, and this mRNA is released by microtubule-disrupting agent colchicine (Pokrywka and Stephenson 1994). The effects of colchicine and cytochalasin on the distribution of specific mRNAs also suggest that microtubules are required for translocation, whereas microfilaments are involved in anchoring the mRNA so as to restrict its localization (Raff et al. 1990; Yisraeli et al. 1990). Recently, the distribution of the protein Staufen (which binds to the localized mRNA bicoid) has been found to be affected by colchicine treatment (Ferrandon et al. 1994). Other observations suggest that microtubules are involved in mRNA transport: confocal microscopy suggests a close apposition of microtubules and particles containing myelin basic protein mRNA (Ainger et al. 1993); in neurons, 55% of mRNA is found close to microtubules by high resolution in situ hybridization (Bassell et al. 1994b); in insect nutritive tubes, there is an association of ribosomes and microtubules (Stebbings 1986). The different effects of colchicine and cytochalasins on mRNA distribution in oocytes are consistent with there being different interactions between different filament systems and mRNAs: available data suggest a model in which microtubules are involved in mRNA/polysome translocation through the cytoplasm and that microfilaments are involved in anchoring mRNAs (Raff et al. 1990; Yisraeli et al. 1990).

2.3 Cytoskeletal-Bound Polysomes Are Enriched in Specific mRNAs Including c-myc and c-fos

As discussed in the previous section, it is now clear that a proportion of polysomes and mRNAs are associated with the cytoskeleton. A number of procedures have been used to produce cell fractions which either are enriched in cytoskeletal components or which are derived by use of cytoskeleton-disruptive agents: such fractions are thought to contain cytoskeletal-bound polysomes (CBP) and they have been found to be enriched in specific mRNAs. In fibroblasts and myoblasts, the majority of β-actin mRNA has been found in cytoskeletal fractions or polysomes isolated from such fractions (Bird and Sells 1986; Hesketh et al. 1991a; Katze et al. 1989; Meadus et al. 1990; Johannessen et al. 1995; Campbell and Hesketh 1996). Treatment of the cell matrix with cytochalasins releases CBPs which are enriched in *c-fos* and histone mRNAs (Bird and Sells 1986; Heuijerjans et al. 1989; Zambetti et al. 1990) and disruption of the cell matrix by exposure to increased hydrostatic pressure also releases histone mRNAs (Symington et al. 1993). More recently, the separation of CBPs from the cell matrix by salt treatment has allowed analysis of a number of mRNAs. As shown in Table 2, it has been found consistently that *c-myc* mRNA is enriched in CBP from a number of cell lines such as 3T3 fibroblasts (Hesketh et al. 1991a), transfected fibroblasts (Hesketh et al. 1994), HepG2 and H4 hepatoma cells (Hovland et al. 1995; Mahon et al. 1995) and Chinese hamster ovary cells (Hesketh et al. 1996). Thus it appears that both *c-myc* and *c-fos* mRNAs are translated on polysomes associated with the cytoskeleton. In addition, the mRNAs for cyclin A and ribosomal protein L4 mRNAs are also recovered in CBPs (Hovland et al. 1995).

The evidence emerging from mRNA analysis of CBP suggests therefore that specific mRNAs are associated with the cytoskeleton. This implies that CBP represent a functionally distinct compartment. However, the nature of CBPs in terms of cellular compartmentation remains to be determined, particularly whether CBPs have a distinct localization in the cytoplasm or whether they are derived from several compartments with different interactions with the cytoskeleton. One common feature of *c-myc, c-fos,* ribosomal proteins and histones, whose mRNAs are found in CBP, is that when newly synthesized they are translocated to the nucleus. It appears therefore that nuclear protein

Table 2. C-myc mRNA is associated with CBP in a variety of cell lines

Cell line	mRNA abundance			
	FP	CBP	MBP	References[a]
3T3 fibroblasts	1	1.8	0.7	[1]
[H4] hepatoma	1	2.00.2	1.4 ± 0.1	[3,7]
Hep G2 hepatoma	1	2.2	0.8	[2]
Ltk- fibroblasts (transfected with exons 2 and 3 of murine *c-myc* gene)	1	2.0 ± 0.3	1.1 ± 0.1	[5]
CHO cells	1	2.9	2.2	[4]

mRNA abundance is shown as relative enrichment of the mRNA compared to FP (= 1). Values shown are means from at least two separate fractionation experiments, and where appropriate ± SEM FP, CBP and MBP were isolated as described in [1–6] and blots hybridized with probes to *c-myc* and 18S rRNA. Specifically bound probe was quantified using a Camberra-Packard Instantimager and c-myc/18S rRNA ratios calculated.

[a] 1, Edwards and Mahadevan (1992); 2, Aziz and Munro (1987); 3, Casey et al. (1988); 4, Mattia et al. (1989); 5, Wisdom (1991); 6, Vakalopoulou et al. (1991); 7, Dolph et al. (1990).

mRNAs may be one class of mRNAs which are translated on CBP. It is not known if the mRNAs for many nuclear proteins are also localized to the perinuclear cytoplasm, but on the basis of the above data and the perinuclear localization of *c-myc* mRNA, it has been proposed (Hesketh et al. 1991a, 1994; Hovland et al. 1995) that the mRNAs for nuclear proteins including those for the transcription factors *c-myc* and *c-fos* are retained on the cytoskeleton in the perinuclear cytoplasm and translated there so as to promote efficient transport of the protein to the nucleus. Such local synthesis could be important for proteins such as *c-myc* and *c-fos*, which are unstable and rapidly degraded in the cytoplasm (Carillo et al. 1994).

2.4 Targeting of c-myc to the Cytoskeleton and the Perinuclear Cytoplasm: Role of the 3′UTR

Details of the precise mechanisms of mRNA localization remain unclear but there is some evidence to suggest that the cytoskeleton may play an important role in the translocation and localization of mRNAs. It is particularly relevant that in some cases localization is destroyed by disruption of the cytoskeleton (Sundell and Singer 1990) and that certain localized mRNAs are associated with the cytoskeleton (Pondel and King 1988; Pokrywka and Stephenson 1991; Forristall et al. 1995). Recent work has also indicated that the localization of mRNAs, including that of *c-myc*, depends upon information which resides in the mRNA itself, particularly in the 3′UTR.

Micro-injection of exogenous mRNA into oocytes or oligodendroglia results in correct localization of the mRNA (Macdonald and Struhl 1988; Yisraeli and Melton 1988; Ainger et al. 1993) and localization of β-actin mRNA in spreading fibroblasts is unaffected by puromycin, and thus independent of nascent polypeptide chains or ribosomes (Sundell and Singer 1990). It would appear therefore that in these particular cases mRNA localization depends upon signals in the mRNA itself. This is also the case in a variety of mRNAs which are localized in oocytes and in a number of these mRNAs the localization signal appears to reside in the 3′UTR. Detailed study of 3′UTR function in mRNA localization has been achieved in a number of instances using molecular approaches to produce, and introduce into appropriate cells or organisms, genes with modified 3′UTRs or chimeric genes in which the 3′UTRs from specific genes are linked to a reporter coding sequence such as β-globin or β-galactosidase. The production of transgenic flies with modified 3′UTRs or with chimeric genes has shown that for a number of localized mRNAs the 3′UTR is either required for localization or is capable of localizing a reporter sequence (Macdonald and Struhl 1988; Davis and Ish 1991; Kim et al. 1993; Macdonald et al. 1993). Similarly, micro-injection of this type of construct has shown that the 3′UTR of the Vg1 mRNA is essential for correct localization of *Xenopus* oocytes (Mowry and Melton 1992).

Such modified genes can be introduced into mammalian cells in culture by transfection, and the expression of chimeric β-globin/*c-myc* gene constructs in which the 3′UTRs were exchanged has shown that removal of the *c-myc* 3′UTR and its replacement by the 3′UTR of β-globin results in loss of perinuclear localization and loss of association with the cytoskeleton (Hesketh et al. 1994). In addition, attachment of the *c-myc* 3′UTR to β-globin coding sequences results in a localization of the transcript in the perinuclear cytoplasm (Fig. 4) and an association with CBP (Hesketh et al. 1994; Veyrune et al. 1996). It would appear therefore that the sorting of *c-myc* mRNA is dependent on some signal within the 3′UTR. This sorting mechanism is not unique to the *c-myc* mRNA. It has also been observed to occur in the case of several mRNA localized in oocytes and with β-actin mRNA in fibroblasts. For example, the 3′UTR from β-actin mRNA directs a reporter sequence to be localized in the cell periphery (Kislauskis et al. 1993, 1994) and the 3′UTRs of vimentin and myosin heavy chain localize β-globin sequence to the perinuclear cytoplasm (Wiseman et al. 1996); in addition, in primary neurons, localization of tau mRNA is dependent on the 3′UTR (Behar et al. 1995). Thus, in mammalian cells, 3′UTR sorting mechanisms appear to be able to sort mRNAs so that some can be retained in the cytoskeleton in the perinuclear cytoplasm (e.g. *c-myc*) and others can be translocated via a cytoskeletal-based mechanism (Sundell and Singer 1990) to the cell periphery (see Fig. 5). Thus, two, or more, types of localization signals must occur in 3′UTR sequences.

The physiological consequences of *c-myc* mRNA localization are as yet unproven since functional studies on cells with altered *c-myc* mRNA localization are very limited. However, immunocytochemistry of fibroblasts transfected with *c-myc* constructs with either the *c-myc* or β-globin 3′UTRs has been used to assess the effects or mRNA relocalization on nuclear localization of the protein (Hesketh et al. 1994). Interpretation is complicated by the very high

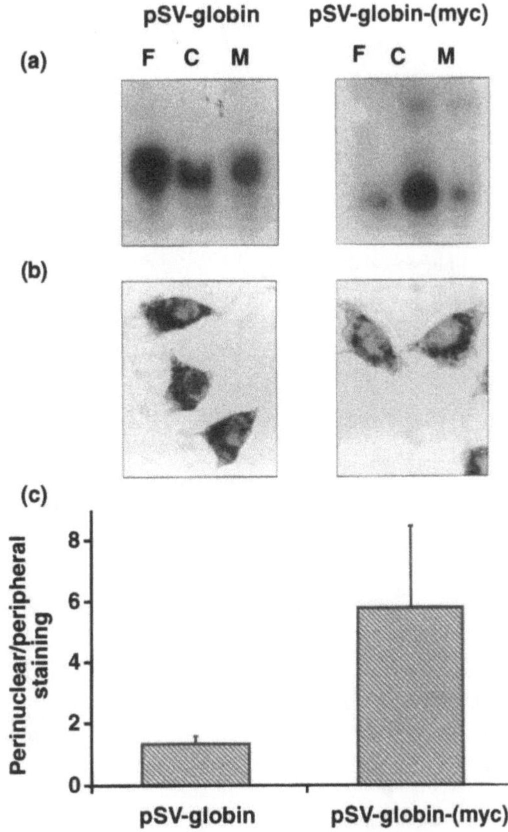

Fig. 4. The c-myc 3′UTR targets *β*-globin coding sequences to the perinuclear cytoplasm and to cytoskeletal-bound polysomes. Ltk⁻ fibroblasts were transfected with a plasmid vector containing either the rabbit *β*-globin gene with the coding sequences linked to the endogenous 3′UTR (pSV-globin) or the *β*-globin coding sequences linked to the c-myc 3′UTR [pSV-globin-(myc)]. Stable transfectants were studied by **a** a fractionation technique to separate free (*F*), cytoskeletal bound (*C*) and membrane-bound (*M*) polysomes and **b** by in situ hybridization. The distribution of transcripts shown in **b** was quantified by microdensitometry and the data are presented **c** as a ratio of staining in the perinuclear cytoplasm to that in the peripheral cytoplasm. Note that the *c-myc* 3′UTR causes both a redistribution of the transcripts from free to cytoskeletal-bound polysomes and a localization of the transcripts in the perinuclear cytoplasm. Data are from experiments described in Hesketh et al. (1994) and Veyrune et al. (1996)

levels of expression in such cells, but it appears (see fig. 6) that there is a smaller proportion of *c-myc* protein in the nucleus in those cells in which the *c-myc* mRNA is no longer localized to the perinuclear cytoplasm or associated with CBP. Such results are compatible with the hypothesis that *c-myc* mRNA is localized on the cytoskeleton around the nucleus to allow efficient targeting of the newly synthesized, unstable *c-myc* protein to the nucleus.

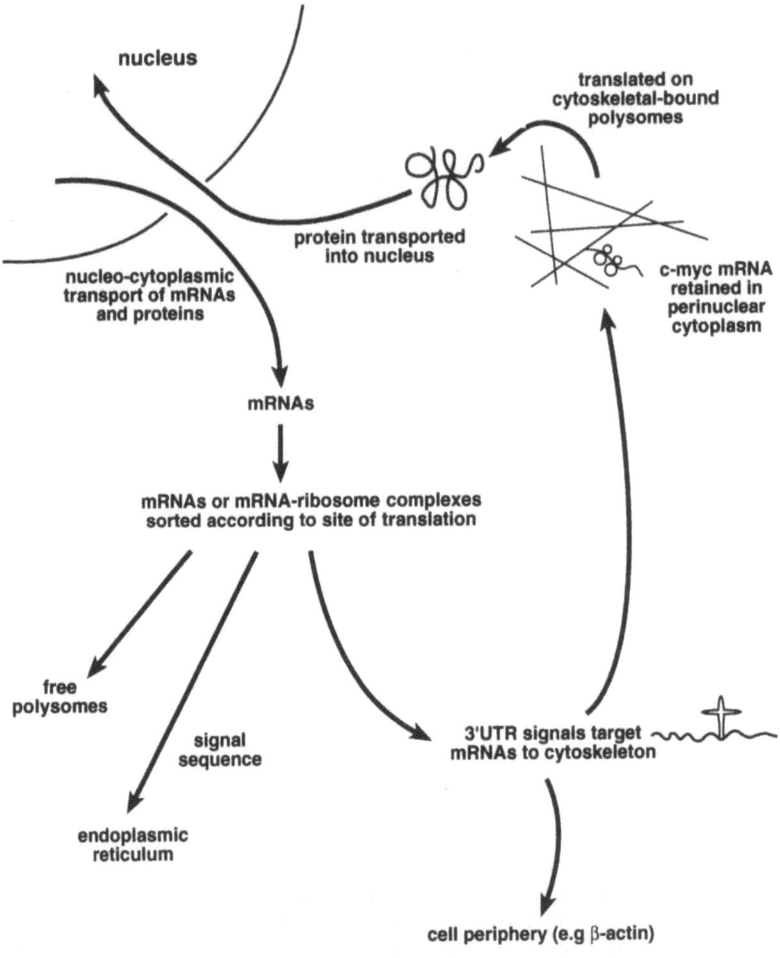

Fig. 5. Sorting of *c-myc* mRNA in the cytoplasm. This hypothetical scheme illustrates key features in mRNA sorting and in the translation of *c-myc* mRNA. Note (1) the targeting of *c-myc* mRNA to the perinuclear cytoplasm via a 3′UTR-dependent mechanism, (2) the need for different 3′UTR signals to sort mRNAs destined for translation at different subcellular sites and (3) the translation of *c-myc* mRNA in the perinuclear cytoplasm would favour efficient import of the protein into the nucleus

In cells transfected with the gene construct in which *c-myc* coding sequences are linked to the *β*-globin 3′UTR, the chimeric *c-myc* mRNA is no longer localized in the perinuclear cytoplasm, but it is present in polysomes; in addition, immunocytochemistry shows the presence of high concentrations of the *c-myc* protein. Both observations indicate that relocalized mRNA is translated and therefore that correct mRNA targeting is not essential for translation to occur.

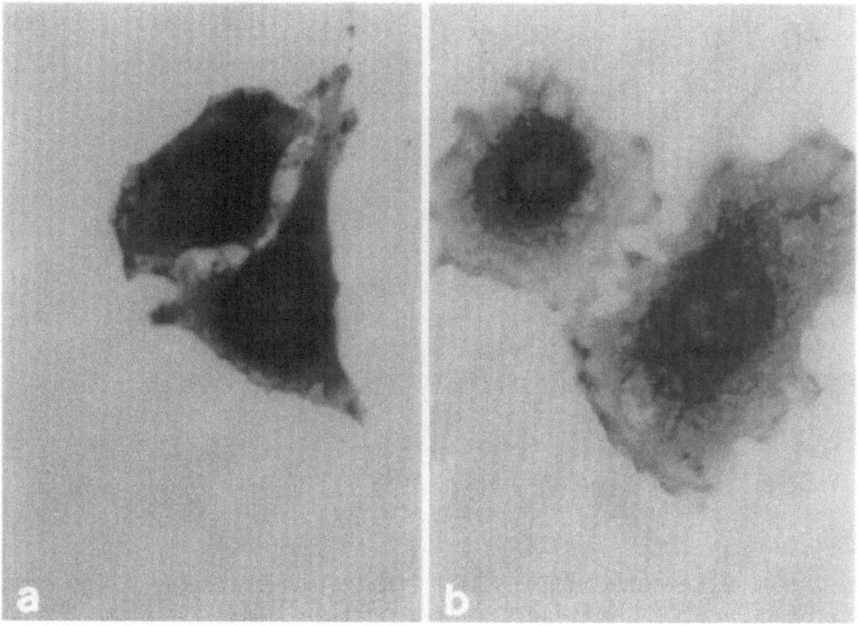

Fig. 6.a, b. Loss of *c-myc* mRNA localization affects distribution of *c-myc* protein. Immunohistochemical detection of *c-myc* protein in either **a** pSV-myc cells transfected with exons 2 and 3 of the *c-myc* gene and in which the mRNA is concentrated in the perinuclear cytoplasm (Wisdom and Lee 1991; Fig. 3) or **b** pSV-myc globin cells transfected with the coding sequences of *c-myc* linked to the β-globin 3′UTR and in which *c-myc* mRNA is found throughout the cytoplasm (Wisdom and Lee 1991). Note the nuclear and perinuclear staining in **a** and the staining throughout the cytoplasm in **b**. The results indicate that a greater proportion of *c-myc* protein is nuclear in the control pSV-myc cells **a**.

2.5 Localization Signals in c-myc mRNA

The construction of a series of gene constructs in which β-globin coding sequences were linked to the *c-myc* 3′UTR with increasingly large deletions has provided one approach to study the part of the 3′UTR responsible for *c-myc* mRNA localization (Veyrune et al. 1996). The positions of these deletions within the 3′UTR are shown in Fig. 7. The data from fractionation and localization studies of fibroblasts transfected with these constructs show that removal of a 86-nt sequence from base 194 to base 280 results in loss of localization and loss of association with the cytoskeleton. Furthermore, this 86-nt sequence, when linked to β-globin sequences, was sufficient to localize the transcripts to the perinuclear cytoplasm and to promote their translation on CBP (Veyrune et al. 1996). It would appear therefore that the localization signal within the 3′UTR lies in the 194- to 280-bp sequence. This part of the 3′UTR contains an AUUUA sequence which is highly conserved between species (Vriz and Mechali 1989) and mutation of this AUUUA causes loss of perinuclear localization and cytoskeletal association of both the chimeric transcript of β-globin linked

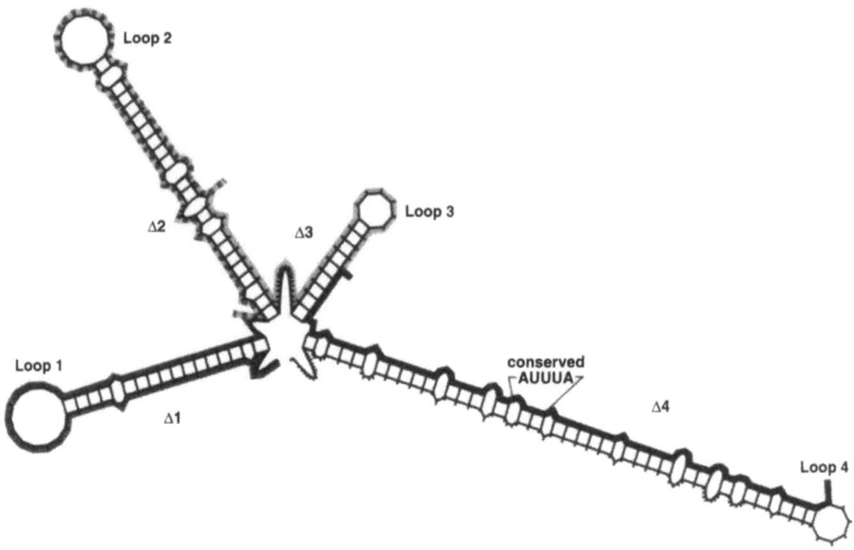

Fig. 7. Predicted secondary structure for *c-myc* 3′UTR. The secondary structure of *c-myc* 3′UTR, as predicted using the RNAFOLD programme, shows four stem-loop structures. Note the positions of the series of deletions (*1, 2 3* and *4*) created in the 3′UTR for studies to define the localization signal (Blobel and Dobberstein 1975. Chimeric constructs in which *β*-globin coding sequences were linked to the *c-myc* 3′UTR were transfected into fibroblasts and those containing deletions *2, 3* and *4* showed an association with the cytoskeleton and perinuclear distribution. Both this distribution and the association with CBP were lost in those with deletion *4*. The sequence lost between deletions *3* and *4* consists of approximately half of the long stem-loop *(loop 4)*

to the 86-nt localization sequence and the *c-myc* mRNA itself (Veyrune et al. 1996). However, this mutation has no effect on mRNA stability (Veyrune et al. 1996), thus supporting previous data showing that this AUUUA is not required for mRNA instability (Bonnieu et al. 1990; Lagnado et al. 1994) and showing that the observed differences in localization are due to changes in targeting of the mRNA rather than differences in stability. It would thus appear both that the localization signal in the *c-myc* 3′UTR is distinct from those elements which regulate mRNA stability, and that the signal involves the 198-280 bp region, including a conserved AUUUA sequence.

The precise extent and nature of the localization signal is not yet known. However, comparison of predicted secondary structures and the effects of deletions and mutations on localization suggests that secondary structure is a vital feature of the localization signal in the *c-myc* 3′UTR. Using RNA-fold programmes, prediction of the secondary structure of the *c-myc* 3′UTR shows four stem-loop structures (see Fig. 7), and one of these loops (loop 4) contains the 194-280 base-pair region defined as important in localization. It is particularly interesting that, as shown in Fig. 8, the mutation of the conserved AUUUA, which causes loss of localization, is predicted to have a dramatic effect on the secondary structure of the 3′UTR with total loss of loop 4, the

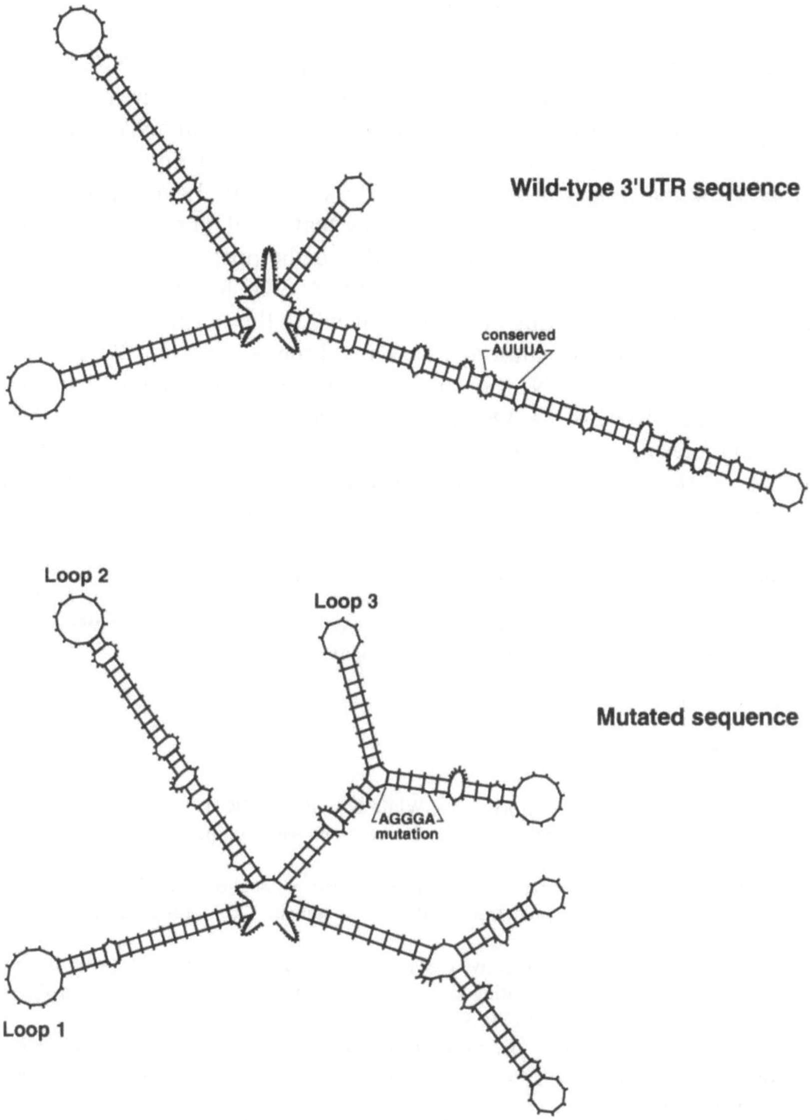

Fig. 8. Effect of mutation of the AUUUA in 194-280 base pair region on predicted secondary structure. Mutation of the conserved AUUUA within the 194-280 base pair region to AGGGA caused loss of ability to target *β*-globin coding sequences to the perinuclear cytoplasm and CBP (Blobel and Dobberstein 1975). Using the RNAFOLD programme, this mutation was predicted to cause a major change in the *c-myc* 3′UTR secondary structure with total loss of a long stem-loop. This suggests that the long stem-loop *(loop 4)* is an important feature of localization signal

long stem-loop structure implicated in localization. It seems likely therefore that, although at present speculative, all or part this long stem-loop structure in the *c-myc* 3′UTR is a major part of the localization signal.

3 Summary and Future Perspectives

It is now emerging that cytoplasmic fate of *c-myc* and *c-fos* mRNAs is complex (Fig. 5). Both mRNAs are translated on polysomes associated with the cytoskeleton and, in the case of *c*-myc, it appears that the mRNA is retained on the cytoskeleton in the perinuclear cytoplasm by a targeting mechanism that involves specific motifs in the 3′UTR (Hesketh et al. 1994). It appears that secondary structure of the 3′UTR is a key feature of this localization signal, although the precise nature of the signal and the proteins with which it presumably interacts remain to be elucidated. A 3′UTR-dependent mechanism also operates in the targeting of β-actin mRNA to the cell periphery (Kislauskis et al. 1993, 1994) and so, although we know little as yet about the localization signals for different mRNAs or the proteins which may bind to them, it is evident that a variety of 3′UTR signals must be used by the cell to sort mRNAs in the cytoplasm. In the case of *c-myc,* it appears that the retention of *c-myc* mRNA on the cytoskeleton in the perinuclear cytoplasm may allow efficient targeting of the newly synthesised protein to the nucleus.

Acknowledgements. JEH is supported by Scottish Office Agriculture, Environment and Fisheries Department (SOAEFD). JLV and JMB are supported by the Centre National de la Recherche Scientifique (CNRS) and by grants from the Ligue Contre le Cancer and ARC. We thank Marc Sitbon for criticism and helpful comments on the manuscript.

References

Adams A, Fey EG, Pike SF, Taylorson CJ, White HA, Rabin BR (1983) Preparation and properties of a complex from rat liver of polyribosomes with components of the cytoskeleton. Biochem J 216:215–226

Aghib DF, Bishop JM, Ottolenghi S, Guerrasio A, Serra A, Saglio G (1990) A 3′ truncation of MYC caused by chromosomal translocation in a human T-cell leukemia increases mRNA stability. Oncogene 5:707–711

Aharon T, Schneider RJ (1993) Selective destabilization of short-lived mRNAs with the granulocyte-macrophage colony-stimulating factor AU-rich 3′ noncoding region is mediated by a cotranslational mechanism. Mol Cell Biol 13:1971–1980

Ainger K, Avossa D, Morgan F, Hill SJ, Barry C, Barbarese E, Carson JH (1993) Transport and localization of exogenous myelin basic protein mRNA microinjected into oligodendrocytes. J Cell Biol 123:431–41

Akashi M, Shaw G, Hachiya M, Elstner E, Suzuki G, Koeffler P 1994) Number and location of AUUUA motifs: role in regulating transiently expressed RNAs. Blood 83:3182–3187

Amur US, Phan T, Campagnoni AT (1993) Myelin basic protein mRNA translocation in oligodendrocytes is inhibited by astrocytes in vitro. J Neurosci Res 36:99–110

Aziz N, Munro HN (1987) Iron regulates ferritin mRNA translation through a segment of its 5′ untranslated region. Proc Natl Acad Sci USA 84:8478–8482

Bassell GJ, Powers CM, Taneja KL, Singer RH (1994a) Single mRNAs visualized by ultrastructural in situ hybridization are principally localized at actin filament intersections in fibroblasts. J Cell Biol 126:863–876

Bassell GJ, Singer RH, Kosik KS (1994b) Association of poly(A) mRNA with microtubules in cultured neurons. Neuron 12:571–582

Behar L, Marx R, Sadot E, Barg J, Ginzburg I (1995) *cis*-acting signals and *trans*-acting proteins are involved in tau mRNA targeting into neurites of differentiating neuronal cells. Int J Dev Neurosci 13:113–127

Bernstein PL, Herrick DJ, Prokipcak RD, Ross J (1992) Control of *c-myc* mRNA half-life in vitro by a protein capable of binding to a coding region stability determinant. Genes Dev 6:642–654

Bickel M, Iwai Y, Pluznik DH, Cohen RB (1992) Binding of sequence-specific proteins to the adenosine-plus uridine-rich sequences of the murine granulocyte/macrophage colony-stimulating factor mRNA. Proc Natl Acad Sci USA 89:10001–10005

Bird RC, Sells BH (1986) Cytoskeleton involvement in the distribution of mRNP complexes and small cytoplasmic RNAs. Biochim Biophys Acta 868:215–225

Blobel G, Dobberstein B (1975) Transfer of proteins across membranes. I. Presence of proteolytically processed and unprocessed nascent immunoglobulin light chains on membrane bound ribosomes of murine myeloma. J Cell Biol 67:835–851

Bohjanen PR, Petryniak B, June CH, Thompson CB, Lindsten T (1991) An inducible cytoplasmic factor (AU-B) binds selectively to AUUUA multimers in the 3′ untranslated region of lymphokine mRNA. Mol Cell Biol 11:3288–3295

Bohjanen PR, Petryniak B, June CH, Thompson CB, Lindsten T (1992) AU RNA-binding factors differ in their binding specificities and affinities. J Biol Chem 267:6302–6309

Bonnieu A, Piechaczyk M, Marty L, Cuny M, Blanchard JM, Fort P, Jeanteur P (1988) Sequence determinants of *c-myc* mRNA turnover: influence of 3′ and 5′ non-coding regions. Oncogene Res 3:155–166

Bonnieu A, Rech J, Jeanteur P, Fort P (1989) Requirements for *c-fos* mRNA down regulation in growth stimulated murine cells. Oncogene 4:881–888

Bonnieu A, Roux P, Marty L, Jeanteur P, Piechaczyk M (1990) AUUUA motifs are dispensable for rapid degradation of the mouse *c-myc* RNA. Oncogene 5:1585–1588

Brewer G (1991) An A+ U-rich element RNA-binding factor regulates *c-myc* mRNA stability in vitro. Mol Cell Biol 11:2460–2466

Brewer G, Ross J (1988) Poly(A) shortening and degradation of the 3′ A+U-rich sequences of human *c-myc* mRNA in a cell-free system. Mol Cell Biol 8:1697–1708

Bruckenstein DA, Lein PJ, Higgins D, Fremeau RJ (1990) Distinct spatial localization of specific mRNAs in cultured sympathetic neurons. Neuron 5:809–819

Campbell GP, Hesketh JE (1996) Distribution of glutathione peroxidase mRNAs between free and cytoskeletal-bound polysomes in H4 hepatoma cells. Biochem Soc Trans (in press)

Caput D, Beutler B, Hartog K, Thayer R, Brown SS, Cerami A (1986) Identification of a common nucleotide sequence in the 3′-untranslated region of mRNA molecules specifying inflammatory mediators. Proc Natl Acad Sci USA 83:1670–1674

Carillo S, Pariat M, Steff AM, Roux P, Etienne JM, Lorca T, Piechaczyk M (1994) Differential sensitivity of FOS and JUN family members to calpains. Oncogene 9:1679–1689

Casey JL, Hentze MW, Koeller DM, Caughman SW, Rouault TA, Klausner RD, Harford JB (1988) Iron-responsive elements: regulatrory RNA sequences that control mRNA levels and translation. Science 240:924–928

Chen CY, Shyu AB (1994) Selective degradation of early-response-gene mRNAs: functional analyses of sequence features of the AU-rich elements. Mol Cell Biol 14:8471–8482

Chen CY, Chen TM, Shyu AB (1994) Interplay of two functionally and structurally distinct domains of the c-fos AU-rich element specifies its mRNA-destabilizing function. Mol Cell Biol 14:416–426

Chen CY, Xu N, Shyu AB (1995) mRNA decay mediated by two distinct AU-rich elements from c-fos and granulocyte-macrophage colony-stimulating factor transcripts: different deadenylation kinetics and uncoupling from translation. Mol Cell Biol 15:5777–5788

Cheng H, Bjerknes M (1989) Asymmetric distribution of actin mRNA and cytoskeletal pattern generation in polarized epithelial cells. J Mol Biol 210:541–549

Cheung HK, Serano TL, Cohen RS (1992) Evidence for a highly selective RNA transport system and its role in establishing the dorsoventral axis of the *Drosophila* egg. Development 114:653–661

Condeelis J (1995) Elongation factor 1 alpha, translation and the cytoskeleton. Trends Biochem Sci 20:169–170

Cripe L, Morris E, Fulton AB (1993) Vimentin mRNA location changes during muscle development. Proc Natl Acad Sci USA 90:2724–2728

Curatola AM, Nadal MS, Schneider RJ (1995) Rapid degradation of AU-rich element (ARE) mRNAs is activated by ribosome transit and blocked by secondary structure at any position 5′ to the ARE. Mol Cell Biol 15:6331–6340

Dalby B, Glover DM (1992) 3′ Non-translated sequences in *Drosophila* cyclin B transcripts direct posterior pole accumulation late in oogenesis and peri-nuclear association in syncytial embryos. Development 115:989–997

Dalby B, Glover DM (1993) Discrete sequence elements control posterior pole accumulation and translational repression of maternal cyclin B RNA in *Drosophila*. Embo J 12:1219–1227

Dang CV, Yang DC, Pollard TD (1983) Association of methionyl-tRNA synthetase with detergent-insoluble components of the rough endoplasmic reticulum. J Cell Biol 96:1138–1147

Dani C, Blanchard JM, Piechaczyk M, El SS, Marty L, Jeanteur P (1984) Extreme instability of myc mRNA in normal and transformed human cells. Proc Natl Acad Sci USA 81:7046–7050

Dani C, Mechti N, Piechaczyk M, Lebleu B, Jeanteur P, Blanchard JM (1985) Increased rate of degradation of *c-myc* mRNA in interferon-treated Daudi cells. Proc Natl Acad Sci USA 82:4896–4899

Davis I, Ish HD (1991) Apical localization of pair-rule transcripts requires 3′ sequences and limits protein diffusion in the *Drosophila* blastoderm embryo. Cell 67:927–940

Dolph PJ, Huang JT, Schneider RJ (1990) Translation by the adenovirus tripartite leader: elements which determine independence from cap-binding protein complex [Published erratum appears in J Virol 1990 Aug; 64(8):4042]. J Virol 64:2669–2677

Edwards DR, Mahadevan LC (1992) Protein synthesis inhibitors differentially superinduce *c-fos* and *c-jun* by three distinct mechanisms: lack of evidence for labile repressors. Embo J 11:2415–2424

Elinson RP, King ML, Forristall C (1993) Isolated vegetal cortex from *Xenopus* oocytes selectively retains localized mRNAs. Dev Biol 160:554–562

Ephrussi A, Dickinson LK, Lehmann R (1991) Oskar organizes the germplasm and directs localization of the posterior determinant nanos. Cell 66:37–50

Ferrandon D, Elphick L, Nusslein VC, St JD (1994) Staufen protein associates with the 3′UTR of bicoid mRNA to form particles that move in a microtubule-dependent manner. Cell 79:1221–1232

Forristall C, Pondel M, Chen L, King ML (1995) Patterns of localization and cytoskeletal association of two vegetally localized RNAs, Vgl and Xcat-2. Development 121:201–208

Fort P, Rech J, Vie A, Piechaczyk M, Bonnieu A, Jeanteur P, Blanchard JM (1987) Regulation of c-fos gene expression in hamster fibroblasts: initiation and elongation of transcription and mRNA degradation. Nucleic Acids Res 15:5657–5667

Garner CC, Tucker RP, Matus A (1988) Selective localization of messenger RNA for cytoskeletal protein MAP2 in dendrites. Nature 336:674–677

Gavis ER, Lehmann R (1992) Localization of nanos RNA controls embryonic polarity. Cell 71:301–313

Gavis ER, Lehmann R (1994) Translational regulation of nanos by RNA localization. Nature 369:315–318

Gavrilova LP, Ruktevitch NM, Gelfand VI, Motuz LP, Stahl J, Bommer UA, Bielka H (1987) Immunofluorescent localization of protein synthesis components in mouse embryo fibroblasts. Cell Biol Int Rep 11:745–753

Glass JR, DeWitt RG, Cress AE (1985) Rapid loss of stress fibers in Chinese hamster ovary cells after hyperthermia. Cancer Res 45:258–262

Greenberg ME, Ziff EB (1984) Stimulation of 3T3 cells induces transcription of the c-fos protooncogene. Nature 311:433–438

Gururajan R, Mathews L, Longo FJ, Weeks DL (1994) An3 mRNA encodes an RNA helicase that colocalizes with nucleoli in *Xenopus* oocytes in a stage-specific manner. Proc Natl Acad Sci USA 91:2056–2060

Gururajan R, Perry OH, Melton DA, Weeks DL (1991) The *Xenopus* localized messenger RNA An3 may encode an ATP-dependent RNA helicase. Nature 349:717–719

Hamill D, Davis J, Drawbridge J, Suprenant KA (1994) Polyribosome targeting to microtubules: enrichment of specific mRNAs in a reconstituted microtubule preparation from sea urchin embryos. J Cell Biol 127:973–984

Hamilton BJ, Nagy E, Malter JS, Arrick BA, Rigby WF (1993) Association of heterogeneous nuclear ribonucleoprotein A1 and C proteins with reiterated AUUUA sequences. J Biol Chem 268:8881–8887

Hentze MW, Caughman SW, Casey JL, Koeller DM, Rouault TA, Harford JB, Klausner RD (1988) A model for the structure and functions of iron-responsive elements. Gene 72:201–208

Hentze MW, Rouault TA, Caughman SW, Dancis A, Harford JB, Klausner RD (1987) A *cis*-acting element is necessary and sufficient for translational regulation of human ferritin expression in response to iron. Proc Natl Acad Sci USA 84:6730–6734

Herrick DJ, Ross J (1994) The half-life of *c-myc* mRNA in growing and serum-stimulated cells: influence of the coding and 3′ untranslated regions and role of ribosome translocation. Mol Cell Biol 14:2119–2128

Hesketh J (1994) Translation and the cytoskeleton: a mechanism for targeted protein synthesis. Mol Biol Rep 19:233–243

Hesketh J, Campbell G, Piechaczyk M, Blanchard JM (1994) Targeting of *c-myc* and beta-globin coding sequences to cytoskeletal-bound polysomes by *c-myc* 3′ untranslated region. Biochem J 298:143–148

Hesketh JE (1996) Sorting of mRNAs in the cytoplasm: mRNA localisation and the cytoskeleton. Exp Cell Res (in press)

Hesketh JE, Pryme IF (1988) Evidence that insulin increases the proportion of polysomes that are bound to the cytoskeleton in 3T3 fibroblasts. Febs Lett 231:62–66

Hesketh JE, Pryme IF (1991) Interaction between mRNA, ribosomes and the cytoskeleton. Biochem J 277:1–10

Hesketh JE, Campbell GP, Whitelaw PF (1991a) *c-myc* mRNA in cytoskeletal-bound polysomes in fibroblasts. Biochem J 274:607–609

Hesketh JE, Horne Z, Campbell GP (1991b) Immunohistochemical evidence for an association of ribosomes with microfilaments in 3T3 fibroblasts. Cell Biol Int Rep 15:141–150

Hesketh JE, Jodar D, Johanessen A, Partridge K, Pryme IF, Tauler A (1996) Enrichment of specific mRNAs in cytoskeletal-bound and membrane-bound polysomes in Chinese hamster ovary cells. Biochem Soc Trans (in press)

Heuijerjans JH, Pieper FR, Ramaekers FC, Timmermans LJ, Kuijpers H, Bloemendal H, Van VW (1989) Association of mRNA and eIF-2 alpha with the cytoskeleton in cells lacking vimentin. Exp Cell Res 181:317–330

Hill MA, Gunning P (1993) Beta and gamma actin mRNAs are differentially located within myoblasts. J Cell Biol 122:825–832

Hill MA, Schedlich L, Gunning P (1994) Serum-induced signal transduction determines the peripheral location of beta-actin mRNA within the cell. J Cell Biol 126:1221–1229

Hovland R, Campbell G, Pryme I, Hesketh J (1995) The mRNAs for cyclin A, c-myc and ribosomal proteins L4 and S6 are associated with cytoskeletal-bound polysomes in HepG2 cells. Biochem J 310:193–196

Johannessen AJ, Pyrme IF, Vedeler A (1995) Changes in distribution of actin mRNA in different polysome fractions following stimulation of MPC-11 cells. Mol Cell Biochem 142:107–115

Jones TR, Cole MD (1987) Rapid cytoplasmic turnover of *c-myc* mRNA: requirement of the 3′ untranslated sequences. Mol Cell Biol 7:4513–4521

Kabnick KS, Housman DE (1988) Determinants that contribute to cytoplasmic stability of human *c-fos* and beta-globin mRNAs are located at several sites in each mRNA. Mol Cell Biol 8:3244–3250

Katze MG, Lara J, Wambach M (1989) Nontranslated cellular mRNAs are associated with the cytoskeletal framework in influenza virus or adenovirus infected cells. Virology 169:312–322

Kim HJ, Smith JL, Macdonald PM (1991) Oskar mRNA is localized to the posterior pole of the *Drosophila* oocyte. Cell 66:23–35

Kim HJ, Webster PJ, Smith JL, Macdonald PM (1993) Multiple RNA regulatory elements mediate distinct steps in localization of oskar mRNA. Development 119:169–178

Kim HJ, Kerr K, Macdonald PM (1995) Translational regulation of oskar mRNA by bruno, an ovarian RNA-binding protein, is essential. Cell 81:403–412

Kislauskis EH, Li Z, Singer RH, Taneja KL (1993) Isoform-specific 3′-untranslated sequences sort alpha-cardiac and beta-cytoplasmic actin messenger RNAs to different cytoplasmic compartments. [Published erratum appears in J Cell Biol 1993 Dec; 123 (6 Pt 2): following 1907]. J Cell Biol 123:165–172

Kislauskis EH, Zhu X, Singer RH (1994) Sequences responsible for intracellular localization of beta-actin messenger RNA also affect cell phenotype. J Cell Biol 127:441–451

Kleiman R, Banker G, Steward O (1990) Differential subcellular localization of particular mRNAs in hippocampal neurons in culture. Neuron 5:821–830

Kloc M, Etkin LD (1994) Delocalization of Vg1 mRNA from the vegetal cortex in *Xenopus* oocytes after destruction of Xlsirt RNA. Science 265:1101–1103

Kloc M, Etkin LD (1995) Two distinct pathways for the localization of RNAs at the vegetal cortex in Xenopus oocytes. Development 121:287–297

Koeller DM, Horowitz JA, Casey JL, Klausner RD, Harford JB (1991) Translation and the stability of mRNAs encoding the transferrin receptor and *c-fos*. Proc Natl Acad Sci USA 88:7778–7782

Kondaiah P, Sands MJ, Smith JM, Fields A, Roberts AB, Sporn MB, Melton DA (1990) Identification of a novel transforming growth factor-beta (TGF-beta 5) mRNA in Xenopus laevis. J Biol Chem 265:1089–1093

Kozak M (1986) Influences of mRNA secondary structure on initiation by eukaryotic ribosomes. Proc Natl Acad Sci USA 83:2850–2854

Kruijer W, Cooper JA, Hunter T, Verma IM (1984) Platelet-derived growth factor induces rapid but transient expression of the *c-fos* gene and protein. Nature 312:711–716

Kruys V, Wathelet M, Poupart P, Contreras R, Fiers W, Content J, Huez G (1987) The 3′ untranslated region of the human interferon-beta mRNA has an inhibitory effect on translation. Proc Natl Acad Sci USA 84:6030–6034

Kruys VI, Wathelet MG, Huez GA (1988) Identification of a translation inhibitory element (TIE) in the 3′ untranslated region of the human interferon-beta mRNA. Gene 72:191–200

Kruys V, Marinx O, Shaw G, Deschamps J, Huez G (1989) Translational blockade imposed by cytokine-derived UA-rich sequences. Science 245:852–855

Ku M, Melton DA (1993) Xwnt-11: a maternally expressed *Xenopus* wnt gene. Development 119:1161–1173

Kuge H, Richter JD (1995) Cytoplasmic 3′ poly(A) addition induces 5′ cap ribose methylation: implications for translational control of maternal mRNA. EMBO J 14:6301–6310

Lagnado CA, Brown CY, Goodall GJ (1994) AUUUA is not sufficient to promote poly(A) shortening and degradation of an mRNA: the functional sequence within AU-rich elements may be UUAUUUA (U/A) (U/A). Mol Cell Biol 14:7984–7995

Laird-Offringa IA, Elfferich P, Van der Eb AJ (1991) Rapid c-myc mRNA degradation does not require (A + U)-rich sequences or complete translation of the mRNA. Nucleic Acids Res 19:2387–2394

Lantz V, Schedl P (1994) Multiple cis-acting targeting sequences are required for orb mRNA localization during *Drosophila* oogenesis. Mol Cell Biol 14:2235–2242

Lantz V, Ambrosio L, Schedl P (1992) The *Drosophila* orb gene is predicted to encode sex-specific germline RNA-binding proteins and has localized transcripts in ovaries and early embryos. Development 115:75–88

Latham VJ, Kislauskis EH, Singer RH, Ross AF (1994) Beta-actin mRNA localization is regulated by signal transduction mechanisms. J Cell Biol 126:1211–1219

Lenk R, Ramsom L, Kaufmann Y, Penman S (1977) A cytoskeletal structure with associated polyribosomes obtained from HeLa cells. Cell 10:67–78

Lindsten T, June CH, Ledbetter JA, Stella G, Thompson CB (1989) Regulation of lymphokine messenger RNA stability by a surface-mediated T cell activation pathway. Science 244:339–343

Linnen JM, Bailey CP, Weeks DL (1993) Two related localized mRNAs from *Xenopus laevis* encode ubiquitin-like fusion proteins. Gene 128:181–188

Litman P, Barg J, Rindzoonski L, Ginzburg I (1993) Subcellular localization of tau mRNA in differentiating neuronal cell culture: implications for neuronal polarity. Neuron 10:627–638

Litman P, Barg J, Ginzburg I (1994) Microtubules are involved in the localization of tau mRNA in primary neuronal cell cultures. Neuron 13:1463–1474

Luk SK, Kilpatrick M, Kerr K, MacDonald PM (1994) Components acting in localization of bicoid mRNA are conserved among *Drosophila* species. Genetics 137:521–530

Macdonald PM (1990) bicoid mRNA localization signal: phylogenetic conservation of function and RNA secondary structure. Development 110:161–171

Macdonald PM, Struhl G (1988) *cis*-acting sequences responsible for anterior localization of bicoid mRNA *Drosophila* embryos. Nature 336:595–598

Macdonald PM, Kerr K, Smith JL, Leask A (1993) RNA regulatory element BLE1 directs the early steps of bicoid mRNA localization. Development 118:1233–1243

Mahon P, Beattie JH, Glover LA, Hesketh JE (1995) Localization of metalothionein isoform mRNAs in rat hepatoma (H4) cells. FEBS Lett 373:76–80

Malter JS (1989) Identification of an AUUUA-specific messenger RNA binding protein. Science 246:664–666

Mattia E, den Blaauwen J, Ashwell G, van Renswoude J (1989) Multiple post-transcriptional regulatory mechanisms in ferritin gene expression. Proc Natl Acad Sci USA 86:1801–1805

Meadus WJ, Pramanik S, Bag J (1990) Cytoskeleton-bound mRNA for a 40-kDa polypeptide in rat L6 cells is not always translated. Exp Cell Res 187:25–32

Mitchell RL, Zokas L, Schreiber RD, Verma IM (1985) Rapid induction of the expression of proto-oncogene fos during human monocytic differentiation. Cell 40:209–217

Morris EJ, Fulton AB (1994) Rearrangement of mRNAs for costamere proteins during costamere development in cultured skeletal muscle from chicken. J Cell Sci 107:377–386

Mosquera L, Forristall C, Zhou Y, King ML (1993) A mRNA localized to the vegetal cortex of *Xenopus* oocytes encodes a protein with a nanos-like zinc finger domain. Development 117:377–386

Mowry KL, Melton DA (1992) Vegetal messenger RNA localization directed by a 340-nt RNA sequence element in *Xenopus* oocytes. Science 255:991–994

Muller R, Bravo R, Burckhardt J, Curran T (1984) Induction of *c-fos* gene and protein by growth factors precedes activation of *c-myc*. Nature 312:716–720

Myer VE, Lee SI, Steitz JA (1992) Viral small nuclear ribonucleoproteins bind a protein implicated in messenger RNA destabilization. Proc Natl Acad Sci USA 89:1296–1300

Ornelles DA, Fey EG, Penman S (1986) Cytochalasin releases mRNA from the cytoskeletal framework and inhibits protein synthesis. Mol Cell Biol 6:1650–1662

Osborn M, Weber K (1977) The detergent-resistant cytoskeleton tissue culture cells includes the nucleus and the microfilament bundles. Exp Cell Res 106:339–349

Pallavicini MG, George T, Deteresa PS, Amendola R, Gray JW (1994) Intracellular dynamic of c-myc mRNA traffic in single cells in situ. J Cell Physiol 158:223–230

Peppel K, Vinci JM, Baglioni C (1991) The AU-rich sequences in the 3′ untranslated region mediate the increased turnover of interferon mRNA induced by glucocorticoids. J Exp Med 173:349–355

Piechaczyk M, Blanchard JM (1994) *c-fos* proto-oncogene regulation and function. Crit Rev Oncol Hematol 17:93–131

Pokrywka NJ, Stephenson EC (1991) Microtubles mediate the localization of bicoid RNA during *Drosophila* oogenesis. Development 113:55–66

Pokrywka NJ, Stephenson EC (1994) Localized RNAs are enriched in cytoskeletal extracts of *Drosophila* oocytes. Dev Biol 166:210–219

Pokrywka NJ, Stephenson EC (1995) Microtubules are a general component of mRNA localization systems in *Drosophila* oocytes. Dev Biol 167:363–370

Pondel MD, King ML (1988) Localized maternal mRNA related to transforming growth factor beta mRNA is concentrated in a cytokeratin-enriched fraction from *Xenopus* oocytes. Proc Natl Acad Sci USA 85:7612–7616

Raff JW, Whitfield WG, Glover DM (1990) Two distinct mechanisms localise cyclin B transcripts in syncytial *Drosophila* embryos. Development 110:1249–1261

Rahmsdorf HJ, Schonthal A, Angel P, Litfin M, Ruther U, Herrlich P (1987) Posttranscriptional regulation of *c-fos* mRNA expression. Nucleic Acids Res 15:1643–1659

Ramaekers F, Benedetti E, Dunia I, Vorstenbosch P, Bloemendal H (1983) Polyribosomes associated with microfilaments in cultured lens cells. Biochim Biophys Acta 740:441–448

Ruther U, Garber C, Komitowski D, Muller R, Wagner EF (1987) Deregulated c-fos expression interferes with normal bone development in transgenic mice. Nature 325:412–416

Sachs AB (1993) Messenger RNA degradation in eukaryotes. Cell 74:413–421

Sachs AB, Deardorff JA (1992) Translation initiation requires the PAB-dependent poly(A) ribonuclease in yeast. Cell 70:961–973

Savant BS, Cleveland DW (1992) Evidence for instability of mRNAs containing AUUUA motifs mediated through translation-dependent assembly of a > 20S degradation complex. Genes Dev 6:1927–1939

Schiavi SC, Wellington CL, Shyu AB, Chen CY, Greenberg ME, Belasco JG (1994) Multiple elements in the *c-fos* protein-coding region facilitate mRNA deadenylation and decay by a mechanism coupled to translation. J Biol Chem 269:3441–3448

Schuler GD, Cole MD (1988) GM-CSF and oncogene mRNA stabilities are independently regulated in trans in a mouse monocytic tumor. Cell 55:1115–1122

Sharpless K, Biegel D, Yang T, Pachter JS (1993) Beta-actin mRNA-binding proteins associated with the cytoskeletal framework. Eur J Biochem 212:217–225

Shaw G, Kamen R (1986) A conserved AU sequence from the 3′ untranslated region of GM-CSF mRNA mediates selective mRNA degradation. Cell 46:659–667

Shestakova EA, Motuz LP, Gavrilova LP (1993a) Co-localization of components of the protein-synthesizing machinery with the cytoskeleton in G0-arrested cells. Cell Biol Int 17:417–424

Shestakova EA, Motuz LP, Minin AA, Gavrilova LP (1993b) Study of localization of the protein-synthesizing machinery along actin filament bundles. Cell Biol Int 17:409–416

Shyu AB, Greenberg ME, Belasco JG (1989) The c-fos transcript is targeted for rapid decay by two distinct mRNA degradation pathways. Genes Dev 3:60–72

Shyu AB, Belasco JG, Greenberg ME (1991) Two distinct destabilizing elements in the c-fos message trigger deadenylation as a first step in rapid mRNA decay. Genes Dev 5:221–231

Singer RH (1993) RNA zipcodes for cytoplasmic addresses. Curr Biol 3:719–721

Singer RH, Langevin GL, Lawrence JB (1989) Ultrastructural visualization of cytoskeletal mRNAs and their associated proteins using double-label in situ hybridization. J Cell Biol 108:2343–2353

St JD (1995) The intracellular localization of messenger RNAs. Cell 81:161–170

Stebbings H (1986) Cytoplasmic transport and microtubules in telotrophic ovarioles of hemipteran insects. Int Rev Cytol 101:101–123

Stephenson EC, Pokrywka NJ (1992) Localization of bicoid message during *Drosophila* oogenesis. Curr Top Dev Biol 26:23–34

Sundell CL, Singer RH (1990) Actin mRNA localizes in the absence of protein synthesis. J Cell Biol 111:2397–3403

Symington AL, Zimmerman S, Zimmerman AM (1993) The influence of hydrostatic pressure on the distribution of histone mRNA in HeLa cells. Biochem Cell Biol 71:150–155

Taneja KL, Lifshitz LM, Fay FS, Singer RH (1992) Poly(A) RNA codistribution with microfilaments: evaluation by in situ hybridization and quantitative digital imaging microscopy. J Cell Biol 119:1245–1260

Tiedge H, Fremeau RJ, Weinstock PH, Arancio O, Brosius J (1991) Dendritic location of neural BC1 RNA. Proc Natl Acad Sci USA 88:2093–2097

Tiedge H, Drager UC, Brosius J (1992) Murine BC1 RNA in dendritic fields of the retinal inner plexiform layer. Neurosci Lett 141:136–138

Tiedge H, Zhou A, Thorn NA, Brosius J (1993) Transport of BC1 RNA in hypothalamo-neuro-hypophyseal axons. J Neurosci 13:4214–4219

Toh B, Lolait S, Mathy J, Baum R (1980) Association of mitochondria with intermediate filaments and of polyribosomes with cytoplasmic actin. Cell Tissue Res 211:163–169

Treisman R (1985) Transient accumulation of c-fos RNA following serum stimulation requires a conserved 5′ element and *c-fos* 3′ sequences. Cell 42:889–902

Trembleau A, Calas A, Fevre MM (1990) Ultrastructural localization of oxytocin mRNA in the rat hypothalamus by in situ hybridization using a synthetic oligonucleotide. Brain Res Mol Brain Res 8:37–45

Trembleau A, Morales M, Bloom FE (1994) Aggregation of vasopressin mRNA in a subset of axonal swellings of the median eminence and posterior pituitary: light and electron microscopic evidence. J Neurosci 14:39–53

Ueno S, Kotani Y, Kondoh K, Sano A, Kakimoto Y, Campagnoni AT (1994) The 3′-Untranslated region of mouse myelin basic protein gene increases the amount of mRNA immortalized mouse oligodendrocytes. Biochem Biophys Res Commun 204:1352–1357

Vakalopoulou E, Schaack J, Shenk T (1991) A 32-kilodalton protein binds to AU-rich domains in the 3′ untranslated regions of rapidly degraded mRNAs. Mol Cell Biol 11:3355–3364

Van Venrooij WJ, Sillekens PTG, Van Eekelen CAG, Reinders RT (1981) On the association of mRNA with the cytoskeleton in uninfected and adenovirus-infected human KB cells. Exp Cell Res 135:79–91

Vedeler A, Pryme IF, Hesketh JE (1991) The characterization of free, cytoskeletal and membrane-bound polysomes in Krebs II ascites and 3T3 cells. Mol Cell Biochem 100:183–193

Veyrune JL, Carillo S, Vie A, Blanchard JM (1995) *c-fos* mRNA instability determinants present within both the coding and the 3′ noncoding region link the degradation of this mRNA to its translation. Oncogene 11; 2127–2134

Veyrune JL, Campbell GP, Wiseman JW, Blanchard JM, Hesketh JE (1996) A localization signal in the 3′ untranslated region of *c-myc* mRNA targets *c-myc* mRNA and *β*-glolbin reporter sequences to the perinuclear cytoplasm and cytoskeletal-bound polysomes. J Cell Sci 109:1185–1194

Vriz S, Mechali M (1989) Analysis of 3′-untranslated regions of seven *c-myc* genes reveals conserved elements prevalent in post-transcriptionally regulated genes. Febs Lett 251:201–206

Watt F, Molloy PL (1993) Specific Clevage of transcription factors by the thiol protease, m-calpain. Nucleic Acids Res 21:5092–5100

Webster PJ, Suen J, Macdonald PM (1994) *Drosphila virilis* oskar transgenes direct body patterning but not pole cell formation or maintenance of mRNA localization in *D. melanogaster*. Development 120:2027–2037

Cell-Free Systems for Analysis of Cytoplasmic mRNA Turnover

C.T. DeMaria and G. Brewer

1 Introduction

The level of mRNA available for protein synthesis is dependent not only on the rates of transcription, processing, and transport, but also on the rates of mRNA turnover. Cytoplasmic mRNA turnover is an important component of gene expression in many different organisms (reviewed in Peltz et al. 1991; Sachs 1993; Beelman and Parker 1995) with abnormal changes in cytoplasmic mRNA turnover potentially leading to developmental defects or various diseases (Carter and Malter 1991; Schiavi et al. 1992). Thus it is important to understand how mRNA decay is regulated and to identify the *cis* and *trans* elements of this regulation. One method for studying cytoplasmic mRNA turnover is the use of cell-free decay systems. The study of mRNA turnover using cell-free systems can be extremely useful for determining mechanisms of mRNA decay, thereby elucidating properties of the factors comprising the decay machinery. Here we present a brief introduction to cell-free mRNA decay, then present an overview of specific uses of cell-free decay systems and recent advances in the study of mRNA turnover for a number of different organisms.

The most important requirement for any cell-free decay system is its fidelity for reconstituting cellular decay processes. Ideally, the system should meet three criteria. First, a cell-free system should reconstitute the rank order of cellular mRNA decay. mRNA degradation in vitro often proceeds more slowly than in cells, so that the half-lives of specific mRNAs may not be identical in vitro and in cells. Nonetheless, a functional cell-free decay system should reproduce the relative half-lives of cellular mRNAs. Secondly, degradation of a particular mRNA should produce the same decay products as are formed in cells. Finally, regulated cellular mRNA turnover rates which result from, for example, viral infection or extracellular stimuli such as hormones, should be reproduced in vitro.

There are many reasons for using cell-free systems to study mRNA turnover. It is often desirable to study mRNA turnover in vitro rather than in intact cells, as the latter can be problematic. For example, it is frequently necessary to

Department of Microbiology and Immunology, Bowman Gray School of Medicine of Wake Forest University, Medical Center Blvd., Winston-Salem, NC 27157-1064, USA

Progress in Molecular and Subcellular Biology, Vol. 18
Ph. Jeanteur (Ed.)

treat cells with transcription-inhibiting drugs to follow decay of cellular mRNAs. Such drugs may affect other cellular processes and have indirect effects on mRNA decay. Also, mRNA decay intermediates may be difficult to observe using intact cells due to their lability. Generally, decay reaction rates are slower in vitro so that normally short-lived decay products can be observed. Other advantages of cell-free decay systems include the ability to directly identify and characterize both *cis*- and *trans*-regulators of decay; the possibility of purifying mRNases and other *trans*-acting degradation factors directly from the system; and the ease of modifying RNA sequence and structure in vitro and studying the effects by adding the exogenous mRNAs to the system. Each of these will be discussed further. Finally, it is generally not possible to study mRNA decay in higher organisms using genetic approaches as is commonly done with bacteria and the yeast *Saccharomyces cerevisiae*. As a system that is easily manipulated, cell-free extracts can facilitate the study of mRNA decay in organisms for which genetic methods are poorly developed.

The cell type used for preparing decay extracts, the extracellular signals that induce decay of certain mRNAs, and the cellular fraction(s) used should all be taken into consideration when designing an in vitro degradation system. Various cell-free systems optimal for the study of specific mRNAs have been used to investigate their turnover in diverse groups of organisms including bacteria, yeast, chicken, frog, plants, mammals and virally infected cells. These cell-free systems typically contain some components found in cell-free translation systems such as GTP, ATP and an ATP-regenerating system, monovalent and divalent cations, and one or more cytoplasmic fractions, such as polysomes, described below.

1. Postnuclear supernatant. Such extracts are easily prepared from cells by lysis followed by removal of cellular debris by low-speed centrifugation. However, further biochemical fractionation by differential centrifugation (Howell et al. 1989) is more useful for identifying the cytoplasmic fractions containing the factors necessary for decay of a particular mRNA.
2. Polysomes. Both ribonucleases and translated mRNAs are present in polysomes, so that turnover of the endogenous (polysome-associated) mRNAs can be studied. Ross and Kobs (1986) showed that in decay reactions utilizing polysomes from K562 human erythroleukemia cells, the rank order of decay of a number of mRNAs is the same as observed in whole cells (e.g., stability of γ-globin mRNA > δ-globin > c-*myc* > H4 histone mRNA). It may also be possible to examine decay of exogenous mRNAs, as was done for H4 histone mRNA, that can be degraded by factors loosely associated with polysomes (Ross and Kobs 1986). The advantage of exogenous mRNAs is that they can be radiolabeled in vitro; this greatly facilitates their detection and obviates the need for Northern analyses or nuclease protection assays.
3. Ribosomal salt wash. Ribosomal salt wash is composed of ribosome-associated proteins solubilized by incubating polysomes with salts (e.g.,

0.5 M KCl); washed polysomes are then removed by centrifugation. It contains RNases and therefore is a potential starting material for the purification of enzymes involved in mRNA decay.
4. mRNP. The decay of some mRNAs has been investigated using polysomal mRNP which are solubilized by incubating polysomes with a low concentration of EDTA; polysomes are then removed by centrifugation. Some mRNA decay enzymes may be present in mRNP and not be directly associated with the ribosome.
5. Reticulocyte lysate. Because efficient translation occurs in this system, several groups have used these extracts to study relationships between translation and mRNA decay.

Despite the numerous advantages of cell-free systems for studying mRNA decay, potential problems do exist. Most importantly, it may not be easy or at all possible to establish a cell-free decay system using extracts prepared from some cell types. This may be particularly true of mammalian primary tissue cells, which are particularly rich in ribonucleases compared to tissue culture cells. In addition, the use of subcellular fractions may destroy some requirements for mRNA decay, such as possible relationships between mRNA localization or the cytoskeleton and mRNA decay. Similarly, a cell-free system may not be useful for studies investigating decay of in vitro-transcribed mRNAs if a ribosome-associated mRNP is required for its decay. In general, even though cell-free systems are often applicable for studying mRNA decay, it is important to be aware of the potential difficulties such systems can present.

The remainder of this chapter will present an overview of cell-free mRNA decay. Extensive descriptions of the various cell-free systems are not presented, in order to allow for ample discussion of the results from each study. For more detailed information on methodology, the reader is directed to the references within the studies described here. Our discussion will include the specific uses for cell-free decay systems and important findings that have been made in recent years by taking advantage of such systems. Cell-free systems from a variety of organisms have been reported and some have been used for studying a specific feature of mRNA decay. Therefore, in vitro decay systems from different organisms will be discussed separately. The description of mammalian systems will be divided into sections, each detailing a particular use of cell-free systems for studying mammalian mRNA decay and/or a noted advancement in this field.

2 mRNA Decay in Lower Eukaryotes

2.1 Tetrahymena thermophila

In *Tetrahymena* cells, the expression of the surface protein SerH3 is controlled mainly by a temperature-dependent change in SerH3 mRNA turnover rates. In cells grown at 30 °C, SerH3 mRNA has a half-life of 60 min, while in cells grown at 40 °C, the half-life decreases to 3 min (Love et al. 1988). McMillan et al. (1993) utilized cytoplasmic extracts from wild-type and mutant *Tetrahymena* cells to study this temperature-dependent change in SerH3 mRNA stability in vitro. In vitro-synthesized ^{32}P-labeled SerH3 RNA was incubated with cytoplasmic extracts from cells grown at different temperatures. In extracts from wild-type cells grown at 30 °C, SerH3 RNA had a half-life of approximately 47 min, and in extracts from cells grown at 40 °C, the SerH3 RNA half-life was 6 min. Thus the relative differential stabilities of SerH3 mRNA observed in cells grown at the two temperatures were reproduced in vitro. Importantly, bulk mRNA was stable in extracts from cells grown at either temperature, indicating that in vitro mRNA decay is selective for SerH3 and perhaps a limited number of other mRNAs. An advantage of the cell-free system used for this study was the ability to distinguish between direct and indirect effects of temperature on the decay process. Performing in vitro decay assays at either 30 or 40 °C indicated that the decay reaction temperature had no effect on decay of SerH3 mRNA. Instead, it was the temperature at which cells were grown prior to extract preparation that affected in vitro mRNA stability. This suggests that SerH3 mRNA stability is affected by a component(s) in the cytoplasmic extract which is itself affected by temperature.

The cell-free system was also used to study a mutant *Tetrahymena* cell line in which SerH3 mRNA is stable at 20 °C but unstable at 30 °C. This cell line contains a temperature-sensitive mutation in the rseB locus, which controls SerH3 expression post-transcriptionally. In cell-free decay reactions with cytoplasmic extracts from rseB cells grown at 20 °C SerH3 mRNA is stable, while in extracts from cells grown at 30 °C SerH3 mRNA is unstable, a reproduction of the in vivo observations. This suggests that the rseB locus encodes a cytoplasmic factor important for the regulation of SerH3 mRNA stability in *Tetrahymena thermophila*.

2.2 Yeast

The yeast *Saccharomyces cerevisiae* has proven useful for identification and purification of factors that are important for mRNA turnover. Stevens (1980a,b) identified a mRNA decapping activity and purified a 5′⇒3′ exoribonuclease (XRN1). Sachs and colleagues have used a cell-free system to study mRNA deadenylation, which is the first step in the degradation of some mRNAs (reviewed in Beelman and Parker 1995). In these studies, cell-free extracts were used to investigate deadenylation of both endogenous and exogenous RNAs as well as the roles of poly(A) binding protein (PAB) and poly(A) nuclease (PAN) (Lowell et al.1992; Sachs and Deardorff 1992).

There have been only a limited number of reports of cell-free mRNA decay experiments using *S. cerevisiae* that examine decay of specific mRNAs. For example, Vreken et al. (1992) reported a system that reproduces the in vivo degradation pathway of phosphoglycerate kinase (PGK) mRNA. Gerstel et al.

(1992) reported a system that mimics turnover of the PGK and heat-shock protein 26 (HSP26) mRNAs. However, in many cases, endogenous mRNAs are very stable in vitro regardless of their degree of instability in intact cells (Stuart Peltz, pers. comm.). Thus it is not yet clear how applicable yeast cell-free systems will be for study of mRNA turnover. Nonetheless, *S. cerevisiae* provides the power of genetics for dissecting mRNA turnover (for reviews see Peltz and Jacobson 1992; Sachs 1993; Beelman and Parker 1995).

3 Viral Induction of mRNA Decay

Several eukaryotic viruses turn off host gene expression by inducing degradation of cellular mRNA, and in some cases viral mRNA stability is also affected (reviewed in Katze and Agy 1990). Cell-free decay systems have been described for the study of mRNA decay induced by Herpes simplex virus type 1 and human papillomavirus type 16.

3.1 Herpes simplex Virus Type 1 (HSV-1)

HSV-1 induces degradation of many host mRNAs soon after infection. This requires the virion host shutoff (vhs) protein, a determinant of viral pathogenesis which induces decay of both cellular and viral mRNAs (Strelow and Leib 1995). To better understand the process of mRNA destabilization induced by vhs, in vitro decay systems were developed. One cell-free system used cytoplasmic in vitro translation extracts prepared from HeLa cells that were either mock-infected or infected with either wild-type HSV-1 or *vhs-1* mutant virions (Krikorian and Read 1991). In this system, endogenous GAPDH mRNA was rapidly degraded in extracts from HSV-1-infected cells but was stable in extracts from both mock-infected and mutant-infected cells. Likewise, viral thymidine kinase mRNA was degraded in extracts from HSV-1-infected cells, while remaining stable in extracts from mutant-infected cells. In addition, the observed in vitro decay was specific for mRNA, since 28S rRNA was stable in all extracts analyzed. Thus, for both cellular and viral mRNAs, the rank order of cell-free decay in the different extracts paralleled that observed in vivo. Biochemical characterization experiments indicated the following: (1) the vhs-dependent mRNA decay activity was inactivated by heat or by proteinase K but not by brief treatment with micrococcal nuclease or RNasin; (2) the activity was dependent on Mg^{2+} ions; and (3) the activity did not require the components of an energy-generating system.

Ross and colleagues also used their cell-free system to investigate the effects of HSV-1 infection on mRNA turnover (Sorenson et al. 1991). In this study, polysomes from uninfected K562 human erythroleukemia cells served as the source of mRNA. Either polysomes or postpolysomal supernatant (S130) from HSV-1-infected, mock-infected or *vhs-1*-infected murine erythroleukemia cells

(MEL) was added to the K562 polysomes. These experiments revealed that only the S130 fraction from HSV-1-infected MEL cells promotes decay of polysome-associated γ-globin and GAPDH mRNAs, which are normally stable; the decay of the normally labile H4 histone and c-*myc* mRNAs was not enhanced further. This suggests that the virus-induced destabilization activity observed in vitro is present in the S130 and requires the vhs gene product. Recent results indicate that the vhs gene product is a mRNase (Jeff Ross, pers. comm.). The finding that the destabilization activity fractionated to the S130 also illustrates the advantage of performing decay reactions with fractionated cytoplasmic extracts. Both cell-free systems described here have provided valuable insights into the process of vhs-dependent mRNA decay.

3.2 Human Papillomavirus Type 16 (HPV-16)

The HPV-16 genome contains a negative regulatory element, located in the L1 3′ UTR, which participates in the control of late viral gene expression (Kennedy et al. 1990). An in vitro mRNA decay assay was devised by Kennedy et al. (1991) to study the regulatory function of this element. In vitro-transcribed and polyadenylated RNAs were incubated in mixtures containing polysomes from HeLa cells; these were then assayed for degradation of the exogenous poly(A)$^+$ RNA. In this system, a mRNA containing the HPV-16 negative regulatory element was rapidly degraded, having a half-life of 30 min. However, a mRNA lacking this element was stable for the duration of the assay (90 min). These results suggest that the role of the negative element is in destabilization of polyadenylated RNA. This study exemplifies the usefulness of a cell-free decay system for studying the role of a *cis*-acting element in mRNA decay.

4 mRNA Stability in Plants

4.1 Spinach Chloroplasts

The expression of chloroplast genes can be developmentally regulated, as in the development of amyloplasts to chloroplasts, and often involves post-transcriptional control (reviewed in Gruissem 1989). An example is *psb*A mRNA which encodes the reaction center protein of photosystem II. *psb*A mRNA is present in low levels in spinach amyloplasts but accumulates to high levels in chloroplasts, despite similar relative transcription rates for the *psb*A gene in the two plastid types (Deng and Gruissem 1987). Like most chloroplast mRNAs, *psb*A mRNA contains an inverted repeat (IR) at the 3′ terminus, which can potentially form a stem/loop structure by posttranscriptional processing. 3′ IRs are necessary for mRNA stability, presumably by blocking a 3′⇒5′ exonuclease(s) (reviewed in Brawerman 1987; Gruissem 1989). To better understand the mechanism(s) of this differential control of mRNA level, degradation of *psbA* mRNA was studied in vitro (Adams and Stern 1990). In this study, mutagenesis of the *psbA* IR was combined with a cell-free degradation assay to investigate the role of IRs in mRNA stability. Decay reactions contained spinach chloroplast soluble protein extracts and in vitro-transcribed radio-labeled RNAs with various IR mutations. These assays revealed that certain point mutations in the *psbA* 3′ IR can result in increased *psbA* mRNA stability. In support of the fidelity of this system, the relative stabilities of the wild-type and mutant mRNAs observed in vitro reflected those in chloroplasts electro-porated with the same mRNAs. These experiments led to identification of an endoribonuclease activity. A mutation in the IR loop region both increased the *psbA* half-life in vitro and led to a qualitatively different endoribonucleolytic cleavage pattern, suggesting that IR sequences may be important for nucleo-lytic cleavage as a part of the mRNA decay process.

This cell-free system was also used to characterize ribonuclease activities in spinach chloroplasts (Chen and Stern 1991). The focus of this study was the stability of *petD* mRNA, which also contains a 3′ IR. In vitro-synthesized RNA corresponding to the precursor *petD* 3′ mRNA was processed in vitro to a mature form ending with a 3′ hairpin structure, as in vivo. Apparently a 33-kDa protein from spinach chloroplasts binds the hairpin structure and decreases the efficiency of 3′ end processing. The authors suggest that this occurs through blockage of a 3′⇒5′ exoribonuclease. The cell-free system also revealed an endoribonuclease activity. Incubation of the *petD* 3′ IR RNA precursor with this activity resulted in two major products that appeared to arise from a single endonucleolytic cleavage. A RNA corresponding to the 5′ endonucleolytic cleavage product was rapidly degraded in vitro, suggesting that the identified endoribonuclease activity determines the decay rate of *petD* mRNA. Thus, experiments using spinach chloroplast extracts have enabled identification of endoribonuclease activities important for both *psbA* and *petD* 3′ IR RNA processing and mRNA degradation in vitro.

4.2 Soybean

mRNA turnover is involved in the regulation of ribulose-1,5-bisphosphate carboxylase (rbcS) synthesis in soybean. In soybean seedlings and transgenic petunia leaves, the rbcS mRNA, SRS4, is degraded to smaller products with varying lengths which all retain intact 5′ ends but lack poly(A) tails (Thompson et al. 1992). To facilitate the study of SRS4 mRNA decay, Tanzer and Meagher (1994) developed a cell-free system using polysomes from either soybean seedlings or mature petunia leaves. In this system, both endogenous SRS4 mRNA and SRS4 mRNA synthesized in vitro were degraded to the same proximal products as in vivo. Addition of plasmid vector sequences to the 5′ and 3′ ends of in vitro-synthesized SRS4 mRNAs or 3′ end truncation of these mRNAs had no effect on degradation in vitro. Moreover, the same 3′-distal fragments were present in vivo and in vitro following mRNA degradation. This suggests that the initial step(s) in SRS4 mRNA decay may be endonucleolytic cleavage.

5 Xenopus

Cell-free decay systems have been developed to study the turnover of mRNAs regulated by hormones or development in the frog *Xenopus laevis.*

5.1 Albumin mRNA Decay in Xenopus Liver

Estrogen induces degradation of mRNAs encoding serum proteins such as albumin in male *Xenopus* liver while not affecting bulk liver mRNAs. The lack of an estrogen-responsive *Xenopus* liver cell line presented the need for an alternative method of studying estrogen-induced liver mRNA decay. Therefore, an in vitro system was designed to investigate albumin mRNA stability (Pastori et al. 1991). Seventy-five percent of polysome-associated albumin mRNA was degraded by 60 min, versus only 10% degraded in post-mitochondrial extracts or mRNPs. To examine the effect of estrogen on polysome-associated decay, an excess of total liver RNA from untreated animals was incubated in vitro with polysomes from either estrogen-treated or untreated animals. Degradation of >90% of the albumin mRNA present in the total RNA population occurred with polysomes from estrogen-treated animals, compared to 10–20% degradation with polysomes from untreated animals. This suggests that estrogen induces a polysome-associated ribonuclease activity involved in albumin mRNA decay. The effect of estrogen on mRNA decay must be due to an increase in levels or activity of the polysomal nuclease, since an inhibitor of the polysomal nuclease was observed in the postmitochondrial extract and estrogen did not suppress the inhibitor activity. Incubation of either purified liver RNA, liver mRNP fractions, or in vitro transcribed mRNA with nuclease activity extracted from polysomes resulted in degradation of albumin mRNA but not ferritin mRNA. The importance of this cell-free system for studying mRNA decay is that it replaces the need for an estrogen-responsive cell line, and it provides a means for adding either extracts containing nuclease activity or specific cellular fractions to substrate RNAs.

5.2 Xenopus Oocyte mRNAs

5.2.1 Maternal mRNA Deadenylation

Some maternally encoded mRNAs are deadenylated and released from polysomes during oocyte maturation (Varnum and Wormington 1990). Using a cell-free system, Varnum et al. (1992) studied this maturation-specific deadenylation. In vitro reactions included whole cell extracts, nuclear extracts, or cytoplasmic extracts and in vitro synthesized RNA substrates. Exogenous RNAs were deadenylated in reactions with whole cell extracts from either immature or mature oocytes. Neither nuclear nor cytoplasmic extracts from immature oocytes could separately induce RNA deadenylation; however, when the two extracts were mixed in vitro, RNA deadenylation activity was observed. These results and those from in vivo studies indicate that factors necessary for deadenylation are present in immature oocytes, but deadenylation is prevented by the retaining of a factor(s) in the nucleus. Upon maturation, nuclear factors would be released to the cytoplasm, in turn allowing deadenylation to occur. The usefulness of this cell-free system for assaying extracts either separately or in combination provided a better understanding of how maturation-specific deadenylation is regulated.

5.2.2 Maternal Xlhbox2 mRNA Degradation

The mRNA encoding the maternal homeo box protein *Xlhbox2* is present at high levels in previtellogenic *Xenopus* oocytes, decreasing to low levels in later developmental stages. *Xlhbox2* mRNA injected into *Xenopus* oocytes is degraded endonucleolytically; a 90-nucleotide region of the *Xlhbox2* 3′ UTR is sufficient for this sequence-specific cleavage (Brown and Harland 1990). Further study of the endonucleolytic cleavage and degradation of *Xlhbox2* mRNA utilized in vitro decay reactions that accurately reproduce in vivo results (Brown et al. 1993). Cell-free reactions contained lysates from *Xenopus* oocytes or embryos and in vitro-synthesized ^{32}P-labeled RNAs. RNA was endonucleolytically cleaved to produce the expected degradation products. When unlabeled competitor RNA containing the endonuclease recognition sites was added to in vitro reactions, cleavage and degradation of the labeled RNA was enhanced rather than inhibited as expected. This suggests that the lysate contains a factor that binds the endonuclease recognition site(s) protecting the RNA from cleavage, and that the unlabeled competitor RNA titrates this factor from the labeled RNA. Lysates from early- and late-stage oocytes were used to identify possible developmental regulation of *Xlhbox2* mRNA decay. In cell-free reactions with lysates from early stage oocytes, increasing competitor RNA had only a slight effect on cleavage of the labeled RNA. By contrast, competitor RNA greatly enhanced cleavage and degradation of the labeled RNA with lysates from late stage oocytes. These results indicate that levels of the protective factor increase between early and late stages of oocyte development.

This cell-free system was also important for defining the RNase cleavage site. In vitro decay reactions with radiolabeled RNAs containing one to six copies of a 19-nucleotide repeat from *Xlhbox2* mRNA revealed that one copy of this sequence is sufficient for endonucleolytic cleavage; each additional copy resulted in a linear decrease in RNA half-life. In this study, therefore, the in vitro system was instrumental in both the identification of an endoribonuclease inhibitor activity in oocytes and the determination of the minimal recognition and cleavage site of the endoribonuclease.

6 Chicken

Vitellogenin, which is the precursor of egg yolk proteins, is encoded by three different genes in chickens. Vitellogenin II gene expression in the liver is activated by estrogen in part through stabilization of the mRNA (Wiskocil et al. 1980). One study of this activation utilized an in vitro decay system (Liang and Jost 1991). UV cross-linking assays identified a polysome-associated, liver-specific, 66-kDa protein (p66) that specifically bound the 5′ UTR of vitellogenin mRNA upon induction by estrogen. Cell-free decay reactions were then used to investigate the role of this protein in vitellogenin mRNA stability. The radiolabeled RNA substrate contained only the p66 binding sites, enabling a study of the effect of presumably just one protein (p66) in the extract. Reactions included the radiolabeled RNA, protein from rooster liver cytoplasmic extract as a source of RNases, and hen polysome extract. The vitellogenin RNA fragment was protected from RNase digestion by the polysome extract, while the corresponding antisense RNA was not protected. p66 partially purified from liver polysomes also had a protective effect on the RNA substrate. These observations suggest that binding of p66 to vitellogenin mRNA upon estrogen induction protects the mRNA from RNase attack and decay. This work provides one example of how a cell-free system can be used to study the effects of extracellular stimuli on mRNA decay.

7 Mammalian mRNA Turnover

In this section we will discuss various cell-free systems that have been used to study the degradation of mRNAs from different mammalian cell types. Because there are many examples of mammalian cell-free decay systems, we have divided this section into groups based on the specific uses and advantages of each system. As a prelude we will briefly summarize the initial development of cell-free systems.

7.1 Development of Functional In Vitro Decay Systems

Some cell-free decay systems include, but are not limited to, the following:

1. K562 cell-derived system (Ross and Kobs 1986; Peltz et al. 1987; Ross et al.1987). This system utilizes human K562 erythroleukemia cell extracts to study the decay of both endogenous polysome-associated mRNA and mRNAs synthesized in vitro. Characterization of the system showed that relative decay rates of the cellular mRNAs examined are reconstituted in

vitro (eg., stability of γ-globin mRNA > δ-globin > c-*myc* > H4 histone mRNA). This system also led to the identification of mRNA decay intermediates that can be detected in whole cells.

2. Friend erythroleukemia (FEL) cell-derived system (Sunitha and Slobin 1987). In this system, which contains 40–80S mRNP from FEL cells, different mRNAs had different half-lives. Most mRNAs were stable, while some, such as eEF-Tu mRNA, were labile. The rapid turnover of eEF-Tu mRNA also occurs in cells. In mRNP treated with 0.5 M NaCl the mRNAs examined were stable, suggesting that a putative mRNase had been removed from the mRNPs.
3. Plasmacytoma cell-derived system (Pei and Calame 1988). This system utilized polysomes from murine plasmacytoma cells and in vitro synthesized radiolabeled RNAs to study c-*myc* mRNA decay. The stabilities of full-length and truncated c-*myc* mRNAs observed in vivo were reproduced in the cell-free system. c-*myc* mRNAs truncated within exon 1, which are often found in tumor cells, are more stable than the full-length mRNA, suggesting that a degradation signal exists in exon 1. In vitro reactions using this system showed instead that c-*myc* exon 1 cannot target a heterologous mRNA for rapid decay. Also, the addition of unlabeled competitor c-*myc* RNAs resulted in stabilization of radiolabeled c-*myc* mRNA in the reactions, suggesting that a titratable factor(s) in the polysome extract is responsible for c-*myc* mRNA degradation.
4. Sarcoma S-180 cell-derived system (Bandyopadhyay et al. 1990). Polysomes from murine sarcoma S-180 cells were used in cell-free decay reactions to study the decay of endogenous mRNA. A ribonuclease activity present in S-180 polysomes was tightly associated with polysomes as well as with free mRNP, and was not associated with free ribosomes. Additional experiments indicated that the ribonuclease activity and the mRNA substrate were also tightly associated in vitro. The system was then used to begin analyzing mRNA cleavage sites and specificity.

7.2 Characterization of Decay Pathways

7.2.1 H4 Histone mRNA Decay

H4 histone mRNA is degraded rapidly in cells and in vitro. A K562 cell-derived decay system was used to study the pathway of H4 histone mRNA decay (Ross and Kobs 1986; Ross et al. 1986). The degradation of both endogenous polysome-associated H4 histone mRNA and radiolabeled H4 histone mRNA synthesized in vitro began at or near the 3′ end. The first detectable intermediate lacked 5 nucleotides from the 3′ end (the -5 intermediate). This was followed by generation of an intermediate lacking 12 nucleotides from the 3′ end (the -12 intermediate). In subsequent steps the mRNA was degraded 3′ to 5′, consistent with exonucleolytic decay. This pathway was observed both in intact cells and in vitro, which suggested that this system would be useful for purification of factors important for histone mRNA turnover (see Sect. 7.3.1 and 7.6.1).

7.2.2 c-myc mRNA Decay

The K562-derived cell-free decay system was also used to investigate the pathway of c-*myc* mRNA decay (Brewer and Ross 1988). The first detectable steps in decay of the endogenous, polysome-associated c-*myc* mRNA in vitro was shortening of the poly(A) tract, resulting in a pool of oligoadenylated mRNA molecules within 30 min of incubation. By contrast, the stable γ-globin mRNA did not undergo significant poly(A) removal in vitro during 3 h of incubation. The next step in the decay pathway was degradation of the oligoadenylated RNA, resulting in products with 3′ termini in the A + U-rich element. These decay products were rapidly degraded. Thus c-*myc* mRNA is degraded by a sequential pathway involving deadenylation followed by degradation of the mRNA body. A similar pathway is observed in cells (GB, unpubl. observ; Chen and Shyu 1994; Swartwout and Kinniburgh 1989).

In conclusion, the examples of H4 histone and c-*myc* mRNAs illustrate how cell-free systems can be used to examine mRNA decay pathways. These studies can then lead to the design of experiments for the purification of the relevant factors.

7.3 Identification/Purification of mRNases

One of the major advantages of cell-free decay systems is the ability to identify and characterize regulatory factors, such as mRNases, present in specific cellular fractions. Often the in vitro system can then serve as an assay for purification of such factors. Although this section focuses solely on mammalian mRNases identified from cell-free systems, we should note that both plant and *Xenopus* mRNases have been similarly identified, as described in previous sections.

7.3.1 Histone mRNA 3′ ⇒ 5′ Exoribonuclease

Histone mRNAs are abundant during S phase of the cell cycle and are rapidly degraded at the end of DNA synthesis (reviewed in Ross 1995). Importantly, the steps in the degradation pathway of histone mRNA are the same in vitro and in intact cells. Studies with this system showed that human histone H4 mRNA is degraded 3′ to 5′ by an exoribonuclease-like activity (Ross and Kobs 1986). Experiments with the cell-free system were then performed to analyze the properties of this exonuclease activity (Ross et al. 1987). First, exonuclease activity in vitro required a divalent cation (Mg^{2+}), did not require exogenously added ATP or GTP, and was functional at monovalent cation concentrations between 0.5 to 200 mM. Secondly, the exonuclease was bound to ribosomes isolated in low salt, and the active enzyme could be eluted from ribosomes treated with 0.5 M KCl.

Caruccio and Ross (1994) took advantage of this property of the activity and the fact that it could also act on radiolabeled H4 histone mRNA synthesized in vitro to purify the $3' \Rightarrow 5'$ exoribonuclease that may be the mRNase responsible for decay of H4 histone mRNA in cells. The RNase had the same properties as the histone exoribonuclease assayed in crude extracts. Poly(A) was also degraded by the RNase in vitro, albeit at a slower rate than that for poly(A)$^-$ RNA. By contrast, poly(A)$^+$ RNA in the presence of poly(A)-binding protein (PAB) was not degraded by the RNase, a finding that suggests PAB protects poly(A) from the RNase. Finally, the effect of the purified enzyme on polysome-associated histone mRNA decay was analyzed in cell-free reactions with salt-washed polysomes which are depleted of exonuclease activity. In control reactions with these polysomes, histone mRNA was 10- to 20-fold more stable than in reactions with untreated polysomes. The addition of partially purified $3' \Rightarrow 5'$ exonuclease to reactions containing salt-washed polysomes accelerated histone mRNA decay, and the resulting decay products were the same as those observed for histone decay with untreated polysomes in vitro and in intact cells. Furthermore, the enzyme did not destabilize γ-globin or c-*myc* mRNA in salt-washed polysomes, consistent with the resistance of poly(A)-PAB to the exonuclease. An advantage of the cell-free system here was the ability to add back purified enzyme to an enzyme-depleted fraction to directly assess its effect on mRNA decay. It will be interesting to see the effects of this ribonuclease on mRNA metabolism in cells.

7.3.2 Poly(A) Exoribonuclease from HeLa Cells

Removal of a 3′ poly(A) tract from many yeast and mammalian mRNAs is an early step in mRNA decay, often required for degradation of the mRNA body to occur (reviewed in Beelman and Parker 1995). It is important, therefore, to understand the enzyme(s) involved in poly(A) removal and how its activity affects overall mRNA decay pathways. A $3' \Rightarrow 5'$ exoribonuclease that deadenylates RNA substrates in vitro was identified in HeLa cell extracts, partially purified, and characterized using a cell-free system (Åström et al. 1991, 1992). Initially, in vitro-synthesized capped RNAs were incubated in vitro with HeLa nuclear or cytoplasmic extracts to identify a poly(A)-degrading activity. Incubation of a polyadenylated RNA substrate with nuclear extract resulted in poly(A) removal and accumulation of a product corresponding to the poly(A)$^-$ RNA body. Nuclear extract was therefore fractionated and the enzymatically active fraction was used for subsequent in vitro reactions. (The authors note that the nuclease activity may be associated with cytoplasmic components which sediment during the nuclear extract preparation. Thus localization of the activity to the nucleus has not been unambiguously proven.) Experiments using the partially purified ribonuclease and various radiolabeled RNA substrates revealed the following properties: (1) deadenylation did not appear to require a specific sequence in the RNA body; (2) the nuclease activity was specific for homopolymeric adenosine tails; (3) the activity gradually degraded poly(A) tracts, leaving the RNA body intact; (4) for efficient deadenylation, the poly(A) had to be located at the 3′ terminus and a 3′ hydroxyl had to be present; (5) the nuclease activity was competed by poly(A), and added poly(U) inhibited deadenylation, so that the exonuclease is specific for single-stranded poly(A) RNA; (6) mononucleotides accumulated during the reaction; and (7) 5′ AMP was released in deadenylation reactions in vitro, and added 5′ AMP inhibited deadenylation. The latter two properties indicate that the enzyme is a 3′ exoribonuclease. Purification of this RNase to homogeneity will undoubtedly provide new insights into deadenylation and its possible control.

7.3.3 5′⇒3′ Exoribonuclease in Mouse Sarcoma Ascites Cells

A cell-free decay system was developed from mouse sarcoma-180 (S-180) ascites cells to better understand how general mRNA degradation is regulated in mammalian cells (Coutts et al. 1993). Ribosomal RNA was rapidly degraded in cytoplasmic extracts from these cells while mRNA was relatively stable. This indicates that the extracts contain high levels of ribonuclease activity and that a protective mechanism(s) exists to prevent nonspecific nuclease degradation of mRNA. In vitro-transcribed RNAs with different 5′ termini were incubated with cytoplasmic extracts and assayed for decay. RNAs with a triphosphate or a cap at the 5′ end were much more stable in vitro than RNAs with a monophosphate at the 5′ end. Cap analogue was also included in reactions as a competitor for cap-binding protein (CBP). The analogue had no effect on the stability of RNAs in the cell-free system, indicating that CBP does not protect the RNAs from nuclease attack. Because protection from RNA degradation in vitro appeared to depend on the presence of a specific 5′ structure, the possibility that the degrading enzyme(s) was an exoribonuclease was tested. Ribosomal RNA was degraded at the same rate by equal amounts of cytoplasmic extract and cytosol; therefore, cytosol was used for characterization of the nuclease activity. In an assay that distinguished between release of radioactive mononucleotides and oligonucleotides from RNA, cytosol from the S-180 cells caused release of mononucleotides from radiolabeled rRNA. Fractionation of cytosol from S-180 cells yielded a fraction that contained an exoribonuclease activity. Subsequent work involved the partial purification and characterization of this exonuclease (Coutts and Brawerman 1993). In vitro enzyme activity was optimal at K^+ concentrations from 100–200 mM and $MgCl_2$ concentrations of 1–2 mM, and activity was increased by the addition of 10 mM phosphate. The activity was not affected by ATP, micrococcal nuclease, or RNasin, was somewhat inhibited by GTP, and was completely inhibited by EDTA. The partially purified enzyme resulted in release of mononucleotides from radiolabeled 5S RNA and in a more rapid release of radiolabel from 5′-

labeled 5S RNA than from 3′-labeled 5S RNA, suggesting that it is a 5′⇒3′ exoribonuclease. In vitro experiments with the partially purified enzyme also revealed that a 5′ cap could not protect RNA from degradation by the exonuclease, in contrast to observations with reactions using crude cytoplasmic extracts. The crude extracts may contain a factor(s) that interacts with the 5′ cap, protecting the 5′ terminus from exonucleolytic attack. For these studies, the cell-free system aided in the identification and characterization of a ribonuclease that discriminates between ribosomal RNA and mRNA in vitro, the mechanism of differential degradation being possibly due to the presence of a 5′ cap. As the authors noted, however, it is not known whether mRNA in cells is normally available for attack by the identified enzyme. Even if cellular mRNAs are exposed to this exonuclease activity, it is unknown if the 5′ cap is protective in vivo.

7.3.4 RNase L

The endoribonuclease RNase L, which mediates the antiviral action of interferon, requires 5′-phosphorylated 2′, 5′-linked oligoadenylates (2-5A) for activity. The production of 2-5A requires double-stranded RNA, and interferon can lead to an increase in RNase L levels (reviewed in Sen and Lengyel 1992). RNase L was molecularly cloned and activity of recombinant purified protein was studied (Dong et al. 1994). In vitro-synthesized radiolabeled RNAs were incubated with RNase L in the presence or absence of 2-5A derivatives for analysis of enzyme activity. Cell-free experiments identified the following characteristics of RNase L activity: (1) one 5′ phosphoryl group linked to at least three 2′-5′-linked adenylyl residues was required; (2) cleavage was specific for poly(rU) and poly(rA), which were each degraded into discrete products; and (3) ATP and either manganese or magnesium were required for optimal enzyme activity.

Cell-free reactions were utilized to study a RNase L inhibitor, RLI (Bisbal et al. 1995). RLI blocked binding of 2-5A to RNase L, and expression of RLI in reticulocyte extracts blocked the ability of RNase L to cleave RNA. RLI did not degrade 2-5A in vitro and, together with the observation that RLI and RNase L coimmunoprecipitate, suggests that RLI's inhibition of RNase L activity resulted from the association of the two proteins. These cell-free experiments were important in that they have permitted characterization of a novel mediator of the 2-5A/RNase L system.

7.4 Identification/Purification of trans-Acting Decay Regulators

7.4.1 c-myc mRNA Decay

The mRNA encoded by the c-*myc* proto-oncogene is rapidly degraded in cells and in vitro (Dani et al. 1984; Brewer and Ross 1988; Pei and Calame 1988). Moreover, the c-*myc* mRNA decay rate can vary, depending on the rate of translation. Brewer and Ross (1989) examined the effects of cytosol on c-*myc* mRNA decay in a K562 cell-based system. In vitro decay reactions included polysomes and postribosomal supernatant (S130). S130 accelerated the decay rate of polysome-bound c-*myc* mRNA six- to eight-fold compared to reactions without S130. By contrast, S130 from cycloheximide-treated cells did not have a destabilizing effect on c-*myc* mRNA, suggesting that continuous protein synthesis is necessary to maintain functional levels of the destabilizing activity. To further examine the cycloheximide effect, mixing experiments were employed. In reactions with a 1:1 mixture of S130s from untreated or cycloheximide-treated cells, c-*myc* mRNA was degraded as rapidly as in reactions with control S130 alone. The mixing experiment demonstrated that S130 does indeed contain a destabilizer that is inactivated by cycloheximide treatment of cells. Additionally, c-*myb* mRNA, which is as unstable as c-*myc* mRNA, was destabilized by S130 to the same extent as c-*myc* mRNA. However, the stabilities of δ-globin, γ-globin, H4 histone, and bulk poly(A)$^+$ were not affected by S130 addition. Thus, the destabilizing activity did not appear to be a general RNase. An important advantage of the cell-free system for this study was the ability to examine the effect of S130 on mRNA decay by adding it back to the polysome fraction. This demonstrated that a factor(s) is present in cytosol that can accelerate decay of specific polysome-associated mRNAs.

Cis-acting instability elements are important for the rapid turnover of many mRNAs (Ross 1995). One *cis*-acting element that can target c-*myc* mRNA for rapid turnover is an A+U-rich element, or ARE, in the 3′ UTR (Shaw and Kamen 1986; Jones and Cole 1987). AREs are present in the 3′ UTRs of many unstable mRNAs, but it is unknown how this sequence element functions in mRNA degradation. S130 from K562 cells was fractionated and the resulting fractions were assayed for c-*myc* ARE-binding activity and c-*myc* mRNA degradation activity (Brewer 1991). By gel mobility-shift analysis and cell-free decay reactions, respectively, fractions with a sedimentation of 7S contained ARE-binding activity and accelerated the rate of c-*myc*, but not γ-globin, mRNA decay. The 7S activity contained major polypeptides of 37 and 40 kDa apparent molecular weights. Proteinase K blocked the mRNA destabilization activity of the 7S fraction, but did not affect its ARE-binding activity, showing that the two activities were separable. Subsequent work led to the purification and cDNA cloning of the ARE-binding proteins referred to as AUF1 (Zhang et al. 1993; Ehrenman et al. 1994). Purified recombinant AUF1 itself does not destabilize c-*myc* mRNA, suggesting that an additional protein(s), perhaps a ribonuclease, associates with or copurifies with AUF1 in the 7S cytosolic fraction to destabilize c-*myc* mRNA in cell-free reactions. Further fractionation of the 7S activity should clarify this.

Another *cis*-acting element involved in c-*myc* mRNA turnover is found in the coding region (Wisdom and Lee 1991; Bernstein et al. 1992). The K562-derived cell-free decay system proved useful for understanding how the c-*myc* coding region may function in mRNA turnover (Bernstein et al. 1992). Addition of competitor RNAs corresponding to different regions of c-*myc* mRNA to cell-free decay reactions indicated that a 182-nucleotide sequence from the 3′ end of the c-*myc* coding region promotes destabilization of the polysome-associated c-*myc* mRNA in vitro. This 182-nucleotide sequence, referred to as the coding region determinant or CRD, was the smallest destabilizing fragment identified. The destabilizing effect of the CRD competitor RNA was restricted, since other endogenous mRNAs were not affected. In addition, polysome-associated β-globin mRNA containing the CRD was destabilized in vitro by the presence of exogenous CRD competitor (Herrick and Ross 1994). This showed that the CRD effect on stability is independent of other sequences in the c-*myc* mRNA. Mobility-shift and UV-crosslinking assays identified a polysome-associated, 75-kDa CRD-binding protein (CRD-BP). Studies with the cell-free system and purified CRD-BP indicated that CRD-BP protects the mRNA from endoribonuclease attack (Prokipcak et al. 1994). The role of CRD-BP in regulation of c-*myc* mRNA turnover is under investigation.

7.4.2 Poly(A)-Binding Protein (PAB) and mRNA Stability

The majority of cellular mRNAs contain a polyadenylate tract at the 3′ terminus which is believed to protect a mRNA from rapid decay (reviewed in Ross 1995). The K562-derived in vitro decay system was used with in vitro-synthesized mRNAs to directly study the effect of the poly(A) tract and PAB on mRNA decay (Bernstein et al. 1989). Polyadenylated mRNAs which are normally stable in decay reactions were destabilized in reactions with excess competitor poly(A); addition of purified yeast PAB stabilized the polyadenylated mRNA. The stability of the nonpolyadenylated H4 histone mRNA was not affected by excess poly(A). These results indicated that PAB binding to polyadenylated mRNA stabilized the mRNA. Two experiments were designed to test whether a PAB interaction with the 3′ terminal poly(A) is important for mRNA stabilization. First, polyadenylated β-globin mRNA in PAB-depleted decay reactions was stabilized by the addition of purified PAB; however, nonadenylated β-globin mRNA was not stabilized by PAB under the same conditions. Secondly, the decay pathway of polyadenylated β-globin mRNA in PAB-depleted extracts revealed that the poly(A) tract was removed before the body of the mRNA was degraded. These data are consistent with PAB protecting mRNA from degradation by interacting with the poly(A) tract. This study exemplifies the advantage of using a cell-free system for studying mRNA decay in factor-depleted extracts.

7.4.3 GM-CSF mRNA Turnover

Like c-*myc* mRNA, GM-CSF mRNA contains an ARE in its 3′ UTR which is bound by cellular factors (see Zhang et al. 1993 and references therein). The role of the adenosine-uridine binding factor AUBF in GM-CSF mRNA decay was examined in a cell-free system using polysomes from human peripheral blood mononuclear cells stimulated with TPA and PHA (Rajagopalan and Malter 1994). GM-CSF mRNA is stabilized by this treatment, and, likewise, polysome-associated GM-CSF mRNA from treated cells is relatively stable in vitro. In cell-free decay reactions, the rate of GM-CSF mRNA decay was accelerated with polysomes depleted of AUBF by preincubation with AUUUA-containing RNA linked to paramagnetic beads; GM-CSF mRNA decay was unaffected by preincubation with AUGUA-containing RNA beads. These results suggested that AUBF binding to the GM-CSF ARE protected the mRNA from rapid degradation, possibly by blocking recognition of the mRNA by a mRNase or other decay factors. The approach used here was valuable in that it enabled assay of GM-CSF mRNA stability in polysomes that had been first depleted of AUBF.

7.4.4 β-Globin mRNA Stability

β-globin mRNA is very stable both in vivo and in vitro (Ross and Kobs 1986; Ross and Pizarro 1983; Ross and Sullivan 1985). Its stability could be due to protection of the mRNA from nucleases by a cellular factor(s). HeLa cell extracts were therefore examined for factors that affect β-globin mRNA stability in a cell-free decay system (Stolle and Benz 1988). Capped and uncapped β-globin mRNAs synthesized in vitro were incubated in either HeLa whole-cell extracts or mouse erythroleukemia cell (MELC) cytoplasmic extracts. In HeLa extracts, capped β-globin mRNA was stable while uncapped β-globin mRNA was rapidly degraded. However, in MELC extracts, both the capped and uncapped mRNAs were rapidly degraded. If capped mRNA was preincubated with HeLa whole-cell extract before addition of MELC extract, the mRNA was stable. This suggested that the HeLa extract, but not the MELC extract, contained a factor(s) that protects capped mRNA from degradation. Addition of competitor RNAs to extracts revealed that the protective activity did not appear to interact directly with the mRNA. Instead, the activity may be a RNase inhibitor that blocks activity of an enzyme that degrades capped mRNAs. Additional experiments showed that the protecting activity shared properties with RNasin, since it inhibited RNase A activity. However, RNasin did not protect mRNA from degradation in MELC extracts, showing that the HeLa cell-protecting factor is distinct from RNasin. Purification of the protecting factor will be required to determine its mode of action.

7.5 Identification/Function of cis-Acting Instability Determinants

In addition to identifying *trans*-acting factors of mRNA turnover, cell-free systems can also aid in the identification and/or characterization of *cis*-acting elements that effect mRNA decay. One example is the identification of the c-*myc* coding region determinant of stability discussed above (Sect. 7.4.1). Other examples are presented here.

7.5.1 *Insulin-Like Growth Factor I (IGF-I) mRNAs*

Two predominant sizes of rat liver IGF-I mRNAs exist, 7.0–7.5-kb and 0.9–1.2-kb. The size differences result from alternate polyadenylation site usage (Lund et al. 1989). The longer 3′ UTR contains sequences potentially important for mRNA decay; therefore, in vitro decay of the two mRNA classes was analyzed using reticulocyte lysate (Hepler et al. 1990). Rat liver poly(A)$^+$, mRNA was isolated and incubated in the cell-free system. The 7.0–7.5-kb mRNA was rapidly degraded while the 0.9–1.2-kb mRNA was stable. To ensure that the smaller mRNA was not simply a degradation product of the larger mRNA, reactions were performed with rat brain poly(A)$^+$ mRNA which contains mainly the 7.0–7.5-kb mRNA. In these reactions, this mRNA was rapidly degraded and no appearance of a smaller product was observed. Similar differences in half-life between the two mRNA classes was subsequently observed in intact cells. Cell-free decay reactions were useful in this instance for first identifying the differential stabilities of the two IGF-I mRNAs, which in turn suggested a role for the longer 3′ UTR in destabilization of the 7.0–7.5-kb mRNA.

7.5.2 *A + U-Rich Sequences*

Reticulocyte lysate was used to investigate the stabilities of different mRNAs (Wreschner and Rechavi 1988). Total cellular RNA from various sources was incubated with reticulocyte lysate, and for all mRNAs analyzed the degree of degradation by reticulocyte RNases corresponded to the extent of $(U)_nA$ sequences present in the 3′ UTR. Three groups of mRNAs were thus classified: (1) $(U)_nA$-poor mRNAs which were stable in the in vitro system. An example is β-globin mRNA; (2) mRNAs with intermediate numbers of $(U)_nA$, which were somewhat unstable in the system. Examples are the Igα and Igμ heavy-chain mRNAs; and (3) $(U)_nA$-rich mRNAs, which were unstable in the system. An example is c-*myc* mRNA. These experiments were among the earliest to suggest that mRNA decay could be controlled biochemically by $(U)_nA$-type sequences in the 3′ UTR using a cell-free system.

A study of interleukin-1α (IL-1α) mRNA turnover also utilized reticulocyte lysate (Gorospe and Baglioni 1994). Incubation of in vitro-synthesized radiolabeled RNAs with reticulocyte lysate revealed that IL-1α mRNA was rapidly degraded while superoxide dismutase mRNA was stable. A RNA corresponding to the 3′ UTR of IL-1α mRNA was also rapidly degraded, but RNAs corresponding to the 5′ UTR or coding region were stable. The addition of excess unlabeled competitor RNA containing four AUUUA motifs stabilized the 3′ UTR-containing RNA in this system. The stabilization effected by a competitor RNA required AUUUA motifs in the competitor sequence, suggesting that AUUUA motifs play a role in the rapid decay of the RNA containing the IL-1α 3′ UTR. Partial fragments of the IL-1α 3′ UTR were then used as RNAs in the cell-free system to localize the region necessary to direct rapid mRNA degradation. The authors concluded that at least three clustered AUUUA motifs were important for mRNA decay, since RNAs with separated AUUUAs were stable. The cell-free decay system used here was helpful for identifying potential sequence requirements for IL-1α mRNA turnover. However, as the authors note, degradation of partial RNAs in vitro may not be a valid representation of the decay of thc full-length mRNA in intact cells.

7.5.3 c-myc 3′ UTR and Coding Region

As mentioned previously, c-*myc* mRNA contains at least two *cis*-acting instability determinants, an A+U-rich element (ARE) in the 3′ UTR and the coding region determinant (CRD). A cell-free decay system was used in conjunction with whole cell experiments to study the function of each *cis* element (Herrick and Ross 1994). One or both sequences was linked to the human β-globin gene, which encodes a stable mRNA. The chimeric constructs were transfected into HeLa cells; polysomes from these cells were used in cell-free decay reactions. A β-globin/CRD chimeric mRNA was destabilized in vitro by the addition of competitor CRD RNA, and an endoribonucleolytic decay product was formed. By contrast, the decay rate of a β-globin/ARE chimeric mRNA was unaffected by the competitor. These results support the existence of alternate decay pathways for c-*myc* mRNA in addition to the deadenylation-dependent pathway.

7.5.4 Ribonucleotide Reductase R2 mRNA 3′ UTR

Ribonucleotide reductase R2 mRNA is stabilized in BALB/c 3T3 mouse fibroblasts treated with either transforming growth factor-β1 (TGF-β1) or 12-*O*-tetradecanoylphorbol-13-acetate (TPA). Cell-free decay reactions and in vitro RNA-binding assays were used to investigate the role of particular RNA sequence elements in the induction of R2 mRNA stability by these stimuli (Amara et al. 1993; Amara et al. 1994). Cell-free reactions, which contained polysomes from untreated cells and postribosomal supernatant (S130) from treated or untreated cells, showed that the effects of TGF-β1 or TPA on R2 mRNA stability were specific, as GAPDH and c-*myc* stabilities did not change with either treatment. A 75-kDa RNA-binding protein (p75) that bound an 83-nucleotide region in the 3′ UTR was present in cells treated with TGF-β1, but not in unstimulated cells. The TGF-β1-induced increase in R2 mRNA stability required the 83-nucleotide 3′ UTR element, and stabilization could be blocked by preincubation of S130 with competitor RNA containing the 83-nucleotide region. R2 mRNA stabilization also correlated with upregulation of p75 by TGF-β1. These data suggested that the 83-nucleotide region contains a *cis*-acting stability determinant that acts via an interaction with p75. mRNA stability did not differ in polysomes from untreated versus TGF-β1-stimulated cells; however, S130 from TGF-β1-treated cells induced stabilization of the polysome-associated mRNA while S130 from untreated cells did not. This suggested that p75 is not polysome-associated.

A 45-kDa protein (p45) that bound a 20-nucleotide region in the R2 3′ UTR was present in unstimulated cells and was downregulated by TPA treatment. S130 from TPA-treated cells stabilized R2 mRNA in vitro. S130 from untreated cells that was preincubated with a competitor RNA containing the 20-nucleotide *cis* element also stabilized R2 mRNA. This suggested that interaction of p45 with the 20-nucleotide element in the endogenous mRNA led to mRNA destabilization, so that downregulation of p45 by TPA stabilizes the mRNA. The cell-free system was essential in determining the roles of both the 83- and 20-nucleotide *cis* elements in ribonucleotide reductase R2 mRNA turnover. Results indicate that for each respective regulatory mechanism, these sequences are necessary; however, it is not known if they are also sufficient.

7.6 Regulation of mRNA Decay

7.6.1 Histone Autoregulation

Histone mRNAs are rapidly degraded at the end of S phase of the cell cycle (reviewed in Ross 1995). In a K562-derived decay system, the four core histone proteins together with cytosol specifically induce a three- to four-fold destabilization of polysome-associated H4 histone mRNA (Peltz and Ross 1987). This is consistent with the hypothesis that increased cytoplasmic pools of histone proteins at the termination of DNA synthesis induces decay of histone mRNAs. Subsequent studies using the cell-free system showed that, in conjunction with cytosol, each purified core histone or purified linker histone H1 could independently induce destabilization of H4 histone mRNA (McLaren and Ross 1993). Furthermore, each separate histone protein was as effective as the four proteins together. Basic polypeptides such as poly-L-lysine could induce an initial step of histone mRNA decay in vitro, but did not lead to full mRNA destabilization. This suggested that a feature of the histone proteins other than their basicity was responsible for the autoregulatory effect.

Fractionation of the S130 using histone-agarose chromatography revealed that both the flow-through fraction (FTS130) and a fraction eluted with high salt (HSWF) were required for histone-induced destabilization in vitro. Experiments with competitor RNAs revealed that a stem-loop-binding protein that appears to be the La protein blocks the $3' \Rightarrow 5'$ exoribonuclease which degrades histone mRNA thus generating one of the decay intermediates (the -12 intermediate; see Sect. 7.2.1; Jeff Ross, pers. comm.). The HSWF fraction contains an activity that acts with histones to relieve this block, thereby accelerating histone mRNA decay. Purification of all the relevant factors will permit experiments to elucidate how this occurs mechanistically.

7.6.2 Stability of Urokinase-Type Plasminogen Activator (uPA) mRNA

A cell-free system was designed to study the relationship between translation and mRNA stability with a focus on uPA mRNA (Altus and Nagamine 1991). In the porcine renal epithelial cell line LLC-PK1, uPA mRNA is stabilized in response to protein synthesis inhibitors (Altus et al. 1987). The use of LLC-PK1 postmitochondrial supernatant (PMS) in cell-free reactions showed that uPA mRNA was unstable in PMS from untreated cells and stable in PMS from cycloheximide-treated cells, corresponding to observations with intact cells. In cells, histone mRNA was also stabilized by cycloheximide treatment; however, histone mRNA remained unstable in PMS from cycloheximide-treated cells. One possible explanation for this discrepancy observed in vitro is that the mechanism by which protein synthesis inhibitors induce uPA mRNA stability is different from that for histone mRNA. Experiments with intact cells revealed that the longer the length of cycloheximide treatment prior to determination of uPA mRNA stability, the more stable the mRNA. This result suggested that uPA mRNA stabilization by protein synthesis inhibition was due to loss of a labile protein decay factor, rather than a lack of ongoing translation of the mRNA. Consistent with this conclusion is the observation that if cycloheximide was added at the start of decay reactions, uPA mRNA was only slightly stabilized. Data obtained from the cell-free system thus support the hypothesis that a labile protein is involved in uPA mRNA decay.

7.6.3 Decay of Manganese Superoxide Dismutase (MnSOD) mRNAs

MnSOD is one enzyme that catalyzes the breakdown of superoxide radicals, thereby protecting cells from free radical damage. Two species of MnSOD mRNAs, 1 kb and 4 kb, arise from alternate polyadenylation site usage and differ only in their 3′ UTRs, the distinction being the presence of an ARE in the 4-kb mRNA. Tumor necrosis factor (TNF) induces high levels of MnSOD, and thus expression of these mRNAs in response to TNF was studied using a cell-free system (Melendez and Baglioni 1993). In vitro reactions contained reticulocyte lysate and poly(A)$^+$ RNA isolated from TNF-treated osteosarcoma HOS cells. In this system, the 4-kb mRNA was degraded rapidly while the 1-kb mRNA was stable, reflecting the relative turnover rates in cells. These observations suggested that MnSOD gene expression was being regulated in part at the level of mRNA turnover. The known role of AREs in mRNA decay supports the idea that rapid decay of the 4-kb mRNA involves the ARE. The use of the cell-free system was important for analyzing MnSOD mRNA decay in the absence of the transcription inhibitors used in experiments with whole cells.

7.6.4 Stabilization of Transforming Growth Factor β1 (TGF-β1) mRNAs

12-tetradecanoylphorbol-13-acetate (TPA) induces differentiation of U937 promonocytes and stabilizes TGF-β1 mRNA. A cell-free system derived from U937 cells was designed to study this effect of TPA on TGF-β1 mRNA stability (Wager and Assoian 1990). In vitro decay reactions contained total deproteinized cellular RNA and U937 cell postnuclear extracts. TGF-β1 mRNA was degraded in reactions with extracts from untreated cells. However, TGF-β1 mRNA was stable in extracts from TPA-treated cells. The degradation rates of 28S rRNA and GAPDH mRNA were the same in extracts from untreated and TPA-treated cells. Thus the cell-free system accurately reproduced the specificity of the TPA effect for TGF-β1 mRNA observed in cells. From both in vivo and in vitro results, the authors concluded that a selective RNase system was selectively degrading TGF-β1 mRNA, and that this system was inhibited by TPA-induced cell differentiation.

The cell-free system was then used to identify the region(s) of TGF-β1 mRNA important for decay and regulation by TPA (Wager et al. 1994). In vitro-synthesized RNAs corresponding to different truncated forms of TGF-β1 mRNA were incubated with U937 postnuclear extracts. The coding region of this mRNA was sufficient to act as a substrate for selective degradation in vitro with extracts from untreated cells. The presence of the 5′ and 3′ UTRs could, however, result in an increased decay rate over that of the coding region alone. By contrast, the 5′ and 3′ UTRs were necessary for mRNA stabilization by TPA. These studies demonstrated both the general usefulness of a cell-free system for complementing in vivo experiments and the specific advantage of the system for investigating the role of different regions of a mRNA in turnover.

8 Conclusion

From the above discussion, it is apparent that cell-free decay systems have contributed to a better understanding of mRNA decay mechanisms. One of the most obvious advantages of cell-free decay systems is the ability to assay effects of specific cellular fractions on mRNA turnover, which can, in turn, lead to the identification of factors that are involved in mRNA decay. Once putative decay factors are identified in vitro, however, it is essential that their functions be tested in vivo. Thus cell-free systems are applicable for the initial identification of in vitro decay factors, with their subsequent purification and analysis in vivo required for determining their precise functions in cellular mRNA decay.

References

Adams CC, Stern DB (1990) Control of mRNA stability in chloroplasts by 3′ inverted repeats: effects of stem and loop mutations on degradation of *psbA* mRNA in vitro. Nucl Acids Res 18:6003–6010

Altus MS, Nagamine Y (1991) Protein synthesis inhibition stabilizes urokinase-type plasminogen activator mRNA. J Biol Chem 266:21190–21196

Altus MS, Pearson D, Horiuchi A, Nagamine Y (1987) Inhibition of protein synthesis in LLC-PK1 cells increases calcitonin-induced plasminogen-activator gene transcription and mRNA stability. Biochem J 242:387–392

Amara FM, Chen FY, Wright JA (1993) A novel transforming growth factor-β_1 responsive cytoplasmic *trans*-acting factor binds selectively to the 3′ -untranslated region of mammalian ribonucleotide reductase R2 mRNA: role in message stability. Nucl Acids Res 21:4803–4809

Amara FM, Chen FY, Wright JA (1994) Phorbol ester modulation of a novel cytoplasmic protein binding activity at the 3′ -untranslated region of mammalian ribonucleotide reductase R2 mRNA and role in message stability. J Biol Chem 269:6709–6715

Åström J, Åström A, Virtanen A (1991) In vitro deadenylation of mammalian mRNA by a HeLa cell 3′ exonuclease. EMBO J 10:3067–3071

Åström J, Åström A, Virtanen A (1992) Properties of a HeLa cell 3′ exonuclease specific for degrading poly(A) tails of mammalian mRNA. J Biol Chem 267:18154–18159

Bandyopadhyay R, Coutts M, Krowczynska A, Brawerman G (1990) Nuclease activity associated with mammalian mRNA in its native state: possible basis for selectivity in mRNA decay. Mol Cell Biol 10:2060–2069

Beelman CA, Parker R (1995) Degradation of mRNA in eukaryotes. Cell 81:179–183

Bernstein P, Peltz SW, Ross J (1989) The poly(A)-Poly(A)-binding protein complex is a major determinant of mRNA stability in vitro. Mol Cell Biol 9:659–670

Bernstein PL, Herrick DJ, Prokipcak RD, Ross J (1992) Control of c-*myc* mRNA half-life in vitro by a protein capable of binding to a coding region stability determinant. Genes Dev 6:642–654

Bisbal C, Martinand C, Silhol M, Lebleu B, Salehzada T (1995) Cloning and characterization of a RNase L inhibitor. J Biol Chem 270:13308–13317

Brawerman G (1987) Determinants of messenger RNA stability. Cell 48:5–6

Brewer G (1991) An A + U-rich element RNA binding factor regulates c-*myc* mRNA stability in vitro. Mol Cell Biol 11:2460–2466

Brewer G, Ross J (1988) Poly(A) shortening and degradation of the 3′ AU-rich sequences of human c-*myc* mRNA in a cell-free system. Mol Cell Biol 8:1697–1708

Brewer G, Ross J (1989) Regulation of c-*myc* mRNA degradation in vitro by a labile destabilizer with an essential nucleic acid component. Mol Cell Biol 9:1996–2006

Brown BD, Harland RM (1990) Endonucleolytic cleavage of a maternal homeo box mRNA in *Xenopus* oocytes. Genes Dev 4:1925–1935

Brown DB, Zipkin ID, Harland RM (1993) Sequence-specific endonucleolytic cleavage and protection of mRNA in *Xenopus* and *Drosophila*. Genes Dev 7:1620–1631

Carter BZ, Malter JS (1991) Regulation of mRNA stability and its relevance to disease. Lab Invest 65:610–621

Caruccio N, Ross J (1994) Purification of a human polyribosome-associated 3′ to 5′ exoribonuclease. J Biol Chem 269:31814–31821

Chen C-YA, Shyu A-B (1994) Selective degradation of early-response-gene mRNAs: functional analyses of sequence features of the AU-rich elements. Mol Cell Biol 14:8471–8482

Chen H-C, Stern DB (1991) Specific ribonuclease activities in spinach chloroplasts promote mRNA maturation and degradation. J Biol Chem 266:24205–24211

Coutts M, Brawerman G (1993) A 5′ exoribonuclease from cytoplasmic extracts of mouse sarcoma 180 ascites cells. Biochim Biophys Acta 1173:57–62

Coutts M, Krowczynska A, Brawerman G (1993) Protection of mRNA against nucleases in cytoplasmic extracts of mouse sarcoma ascites cells. Biochim Biophys Acta 1173:49–56

Dani C, Blanchard JM, Piechaczyk M, El Sabouty S, Marty L, Jeanteur P (1984) Extreme instability of myc mRNA in normal and transformed human cells. Proc Natl Acad Sci USA 81:7046–7050

Deng XW, Gruissem W (1987) Control of plastid gene expression during development: the limited role of transcriptional regulation. Cell 49:379–387

Dong B, Xu L, Zhou A, Hassel BA, Lee X, Torrence PF, Silverman RH (1994) Intrinsic molecular activities of the interferon-induced 2-5A-dependent RNase. J Biol Chem 269:14153–14158

Ehrenman K, Long L, Wagner BJ, Brewer G (1994) Characterization of cDNAs encoding the murine A + U-rich RNA-binding protein AUF1. Gene 149:315–319

Gerstel B, Tuite MF, McCarthy JEG (1992) The effects of 5′ -capping, 3′ -polyadenylation and leader composition upon the translation and stability of mRNA in a cell-free extract from the yeast *Saccharomyces cerevisiae*. Mol Microbiol 6:2339–2348

Gorospe M, Baglioni C (1994) Degradation of unstable interleukin-1α mRNA in a rabbit reticulocyte cell-free system. J Biol Chem 269:11845–11851

Gruissem W (1989) Chloroplast gene expression: how plants turn their plastids on. Cell 56:161–170

Hepler JE, Van Wyk JJ, Lund PK (1990) Different half-lives of insulin-like growth factor I mRNAs that differ in length of 3′ untranslated sequence. Endocrinology 127:1550–1552

Herrick DJ, Ross J (1994) The half-life of c-*myc* mRNA in growing and serum-stimulated cells: influence of the coding region and 3′ untranslated regions and role of ribosome translocation. Mol Cell Biol 14:2119–2128

Howell EK, Devaney E, Gruenberg J (1989) Subcellular fractionation of tissue culture cells. Trends Biochem Sci 14:44–47

Jones TR, Cole MD (1987) Rapid cytoplasmic turnover of c-*myc* mRNAs: requirement of the 3′ untranslated sequences. Mol Cell Biol 7:4513–4521

Katze MG, Agy MB (1990) Regulation of viral and cellular RNA turnover in cells infected by eukaryotic viruses including HIV-1. Enzyme 44:332–346

Kennedy IM, Haddow JK, Clements JB (1990) Analysis of human papillomavirus type 16 late mRNA 3′ processing signals in vitro and in vivo. J Virol 64:1825–1829

Kennedy IM, Haddow JK, Clements JB (1991) A negative regulatory element in the human papillomavirus type 16 genome acts at the level of late mRNA stability. J Virol 65:2093–2097

Krikorian CR, Read GS (1991) In vitro mRNA degradation system to study the virion host shutoff function of herpes simplex virus. J Virol 65:112–122

Liang H, Jost J-P (1991) An estrogen-dependent polysomal protein binds to the 5′ untranslated region of the chicken vitellogenin mRNA. Nucl Acids Res 19:2289–2294

Love Jr HD, Allen-Nash A, Zhao Q, Bannon GA (1988) mRNA stability plays a major role in regulating the temperature-specific expression of a *Tetrahymena thermophila* surface protein. Mol Cell Biol 8:427–432

Lowell JE, Rudner DZ, Sachs AB (1992) 3′-UTR-dependent deadenylation by the yeast poly(A) nuclease. Genes Dev 6:2088–2099

Lund P, Hoyt E, Van Wyk J (1989) The size heterogeneity of rat insulin-like growth factor-I mRNAs is due primarily to differences in the length of 3′ -untranslated sequence. Mol Endocrinol 3:2054–2061

McLaren RS, Ross J (1993) Individual purified core and linker histones induce histone H4 mRNA destabilization in vitro. J Biol Chem 268:14637–14664

McMillan PJ, Tondravi MM, Bannon GA (1993) *rseB*, a chromosomal locus that affects the stability of a temperature-specific surface protein mRNA in *Tetrahymena thermophila*. Nucl Acids Res 21:4356–4362

Melendez JA, Baglioni C (1993) Differential induction and decay of manganese superoxide dismutase mRNAs. Free Rad Biol Med 14:601–608

Pastori RL, Moskaitis JE, Schoenberg DR (1991) Estrogen-induced ribonuclease activity in *Xenopus* liver. Biochemistry 30:10490–10498

Pei R, Calame K (1988) Differential stability of c-*myc* mRNAs in a cell-free system. Mol Cell Biol 8:2860–2868

Peltz SW, Jacobson A (1992) mRNA stability: in *trans*-it. Curr Opin Cell Biol 4:979–983

Peltz SW, Ross J (1987) Autogenous regulation of histone mRNA decay by histone proteins in a cell-free system. Mol Cell Biol 7:4345–4356

Peltz SW, Brewer G, Kobs G, Ross J (1987) Substrate specificity of the exonuclease activity that degrades H4 histone mRNA. J Biol Chem 262:9382–9388

Peltz SW, Brewer G, Bernstein P, Hart PA, Ross J (1991) Regulation of mRNA turnover in eukaryotic cells. Crit Rev Euk Gene Expr 1:99–126

Prokipcak RD, Herrick DJ, Ross J (1994) Purification and properties of a protein that binds to the C-terminal coding region of human c-*myc* mRNA. J Biol Chem 269:9261–9269

Rajagopalan LE, Malter JS (1994) Modulation of granulocyte-macrophage colony-stimulating factor mRNA stability in vitro by the adenosine-uridine binding factor. J Biol Chem 269:23882–23888

Ross J (1995) mRNA stability in mammalian cells. Microbiol Rev 59:16–95

Ross J, Kobs G (1986) H4 histone messenger RNA decay in cell-free extracts initiates at or near the 3′ terminus and proceeds 3′ to 5′. J Mol Biol 188:579–593

Ross J, Pizarro A (1983) Human beta and delta globin messenger RNAs turn over at different rates. J Mol Biol 167:607–617

Ross J, Sullivan TD (1985) Half-lives of beta and gamma globin messenger RNAs and of protein synthetic capacity in cultured human reticulocytes. Blood 66:1149–1154

Ross J, Peltz SW, Kobs G, Brewer G (1986) Histone mRNA degradation in vivo: the first detectable step occurs at or near the 3′ terminus. Mol Cell Biol 6:4362–4371

Ross J, Kobs G, Brewer G, Peltz SW (1987) Properties of the exonuclease activity that degrades H4 histone mRNA. J Biol Chem 262:9374–9381

Sachs A (1993) Messenger RNA degradation in eukaryotes. Cell 74:413–421

Sachs AB, Deardorff JA (1992) Translation initiation requires the PAB-dependent poly(A) ribonuclease in yeast. Cell 70:961–973

Schiavi SC, Belasco JG, Greenberg ME (1992) Regulation of proto-oncogene mRNA stability. Biochim Biophys Acta 1114:95–106

Sen G, Lengyel P (1992) The interferon system. A bird's eye view of its biochemistry. J Biol Chem 267:5017–5020

Shaw G, Kamen R (1986) A conserved AU sequence from the 3′ untranslated region of GM-CSF mRNA mediates selective mRNA degradation. Cell 46:659–667

Sorenson CM, Hart PA, Ross J (1991) Analysis of Herpes simplex virus-induced mRNA destabilizing activity using an in vitro mRNA decay system. Nucl Acids Res 19:4459–4465

Stevens A (1980a) An mRNA decapping enzyme from ribosomes of *Saccharomyces cerevisiae*. Biochem Biophys Res Commun 96:1150–1155

Stevens A (1980b) Purification and characterization of an exoribonuclease with yields 5′ -mononucleotides by a $5' \Rightarrow 3'$ mode of action. J Biol Chem 255:3080–3085

Stolle CA, Benz EJ (1988) Cellular factor affecting the stability of beta-globin mRNA. Gene 62:65–74

Strelow LI, Leib DA (1995) Role of the virion host shutoff (*vhs*) of herpes simplex virus type 1 in latency and pathogenesis. J Virol 69:6779–6786

Sunitha I, Slobin LI (1987) An in vitro system derived from Friend erythroleukemia cells to study messenger RNA stability. Biochem Biophys Res Commun 144:560–568

Swartwout SG, Kinniburgh AJ (1989) c-*myc* mRNA degradation in growing and differentiating cells: Possible alternate pathways. Mol Cell Biol 9:288–295

Tanzer MM, Meagher RB (1994) Faithful degradation of soybean rbcS mRNA in vitro. Mol Cell Biol 14:2640–2650

Thompson DM, Tanzer MM, Meagher RB (1992) Degradation products of the mRNA encoding the small subunit of ribulose-1,5-bisphosphate carboxylase in soybean and transgenic petunia. Plant Cell 4:47–58

Varnum SM, Wormington WM (1990) Deadenylation of maternal mRNAs during *Xenopus* oocyte maturation does not require specific *cis*-acting sequences: a default mechanism for translational control. Genes Dev 4:2278–2286

Varnum SM, Hurney CA, Wormington WM (1992) Maturation-specific deadenylation in *Xenopus* oocytes requires nuclear and cytoplasmic factors. Dev Biol 153:283–290

Vreken P, Buddelmeijer N, Raué HA (1992) A cell-free extract from yeast cells for studying mRNA turnover. Nucl Acids Res 20:2503–2510

Wager RE, Assoian RK (1990) A phorbol ester-regulated ribonuclease system controlling transforming growth factor $\beta 1$ gene expression in hematopoietic cells. Mol Cell Biol 10:5983–5990

Wager RE, Scotto L, Assoian RK (1994) Analysis of transforming growth factor β_1 messenger RNA degradation by the transcript-selective, 12-*O*-tetradecanoylphorbol-13-acetate-regulated ribonuclease system from U937 promonocytes. Cell Growth Diff 5:117–124

Wisdom R, Lee W (1991) The protein-coding region of c-*myc* mRNA contains a sequence that specifies rapid mRNA turnover and induction by protein synthesis inhibitors. Genes Dev 5:232–243

Wiskocil R, Bensky P, Dower W, Goldberger RF, Gordon JI, Deeley RG (1980) Coordinate regulation of two estrogen-dependent genes in avian liver. Proc Natl Acad Sci USA 77:4474–4478

Wreschner DH, Rechavi G (1988) Differential mRNA stability in reticulocyte ribonucleases correlates with 3′ non-coding sequences. Eur J Biochem 172:333–344

Zhang W, Wagner BJ, Ehrenman K, Schaefer AW, DeMaria CT, Crater D, DeHaven K, Long L, Brewer G (1993) Purification, characterization and cDNA cloning of an AU-rich element RNA-binding protein, AUF1. Mol Cell Biol 13:7652–7665

Mechanisms for Posttranscriptional Regulation by Iron-Responsive Elements and Iron Regulatory Proteins

M. MUCKENTHALER and M.W. HENTZE

1 Introduction

Iron deficiency is a wide-spread nutritional problem amongst the world's population. Often the symptoms like pallor, fatigue, and anemia yield to treatment with iron, but irreversible harmful effects of iron deficiency in the development of children have also been observed. On the other extreme, primary or secondary iron overload can lead to hemochromatosis or hemosiderosis, which are characterized by iron deposits and multiorgan pathology including liver cirrhosis, cardiac dysfunction, and arthritis.

Iron is best known for its role in oxygen transport in the blood. More than 80% of the iron in the human body is found in red blood cells as part of the hemoglobin molecule. In addition to its role in oxygen transport, iron is utilized as a cofactor by a variety of proteins, which fulfill a wide range of metabolic functions. In enzymes like oxidases, reductases, or dehydrogenases, iron is mostly involved as a reversible donor or acceptor of electrons.

Iron is the second most abundant metal (after aluminum) in the earth's crust, and it is essential for growth of almost all cells, with the exception of some bacilli and lactobacilli. In spite of its abundance, iron metabolism in an oxygen atmosphere poses several problems to living cells. In the presence of oxygen, iron is oxidized from the soluble ferrous (Fe^{2+}) to the insoluble ferric state (Fe^{3+}). Moreover, iron toxicity can result from the generation of $OH^{\bullet}$-radicals (Gutteridge 1989). As a consequence, cells developed means to take up iron in a soluble form and to counteract its potential toxicity by storage in a nontoxic form (Britton et al. 1994; Halliday et al. 1994). Both iron uptake and iron storage are regulated in response to iron requirements and iron availability.

The regulation of iron homeostasis has evolved as an inevitable consequence of life with oxygen. Throughout evolution, the regulation of iron metabolism has been solved in different ways. In *E.coli*, ferric iron forms a soluble chelate with a siderophore that is secreted into the environment. The siderophore/iron complex is then taken up into the cell via a high affinity receptor. The expression of both the siderophore and its receptor is controlled transcriptionally

Gene Expression Programme, European Molecular Biology Laboratory, Meyerhofstrasse 1, 69117 Heidelberg, Germany

Progress in Molecular and Subcellular Biology, Vol. 18
Ph. Jeanteur (Ed.)

by the DNA-binding protein, FUR. Following the binding of iron, FUR represses transcription of the siderophore and the siderophore receptor genes, so that increasing amounts of intracellular iron reduce further iron uptake (Neilands and Nakamura 1985; Braun et al. 1991). Iron uptake in *Saccaromyces cerevisiae* is mediated by a ferric reductase/ferrous iron transporter system, that is located on the cell surface. Its expression is also regulated transcriptionally by iron (Dancis et al. 1992; Eide et al. 1992).

In mammals, the absorption of iron from the small intestine is the main regulatory step for iron homeostasis (Skikne and Baynes 1994). However, the molecular details of this process are not yet well understood. Once iron has entered into the circulation, ferric iron is bound to the serum protein transferrin. Virtually all cells express transferrin receptors on their surface to mediate iron acquisition. Diferric transferrin binds to the receptor and the complex is internalized into the cell in a clathrin-coated vesicle. Acidification of the vesicle leads to the release of the transferrin-bound iron into the fluid phase of the endosome. Apotransferrin and the transferrin receptor are recycled to the cell surface. How iron is transported from the endosome into the cytosol is still unresolved. The iron released into the cytosol can be bioactive in different ways: (1) it can be utilized for iron containing proteins in cytoplasm or after transport into the mitochondria; (2) it can be sequestered by the iron storage protein ferritin; or (3) it can function in a regulatory capacity.

Free iron is required for the synthesis of heme and hemoproteins, iron-sulfur (Fe-S) proteins and a number of nonheme, non-Fe-S proteins in all cells and tissues. Most iron is utilized in erythroid cells for hemoglobin synthesis (Bothwell et al. 1983). Interestingly, the biosynthetic pathway for heme is controlled by iron. If intracellular iron levels are low, the synthesis of 5-aminolevulinate (eALAS) synthase, the first enzyme in the hemoglobin biosynthesis pathway, is decreased (Melefors et al. 1993), whereas it is augmented when iron levels are high, suggesting that the heme precursor protophorin IX is synthesized more efficiently when the iron supply is sufficient.

If iron is not used immediately in the cell it can be stored. The intracellular iron storage protein ferritin forms a hollow shell consisting of 24 subunits of heavy and light chains. As many as 4500 iron atoms can be sequestered within one molecule of ferritin. Ferrous iron taken up into ferritin is subsequently oxidized on the surface of the molecule and deposited into the ferrihydrite core of the protein (Bakker and Boyer 1986).

In addition to eALAS, the expression of ferritin and the transferrin receptor is tightly controlled by the intracellular iron content. When the iron level is low, the expression of the transferrin receptor is increased to insure a sufficient iron supply, while the expression of the iron storage protein ferritin is decreased to avoid intracellular iron sequestration. The opposite scenario takes place when iron is abundant. The mRNAs that encode these proteins have been found to be regulated posttranscriptionally in response to iron. The physiological mechanisms of how ferritin, eALAS and the transferrin receptor are regulated by the intracellular iron level will be the topic of this review. The components of this posttranscriptional regulatory system, although outlined here, is discussed in detail in Henderson and Kühn (this vol.).

2 The Iron-Responsive Element (IRE): A *Cis*-Regulatory RNA Element

Early studies indicated that the stimulatory effect of iron on ferritin synthesis was likely to occur posttranscriptionally (Drysdale and Munro 1965; Zähringer et al. 1976). When cellular iron levels are high, ferritin mRNAs predominantly associate with polysomes, while they are found in the mRNP fraction in iron-deficient cells, suggesting a translational control mechanism (Aziz and Munro 1986; Rogers and Munro 1987). In rats, iron administration increases ferritin H- and L-chain translation. In addition to the translational effect, L-chain expression is stimulated transcriptionally about twofold (White and Munro 1988). Deletion analyses and the construction of gene chimeras allowed the definition of a *cis*-acting element in the nontranslated 5′leader sequence of ferritin mRNAs, which is able to mediate translational control in response to altered iron levels. A minimal region of about 35 nucleotides suffices to confer iron regulation to reporter mRNAs (Aziz and Munro 1987; Hentze et al. 1987a,b) and was called iron-responsive element (IRE; Fig. 1). Deletion of the IRE from human ferritin H-chain mRNA leads to constitutive, iron-independent ferritin synthesis in transfected murine fibroblasts (Hentze et al. 1987b). Thus, a 5′UTR IRE is necessary and sufficient for the translational regulation of ferritin synthesis by iron. Computer modeling predicts this IRE sequence to form a moderately stable RNA hairpin (~-5.0 kcal/mol; Fig. 1). The IRE is conserved in H- as well as L-chain ferritin mRNAs and was subsequently identified in the 5′UTRs of other iron-regulated messages (see below). Comparison of the IREs from different mRNAs and species reveals a striking conservation of the primary sequence of the 6-nucleotide loop (5′-CAGUGN-3′; N = any nucleotide except G) and an unpaired "bulge" cytosine five nucleotides 5′ from the loop, which divides the stem of the hairpin into an upper and a lower region (Fig. 1). The nucleotide sequence of the lower and upper stem region seems functionally less important than its ability to form appropriate hydrogen bonds (Hentze et al. 1988). The structure of the IRE was probed chemically and enzymatically (Barton et al. 1990; Leibold et al. 1990; Jaffrey et al. 1993) and is described in detail in Henderson and Kühn (this vol.).

3 Iron-Regulatory Proteins (IRPs)

Binding studies [electrophoretic mobility shift assays (EMSA)] using lysates from cultured rat cells initially identified a cytoplasmic protein that interacts with the ferritin IRE (Leibold and Munro 1988). Complex formation is diminished following treatment of rats or tissue culture cells with iron. (Rouault et al. 1988). This IRE-binding protein, now commonly referred to as iron

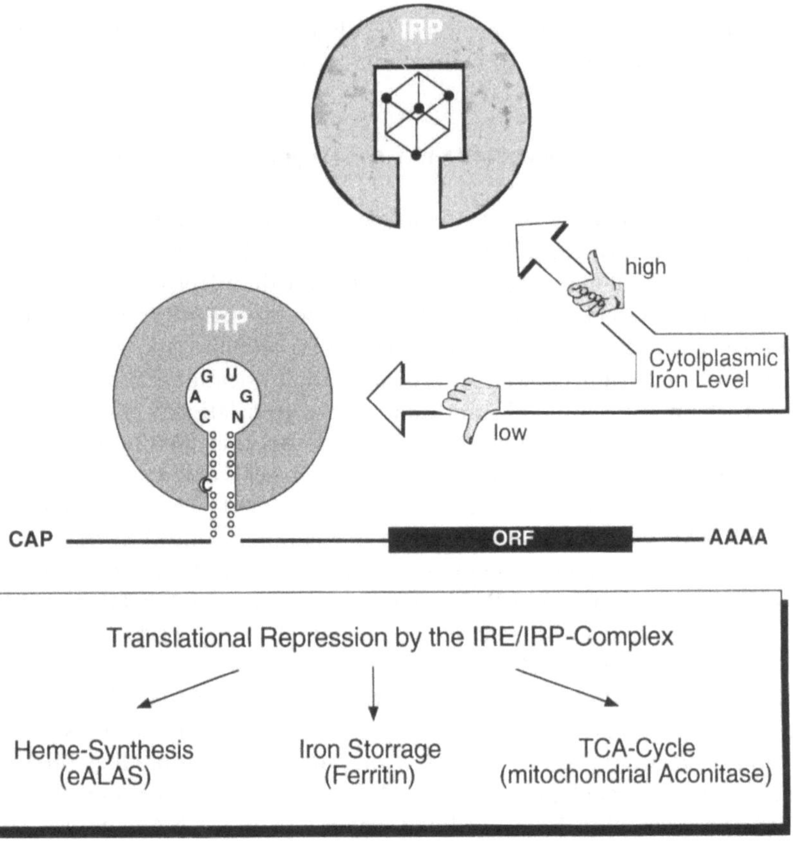

Fig. 1. Translational repression by IRE/IRP complexes. Reduction of cytoplasmic iron levels induces the conversion of the aconitase form of IRP-1 (that contains a 4Fe-4S cluster) to the apoprotein form, which binds to the IRE and acts as a translational repressor. IREs exist in the 5′UTR of ferritin, eALAS, and mitochondrial aconitase mRNAs. The expression of these IRE-containing mRNAs is therefore directly controlled by the IRE/IRP interaction and thus by signals like the intracellular iron level, nitric oxide, or oxidative stress

regulatory protein or IRP, has previously been called IRE-BP (Rouault et al. 1988), IRF (Müllner et al. 1989), FRP (Walden et al. 1988), or p90 (Harrell et al. 1991). IRP is a ubiquitous cytoplasmic protein of ~98 kDa and has been cloned from several species, including human (Rouault et al. 1989; Neupert et al. 1990), mouse (Philpott et al. 1991), rabbit (Patino and Walden 1992), and rat (Yu et al. 1992). IRE/IRP complexes that form in an iron-regulated manner are also found in chicken, frog, and fly. Moreover, IRP activity was demonstrated in fish and worm, while it could not be detected in extracts from plants, yeast, or bacteria (Rothenberger et al. 1990). In organisms where IRP could not be detected, different iron control mechanisms appear to be operative (see above).

Iron salts or hemin as iron donors, or desferrioxamine as an iron chelator, influence the binding of IRP to IREs, as determined by EMSA of lysates prepared from pretreated cells. In comparison to untreated cells, iron starvation increases the IRE-binding activity, whereas iron administration reduces RNA binding (Rouault et al. 1988; Fig. 1). First indications of how cellular iron levels may control IRP came from the observation of a striking homology between the amino acid sequences of IRP and the mitochondrial Krebs cycle enzyme aconitase (Hentze and Argos 1991; Rouault et al. 1991). The catalytic center of mitochondrial aconitase is formed by a 4Fe-4S cluster. Amino acid residues that contribute to the formation of the active site in mitochondrial aconitase are the most highly conserved residues in IRP, suggesting that IRP might be an iron-sulfur protein itself. Conditions which favor 4Fe-4S cluster formation of mitochondrial aconitase in vitro specifically abolish IRE-binding by IRP (Constable et al. 1992). It was subsequently demonstrated that IRP has aconitase activity (Haile et al. 1992a; Emery-Goodman et al. 1993; Gray et al. 1993). Whether IRP exerts aconitase activity or is able to interact with iron-responsive elements was hypothesized to depend on the iron-sulfur cluster in the protein. IRP from iron-replete cells contains an intact 4Fe-4S cluster and exerts aconitase activity. In contrast, IRP from iron-deprived cells is predominantly found in the apoprotein form and displays IRE-binding activity (Haile et al. 1992b; Kennedy et al. 1992; Emery-Goodman et al. 1993). The model arising from these studies suggests that IRP is a bifunctional protein that responds to cellular iron levels by posttranslational assembly or disassembly of its iron-sulfur cluster (Haile et al. 1992a; Kennedy et al. 1992; Pantopoulos et al. 1995). Consistent with this model, mutations of the cysteine residues that are required for the coordination of the 4Fe-4S cluster result in the loss of aconitase activity while the mutant proteins exhibit high IRE-binding activity (Philpott et al. 1993; Hirling et al. 1994). The mechanism by which the two activities of IRP are regulated by iron is discussed in more detail in Henderson and Kühn (this vol.).

A second protein that interacts with IREs in EMS assays was initially identified in extracts from rodent cells (Leibold and Munro 1988; Müllner et al. 1989; Rothenberger et al. 1990; Henderson et al. 1993), but it is also present in human cells (Rouault et al. 1990; Samaniego et al. 1994). This protein, termed IRP-2 (Samaniego et al. 1994); IRE-BP2 (Leibold and Munro 1988) or IRP_B (Henderson et al. 1993) is ~105 kDa in size (Rouault et al. 1990; Henderson et al. 1993; Guo et al. 1994). It also binds to IREs in an iron-dependent manner and with affinity and specificity similar to IRP-1, the protein described above (Henderson et al. 1993; Guo et al. 1994; Samaniego et al. 1994). Despite the similarities between IRP-1 and IRP-2, they differ in two significant ways: unlike IRP-1, IRP-2 protein levels decrease in iron-replete cells due to degradation of the protein. Furthermore, IRP-2 does not seem to exert aconitase activity (Guo et al. 1994; Samaniego et al. 1994). For further details see Henderson and Kühn (this vol.).

4 Translational Control of IRE Containing RNAs by IRE/IRP Interactions in the 5′UTR

In iron-deficient cells, IRPs bind to the IRE in the 5′UTR of ferritin mRNA, suggesting that they act as translational repressors (Fig. 1). Walden et al. (1989) showed that translation of mouse ferritin mRNA was specifically repressed in a cell-free wheat germ translation system when IRP-1 purified from rabbit liver was added. IRE-containing reporter constructs are responsive to changes in cellular iron levels following transfection into mammalian cells (Aziz and Munro 1987; Hentze et al. 1987a). The ferritin promoter, coding region, IRE-flanking sequences, or the 3′UTR do not seem to contribute significantly to the translational regulation observed in transfected cells (Caughmann et al. 1988; Goossen and Hentze 1992), although a role for ferritin IRE-flanking sequences has been reported on the basis of in vitro translation studies (Harrell et al. 1991). Quantitative differences in the range of iron responsiveness have, however, been observed when ferritin translation was directly compared to that of IRE-containing indicator constructs (Hentze et al. 1987a,b; Hentze et al. 1988; Goossen and Hentze 1992; Coulson and Cleveland 1993). This quantitative aspect of ferritin regulation may be influenced by the experimental conditions and the details of iron perturbations (Caughmann et al. 1988), but it may also relate to contributions by additional transcriptional, posttranscriptional or posttranslational mechanisms. For example, one report suggests that the 3′UTR of the ferritin mRNA may play a role to obtain full translational iron regulation (Dickey et al. 1998).

Translational control of IRE-containing messages occurs in tissue culture cells treated with the transcription inhibitors actinomycin D or cordycepin (Zähringer et al. 1976) as well as in cell-free translation systems from rabbit reticulocytes or wheat germ (Brown et al. 1989; Walden et al. 1989; Gray et al. 1993; Fig. 2). These experiments confirm that IRE/IRP-mediated translational repression can be fully uncoupled from transcription. Translational repression by the IRE/IRP-1 interaction was also demonstrated in yeast, where a constitutively transcribed IRE-containing luciferase reporter mRNA was translationally repressed when IRP-1 was coexpressed from an inducible IRP expression plasmid (Oliveira et al. 1993; Fig. 2). Experiments in yeast as well as in wheat germ extract indicate that cells that have not evolved to regulate endogenous IRE-containing mRNAs are able to mediate translational control exerted by the IRE/IRP interaction. These experiments show that the translation initiation apparatus of these cells is receptive to the repressor and that IRE/IRP are sufficient for translational repression. Further cofactors do not seem to be required in the process.

In addition to ferritin mRNAs, IREs have also been identified in the 5′UTR of eALAS mRNAs (Dierks 1990; Cox et al. 1991; Dandekar et al. 1991) and mitochondrial aconitase mRNA (Dandekar et al. 1991; Fig. 1). The eALAS IRE binds IRPs contained in murine B6 cells, MEL cells (Dandekar et al. 1991), and K562 cells (Cox et al. 1991), and represses translation in a cell-free

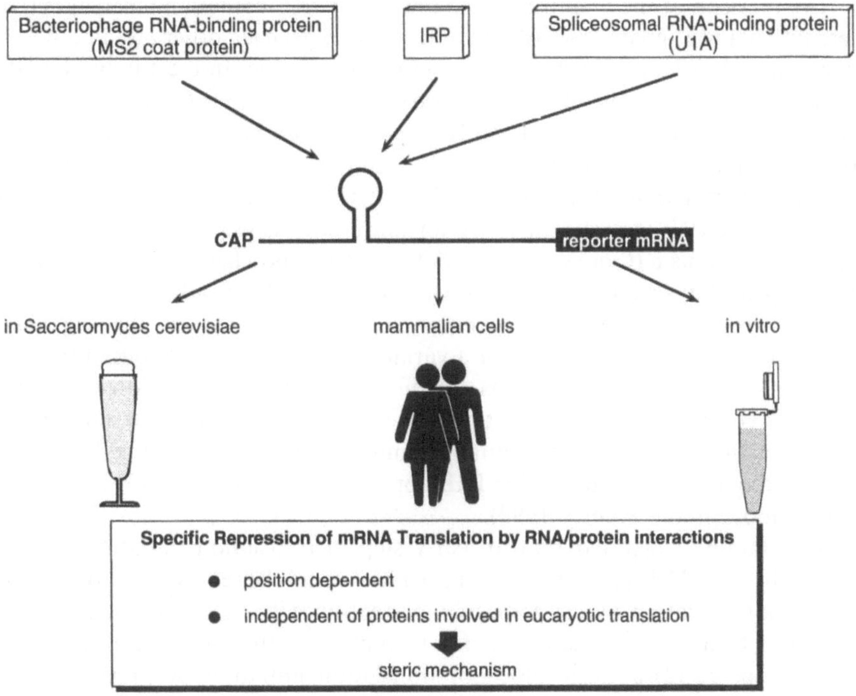

Fig. 2. Specific repression of mRNA translation by RNA/protein interactions. The bacteriophage MS2 coat protein, the spliceosomal RNA-binding protein U1A and IRP can specifically repress the translation of mRNAs in mammalian cells, *Saccharomyces cerevisiae* and in vitro, if their respective RNA target sequences are located in a cap-proximal position in the 5′UTR. As translational repression of these RNA/protein interactions depends on the position within the 5′UTR, and MS2 coat protein, as well as U1A, are not directly involved in eukaryotic translation, it seems likely that an early step in translation initiation is blocked by steric inhibition of the translation apparatus

translation system containing IRP-1 (Dandekar et al. 1991; Bhasker et al. 1993). eALAS synthesis increases when erythroid cells (MEL cells) are treated with iron and decreases when the cells are iron-deficient (Melefors et al. 1993). While eALAS regulation qualitatively resembles that of ferritin, the range of iron control of eALAS mRNA translation is smaller (Melefors et al. 1993). This less extensive range of regulation appears to originate from a lower affinity of IRPs for the eALAS IRE. Regulation of eALAS, therefore, depends on the intracellular iron level, linking the first step in heme biosynthesis, the major iron utilization pathway, with iron availability in the cell (Fig. 1).

An IRE motif contained in mitochondrial aconitase mRNA (which is encoded in the nucleus and translated in the cytoplasm) also binds IRP-1 in vitro (Zheng et al. 1992; Fig. 1) and mediates translational control in wheat germ extract (Gray et al. 1996). The biological role of iron-dependent regulation of this Krebs cycle enzyme is currently less well understood.

Comparison of the 5′ untranslated regions of IRE-containing messages (i.e., ferritin H- and L-chain, eALAS, aconitase) amongst each other and in different organisms shows that their overall lengths are variable. Interestingly, the IRE is always located within 15–40 nucleotides from the transcription start site. By contrast, the distance between the IRE and the translation start codon varies from 150 nucleotides in ferritin mRNA to zero in aconitase mRNA, where the translation initiation codon is part of the IRE. This conservation of the position of the IRE was shown to be an important aspect of the translational control mechanism (Goossen et al. 1990; Goossen and Hentze 1992). When the IRE is placed at its natural position within the 5′UTR of indicator constructs, efficient iron regulation can be observed. An alteration of the sequences flanking an IRE in this position or a variation of the distance between the IRE and the AUG initiator codon does not affect the regulatory function of the IRE in vivo. However, a spacer that places the IRE more than 70 nucleotides downstream of the cap substantially diminishes translational regulation, although the binding of IRP to the IRE appears to be unaffected (Goossen et al. 1990; Goossen and Hentze 1992).

This finding suggested that an early step in translation initiation was affected by the IRE/IRP interaction, which could be envisaged to prevent translation initiation by (1) a specific interaction with or enzymatic modification of a translation initiation factor to impair its function and thus inhibit translation, or (2) a steric block of translation initiation by preventing the binding of components of the translation machinery to the mRNA. In an experimental approach devised to distinguish between these two possibilities, the IRE/IRP interaction in the 5′leader was replaced by two heterologous RNA/protein interactions, those of the spliceosomal protein U1A or the bacteriophage MS2 coat protein with their respective RNA target sequences. Translational repression of mRNAs containing either one of these heterologous RNA target sequences occurs when the appropriate RNA-binding protein is added with the indicator mRNA to a cell-free translation system (Stripecke and Hentze 1992) or when both the reporter construct and the expression vector for the binding protein are cotransfected into yeast (Stripecke et al. 1994; Fig. 2). In mammalian cells, only the interaction of the MS2 coat protein with its RNA target site repressed translation. U1A did not display its ability to block translation in mammalian cells. This unexpected result may be explained by the existence of complex negative autoregulatory feedback mechanisms, which reduce U1A expression (Boelens et al. 1993). Furthermore, U1A predominantly accumulates in the nucleus and not in the cytoplasm, where translational control could be exerted.

The inhibition of indicator protein expression by these RNA/protein interactions is translational, because cytoplasmic mRNA levels were similar in the presence and the absence of the repressor proteins. Analogous to the situation with the IRE/IRP interaction, these alternative RNA/protein interactions mediated efficient translational repression only from a cap-proximal position (Stripecke and Hentze 1992; Stripecke 1993). Thus, RNA-binding proteins with physiological functions unrelated to the regulation of eukaryotic translation are able to act as translational repressor proteins and, therefore, appear to block translation initiation by a steric mechanism rather than by a specific modification of the translation apparatus (Fig. 2). By extrapolation of these results, an IRE/IRP complex may act in a similar way.

Translation initiation is a multistep process that leads to the assembly of an 80S ribosome at the initiation codon (Fig. 3). In a simplified model, translation initiation can be subdivided into the following steps (Hershey 1991): (1) cap recognition by the eIF-4F complex (consisting of the cap recognition factor eIF-4E, the 220-kDa protein, eIF-4G and eIF-4A, a protein with helicase activity), (2) unwinding of cap-proximal secondary structure, probably catalyzed by eIF-4A and 4B, (3) binding of the 43S preinitiation complex (con-

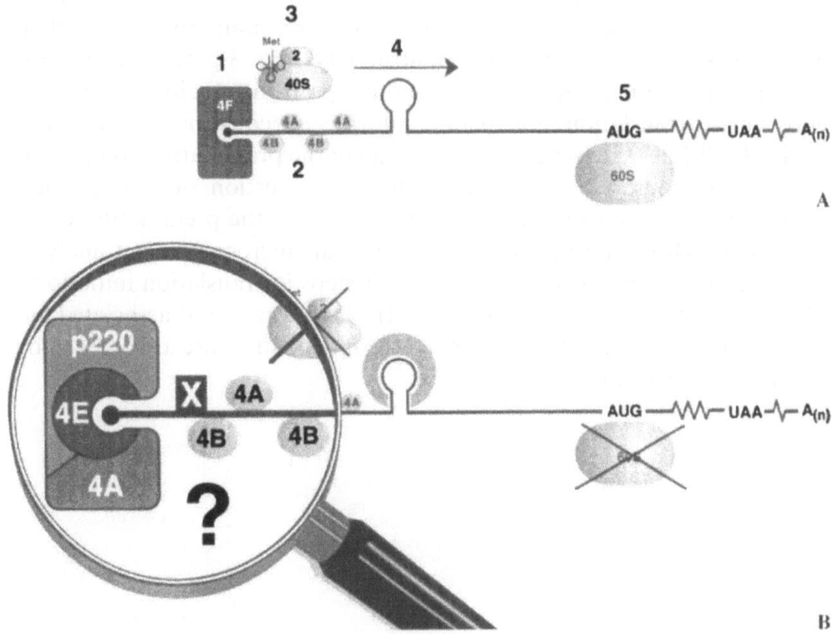

Fig. 3A, B. Mechanism of translational repression by IRP/IRE interactions. **A** Simplified scheme of the cap-dependent translation initiation pathway. The cap-binding complex eIF-4F first binds to the cap structure *1*, facilitating the binding of initiation factors 4A and 4B to the 5′end of the mRNA *2* which probably catalyzes the unwinding of cap-proximal secondary structure. This process is believed to aid the association of the 43S preinitiation complex (consisting of the small ribosomal subunit (40S), eIF-3, initiator ^{met}t-RNA, eIF-2 and GTP) to the mRNA *3*, which will subsequently scan along the mRNA until a suitable AUG start codon is reached *4*. At the AUG, the 60S ribosomal subunit binds and polypeptide synthesis begins *5*. **B** IRP inhibits the association of the 43S preinitiation complex with the mRNA. The inhibition of 43S binding to the mRNA is indicated, as is the consequential lack of 60S joining. Future experiments have to determine whether the block of 43S association is primary or whether even earlier steps of translation initiation are inhibited, e.g., the binding of the 4F complex (consisting of p220, eIF-4E, and eIF-4A) or the unwinding of cap-proximal RNA secondary structure

sisting of the small ribosomal subunit (40S), eIF-3, initiator t-RNA, eIF-2 and GTP) to the mRNA, (4) scanning of the 43S preinitiation complex along the mRNA to select the initiator codon, and (5) joining of the 60S large ribosomal subunit to form a translation competent 80S ribosome. During this latter process, GTP is hydrolyzed and 43S-associated translation initiation factors are released (Fig. 3).

Which step(s) in this pathway is inhibited by the IRE/IRP complex? Using a cell-free translation system and sucrose gradient centrifugation, different intermediates of the translation initiation process can be resolved. This approach revealed that the IRE/IRP complex prevents the stable association of the 43S preinitiation complex with the mRNA and, as a consequence, joining of the 60S ribosomal subunit at the start codon is inhibited (Gray and Hentze 1994). Likewise, heterologous RNA binding proteins (e.g., the spliceosomal factor U1A) bound to their target mRNAs are able to prevent the binding of the small ribosomal subunit to the mRNA when their binding sites are located in a cap-proximal position (Gray and Hentze 1994; Fig. 3). As predicted by the result that an RNA/protein complex in a cap-distant position does not efficiently impair translation in vivo and in vitro, such constructs examined in the described cell-free system permit the binding of the preinitiation complex (Gray et al. unpubl. observ.), raising the intriguing question of how a potential translational repressor protein can be overcome by the preinitiation complex, once bound. More discriminating methods than sucrose gradient analysis are required to investigate whether even earlier steps in translation initiation than 43S association, like the binding of the eIF-4F complex and associated factors or the unwinding of cap-proximal RNA secondary structure are blocked by the IRE/IRP complex (Fig. 3).

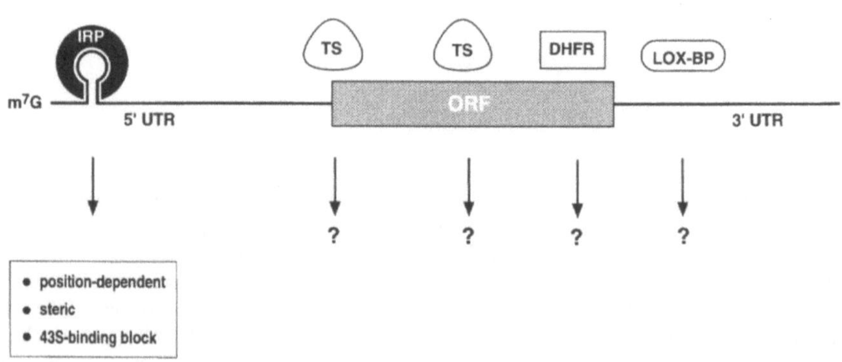

Fig. 4. How do mRNA-binding proteins regulate translation? The mechanism of translational repression by IRP has been intensively characterized. It appears to be positionally restricted to the cap-proximal portion of the 5′UTR. In contrast, very limited information is currently available for RNA/protein interactions that inhibit translation from cap-distant binding sites within the open reading frame [e.g., dihydrofolate reductase (DHFR) or thymidylate synthase (TS)], or within the 3′UTR [e.g., 15-lipoxygenase binding protein (LOX-BP)]

In summary, the mechanism of translational control of IRE-containing messages, like ferritin or eALAS, by IRPs involves the (sterical) blockage of the association of the small ribosomal subunit. Likewise, other RNA/protein complexes located in a cap-proximal position prevent the binding of the small ribosomal subunit. It will be interesting to compare this mechanism for translational regulation with those involving RNA/protein interactions that inhibit translation from cap-distant binding sites within the open reading frame, e.g., dihydrofolate reductase (Chu et al. 1993a), thymidilate synthetase (TS; Chu et al. 1991, 1993b) or even in the 3′UTR, e.g., LOX-BP (Ostareck-Lederer et al. 1994) of a message (Fig. 4).

5 Control of Transferrin Receptor mRNA Stability by IRE/IRP Interactions in the 3′UTR

In addition to ferritin and eALAS, the expression of the cell surface receptor responsible for iron uptake, the transferrin receptor (TfR), is also controlled by iron. Abundance of iron reduces the expression of transferrin receptors on the cell surface (Pelicci et al. 1982; Ward et al. 1982; Mattia et al. 1984). First results indicated that the iron-dependent expression of the transferrin receptor was reflected in the steady-state level of its mRNA and that the regulatory mechanism was thus not translational (Mattia et al. 1984; McClelland et al. 1984; Rao et al. 1985). Treatment of cells with an iron chelator increases TfR mRNA levels and TfR protein expression. Although TfR promoter activity is stimulated ~three-fold in cells treated with an iron chelator in comparison to cells treated with iron, this effect does not explain the profound difference in TfR mRNA expression observed (Rao et al. 1986; Casey et al. 1988). An iron-responsive region that mediates a large iron-dependent effect on TfR mRNA levels was subsequently found in the 3′UTR of the TfR mRNA (Owen and Kühn 1987; Casey et al. 1988). When this region was transferred to a reporter gene, the construct was responsive to iron (Casey et al. 1988; Müllner and Kühn 1988). Based on run-on transcription assays and measurements of TfR mRNA levels in the nucleus and the cytoplasm, it was concluded that the regulation of TfR as well as the chimeric reporter mRNAs occurs post-transcriptionally in the cytoplasm at the level of mRNA stability (Casey et al. 1988; Müllner and Kühn 1998).

The 3′UTR of the TfR mRNA is highly structured and contains sequences that are conserved between human, rat, and chicken. Mapping of functionally necessary domains in the 3′UTR by deletion and mutational analysis defined two conserved regions of approximately 150 nucleotides each, which are separated by a less conserved spacer element of approx. 250 nucleotides. Strikingly, the two domains harbor five stem-loop structures (A-E) similar to the IRE located in the 5′UTR of the ferritin message (Casey et al. 1988; Fig. 5). Every single one of these motifs was shown to bind IRP-1 in vitro (Koeller et al. 1989). Moreover, IRE(B) and IRE(C) can substitute for the native ferritin

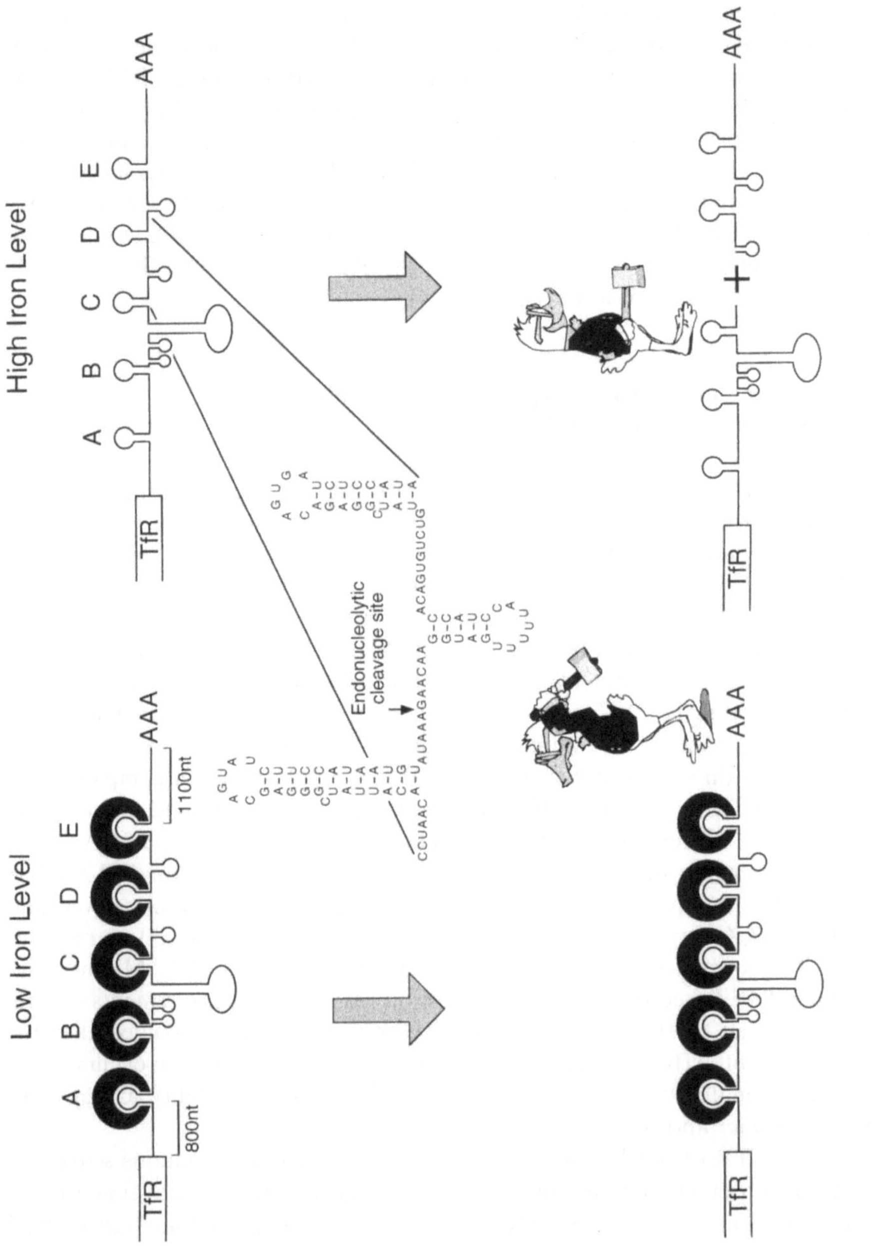
Low Iron Level
High Iron Level
A
B
C
D
E
TfR
AAA
800nt
1100nt
Endonucleolytic cleavage site

IRE and mediate translational control in tissue culture cells when cloned into the 5′UTR of reporter constructs, implying that they also bind IRP in vivo (Casey et al. 1988). In the natural context of the TfR 3′UTR, up to four molecules of IRP have been shown to bind to the TfR regulatory region simultaneously (Müllner et al. 1989). How many molecules of IRP are required to interact with the TfR IREs to confer iron regulation is not known, but IRE (A) and IRE (E) can be removed from the TfR mRNA without an apparent effect on iron regulation in vivo (Casey et al. 1989).

A more detailed analysis of the TfR regulatory region led to the definition of a synthetic 250 nucleotide minimal regulatory region (TRS-1), that contains three IREs (B,C,D) and lacks most of the nonconserved spacer region (Casey et al. 1989). A TfR mRNA containing TRS-1 is functionally indistinguishable from the wild-type TfR message. When the three IREs are deleted from this construct, iron regulation is abolished, and TfR expression obtained from this construct (TRS-3) is high. Point mutations in the three IRE structures that abolish IRP binding (TRS-4) also eliminate iron regulation. However, in contrast to TRS-3, this construct yields an mRNA that is intrinsically unstable (Casey et al. 1989; Koeller et al. 1991). A phenotype similar to that of TRS-3 was obtained when non-IRE stem-loop structures contained in the regulatory region were deleted (Casey et al. 1989). Consistent with these results, Müllner and Kühn (1988) found that mutations outside of the IREs could abolish iron regulation.

These studies suggest the presence of two functionally interdependent components within the minimal regulatory region: the three IRE structures (that confer iron-mediated regulation) and an instability element that is functionally distinct from but appears to be physically overlapping with the IREs (Casey et al. 1989; Koeller et al. 1991). In TRS-4, iron regulation is lost by selective interference with IRP binding while preserving the instability element, which results in an unstable mRNA. As the non-IRE secondary structures that seem to be part of the instability element are separated by more than 300 nucleotides in natural TfR mRNAs, a higher-order RNA structure may have to form to constitute the turnover element. In contrast to the point deletions in TRS-4, the complete deletion of the IREs removes a necessary part of the instability determinant, resulting in the loss of iron regulation of a relatively stable mRNA. Thus, the role of the IRE/IRP interaction in the regulation of

◄

Fig. 5. Control of transferrin receptor (TfR) mRNA stability by IRE/IRP interactions in the 3′UTR. A regulatory region that mediates a large iron-dependent effect on TfR mRNA stability is located in the 3′UTR [approx. 800 nt from the stop codon and 1100 nt from the poly(A) tail] of TfR mRNAs. It contains five IREs (*A-E*), several conserved non-IRE stem-loop structures and a sequence, located between IREs *C* and *D*, that is referred to as the endonucleolytic cleavage site. When cellular iron is low, IRP binds to the IREs and protects the TfR mRNA from an endonucleolytic attack by a so far unidentified nuclease. When the iron level is high, IRP does not bind to the IREs, and the TfR mRNA is cleaved into two fragments at the endonucleolytic cleavage site, initiating further degradation of the TfR mRNA.

TfR mRNA stability is best reflected in the following model: the binding of IRP to the IREs protects the mRNA against degradation. In the absence of IRP binding, as is the case in iron-loaded cells, the instability determinants confer targeted degradation of the TfR mRNA, and TfR expression on the cell surface is reduced (Fig. 5).

In the model outlined above, IRP is proposed to protect the TfR mRNA in an iron-regulated way. A nuclease involved in the degradation would therefore not necessarily have to be iron-regulated itself, nor would the TfR mRNA necessarily have to be its only substrate. However, this nuclease has not yet been identified.

Interestingly, truncated TfR mRNAs have been identified in cell lines under conditions where the TfR mRNA turns over rapidly (Binder et al. 1994). In cells stably transfected with TRS-1, the appearance of truncated TfR mRNAs correlated with iron treatment of the cells. Conversely, the truncated RNAs are constitutively expressed in cell lines transfected with TRS-4, suggesting that the truncated RNAs are bona fide intermediates of TfR mRNA breakdown. Further experiments, using a construct in which the TRS-4 region was placed upstream of the TRS-1 region suggested that an endoribonuclease is involved in the degradation of TfR mRNA, as the binding of IRP to the IREs in the downstream TRS-1 cassette did not prevent the cleavage within the upstream TRS-4 cassette, arguing against the involvement of a 3′ to 5′ exonuclease. Consistent with the idea of an endonucleolytic cleavage, two fragments were identified: one that entails the main body of the message and extends into the 3′UTR regulatory region, and a second, polyadenylated fragment that corresponds to the remaining 3′ end of the TfR mRNA (Binder et al. 1994). The suggested cleavage site maps just 3′ of IRE(C) within a single stranded A-rich region (Binder et al. 1994; Fig. 5). The sequence at the cleavage site is conserved in human and chicken (Koeller et al. 1989) as well as in rat TfR mRNAs (Roberts and Griswold 1990), and its mutation causes lack of cleavage and mRNA stabilization (Binder et al. 1994). The endonucleolytic cleavage does not require prior shortening of the poly (A) tail (Binder et al. 1994). The location of the cleavage site in the vicinity of an IRE is consistent with the model that an IRE/IRP complex protects the TfR mRNA against a nucleolytic attack (Fig. 5). The IRE/IRP complex could either sterically block the access of the endonuclease or, alternatively, induce an alteration in the RNA structure in a way that prevents recognition of the endonucleolytic cleavage site by the nuclease.

Several studies addressed the possibility that ongoing translation might be required for the degradation of the TfR mRNA. In addition to the TfR coding region itself, reporter constructs encoding either HLA membrane proteins or the secretory hormone hGH have been used. Recently, luciferase was also employed successfully as a reporter (Pantopoulos et al. 1995b), showing that translation on the rough endoplasmatic reticulum was not required for the degradation process. Human TfR mRNA expressed in mouse L cells is stabilized by cycloheximide (Müllner and Kühn 1988), suggesting that inhibition of translation stabilizes the TfR mRNA. Likewise, the stability of TRS-1 mRNA in cells treated with hemin (but not desferrioxamine) is markedly increased in the presence of the translation inhibitor cycloheximide (Koeller et al. 1991) and TRS-4 mRNA levels increase when translation is pharmacologically inhibited (Koeller et al. 1991). Because cycloheximide inhibits the translation of all cellular mRNAs, these experiments cannot distinguish between a direct effect of translation on TfR mRNA stability or the involvement of a highly unstable protein that is required for degradation.

In an alternative approach, constructs that contain an IRE in the 5′UTR and either the instability element of c-fos mRNA or the intrinsically unstable TRS-4 element in the 3′UTR of a reporter gene were used (Koeller et al. 1991). Translation of these reporter mRNAs can be switched off specifically in iron-deficient cells by binding of IRP to the 5′ IRE without affecting the translation of other (non-IRE containing) cellular mRNAs. Inhibition of the translation of these constructs did not change the levels of these transcripts (Koeller et al. 1991). It was concluded that the TfR (and c-fos 3′UTR) mRNA degradation pathways are independent of the polysomal association of the message and that cycloheximide most likely prevents the synthesis of a short-lived protein that is required for degradation. However, the results reported by Koeller et al. (1991) have, at least with regard to c-fos mRNA turnover, been challenged by three studies (Aharon and Schneider 1993; Winstall et al. 1995; Veyrune et al. 1995), who have argued that the degradation of c-fos mRNA mediated by an $(AUUUA)_n$ sequence in its 3′UTR is dependent on ongoing translation. For future studies, a cell-free system that faithfully reproduces the regulated degradation of TfR mRNA would add an important experimental tool to identify the nuclease(s) or possible additional factors that are involved in the degradation of the TfR mRNA.

6 "Iron-Responsive" Elements Also Respond to Other Cellular Signals

As described, the expression of ferritin, eALAS and TfR mRNA is regulated posttranscriptionally by IRE/IRP interactions. Studies investigating the regulation of IRP binding to IREs have initially focused on the role of iron as the agent to steer the IRE/IRP interaction (see above). Recent findings suggest that two micromolecular transmitter molecules, nitric oxide (NO), and H_2O_2, can also do so and regulate the expression of IRE-containing mRNAs.

Nitric oxide (NO) controls diverse processes including the regulation of vascular tone, neurotransmission or macrophage-mediated cytotoxicity towards tumor cells and microbes (Lowenstein and Snyder 1992; Nathan 1992; Schmidt and Walter 1994). Its broad spectrum of biological effects largely arises from the chemical reactivity of NO towards metals like iron (Stamler 1994). One iron-containing protein that represents a target for NO is mitochondrial aconitase, which has been shown to be inactivated by NO (Drapier et al. 1991). The loss of aconitase activity coincides with the formation of iron-nitrosyl complexes, suggesting that NO targets the iron-sulfur cluster of aconitase. In the light of this, the response of IRP to NO was investigated. Induction of NO synthesis by stimulating mouse macrophages or macrophage cell lines with γ-interferon (γ-IFN) and lipopolysaccharide (LPS) diminished the cytoplasmic aconitase activity and activated the IRE-binding activity of IRP-1 and IRP-2 (Drapier et al. 1993; Weiss et al. 1993). The NO-dependent increase in IRE-binding is comparable to the effect of desferrioxamine treatment and results in the translational repression of ferritin mRNA (Weiss et al. 1993). Exposure of recombinant IRP-1 to NO gas in vitro also increases the IRE-binding activity (Drapier et al. 1993), but this response is incomplete and it thus remains to be established whether the effect of NO on IRP is a direct one.

In a subsequent study, fibroblasts were stably transfected with a murine macrophage NO synthase cDNA under the control of the SV40 promoter, thus producing NO constitutively (Pantopoulos and Hentze 1995a) In these cells, NO synthesis represses ferritin mRNA translation and stabilizes transferrin receptor mRNA against targeted degradation (Pantopoulos and Hentze 1995a), consistent with the idea that NO has an equivalent effect on IRP as desferioxamine. The stabilization of TfR mRNA was not observed in macrophages that were stimulated to produce NO with γ-IFN and LPS (Drapier et al. 1993). These results suggest that the regulation of TfR expression by γ-IFN/LPS in murine macrophages is controlled at multiple levels, possibly involving additional nuclear mechanisms (Bourgeade et al. 1992).

Desferrioxamine treatment of cells induces the complete disassembly of the [4Fe-4S] cluster of IRP-1 (Haile et al. 1992b; Emery-Goodman et al. 1993) and IRE-binding in the apoprotein form. Whether NO also induces apoprotein formation of IRP is currently not known, but biochemical studies suggest that NO also induces cluster disassembly beyond the 3Fe-4S stage (Pantopoulos and Hentze 1995a). This disassembly could either be a direct reaction of NO with the Fe-S cluster, a response to NO-induced second messengers that act on IRP, or NO could exert its function by reducing the level of iron available for cluster formation and maintenance. Recent data suggest that NO does not directly act upon the [4Fe4S] cluster but that a more oxidized form, $OONO^-$, does (Castro et al. 1994; Hausladen and Fridovich 1994). Further studies need to address this question and probe the (patho-) physiological role(s) of the regulatory interconnection between iron metabolism and NO. The finding that cellular iron levels regulate the transcription of the inducible nitric oxide synthase gene in the murine macrophage cell line J774 further substantiates the link between NO and iron metabolism (Weiss et al. 1994). However, as far as the signaling of iron or NO to IRP is concerned, the two seem to occur independently from each other, as inhibitors of NO synthesis are not able to block the effect of the iron chelator desferrioxamine in inducing the IRE-binding activity of IRP (Pantopoulos and Hentze 1995a).

Within the last few months, a third cellular signal has been discovered to induce the conversion of IRP-1 into its RNA binding form: H_2O_2 (Pantopoulos and Hentze 1995b). Reactive oxygen intermediates (ROIs), including the superoxide anion ($O_2^{\bullet -}$) and hydrogen peroxide (H_2O_2) are inevitable byproducts of aerobic metabolism and physiological mediators of the antimicrobial response of phagocytic cells (Baggiolini and Thelan 1991). However, H_2O_2 can react with ferrous iron to yield the aggressive hydroxyl radical ($OH^\bullet$; Fenton reaction), which attacks macromolecules like lipids, proteins or DNA.

Treatment of murine B6 fibroblasts with H_2O_2 reduces ferritin translation and increases TfR mRNA levels (Pantopoulos and Hentze 1995b). Both effects are antagonized by the inclusion of the antioxidant N-acetyl-L-cysteine (NAC). The effect of H_2O_2 on the expression of ferritin and TfR is mediated by IRP-1, as the mechanism of this novel oxidative stress response pathway is posttranscriptional and the exposure of cells to H_2O_2 increase the IRE-binding activity in a dose- and time-dependent fashion that corresponds to the changes in ferritin and TfR expression. Moreover, H_2O_2-induced IRE-binding by IRP-1 coincides with the loss of its aconitase activity and probably with a disassembly of the 4Fe4S cluster beyond the 3Fe-4S state, possibly into the apoIRP form. Thus, the regulation of IRP-1 by H_2O_2 provides an intriguing link between the control of cellular iron homeostasis and the levels of H_2O_2, the substrate of the Fenton reaction. The kinetics with which the RNA-binding activity of IRP is induced differs significantly between iron starvation and H_2O_2. While iron starvation requires 8–12 h for maximal activation of IRP-1, H_2O_2 induces maximal activation already 30–60 min after its addition to the cells. This result may be explained by different pathways that could be involved in the responses to iron and H_2O_2. The rapid induction by H_2O_2 can be blocked by ocadaic acid, an inhibitor of type I/IIa protein phosphatases, which interestingly does not affect activation of IRP-1 by iron depletion, suggesting the involvement of stress-induced kinase/phosphatase pathways during oxidative stress signaling and permitting the biochemical uncoupling of the iron and oxidative stress responses (Pantopoulos and Hentze 1995b). In contrast to IRP-1, oxidative stress exerts only a minor (if any) effect on IRP-2 (Pantopoulos and Hentze, unpubl. results).

7 Summary and Perspectives

The study of iron metabolism in mammalian cells has led to the characterization of two posttranscriptional control mechanisms that both involve an interaction of iron-responsive elements (IREs) with iron-regulatory proteins (IRPs): translational regulation and the control of mRNA stability.

An IRE/IRP complex that forms in the cap-proximal part of the 5′UTR of ferritin, eALAS and, most likely, mitochondrial aconitase mRNAs in iron-deficient cells inhibits the translation of these messages by blocking the stable association of the 43S preinitiation complex with these mRNAs. Thus, iron metabolism provides the first example of a mechanism by which an RNA/protein complex can block eukaryotic translation. The association of the 43S preinitiation complex with the mRNA is also blocked when the IRE/IRP complex is substituted by heterologous RNA/protein complexes, suggesting that this mechanism reflects a more general property of RNA/protein complexes, located in a cap-proximal position to repress translation. Such a translational control mechanism may, therefore, not be unique to iron metabolism but also be involved, for example, during early development.

Future studies will have to further dissect the mechanism of translational control by such RNA/protein interactions. The focus will be on those initiation factors whose interaction with the mRNA precedes the binding of the 43S preinitiation complex, such as the eIF-4A/4B helicase or the eIF-4F cap-binding complex that sets the way for the rest of the translation initiation apparatus to assemble. A further interesting point that arises from the position dependence of translational inhibition by the IRE/IRP complex addresses the question of how the translation initiation machinery can overcome IRE/IRP complexes located more distally from the cap.

In addition to translation, IRE/IRP interactions regulate mRNA stability. Five IREs are found in the 3′UTR of the TfR message. When IRP is not bound, it appears as if an endonucleolytic cleavage occurs at a site that is located within the IRE-containing region and initiates further mRNA degradation. An IRE/IRP complex prevents the endonucleolytic attack and thus mRNA degradation. The degradation pathway followed by the TfR mRNA degradation. The degradation pathway followed by the TfR mRNA seems to be independent of prior deadenylation.

The identification and characterization of the endonuclease that is involved in the degradation process will be a future key issue in the studies concerning the mechanism of TfR mRNA degradation. The establishment of a suitable cell-free system would constitute a major step forward in this undertaking. Since components of the degradation pathway may be short-lived or cytoskeletal integrity be required, such an undertaking could meet with substantial technical difficulties.

Most IRE-controlled mRNAs that have been identified so far seem to play a direct role in the regulation of iron metabolism, consistent with the finding that iron is one of the agents that controls the switch from the IRE-binding to the aconitase form. In addition to iron, other intracellular signals have been found more recently to control the IRP switch: nitric oxide (NO) and oxidative stress induced by H_2O_2. The physiological role(s) for NO or H_2O_2-mediated regulation of ferritin or TfR mRNAs requires further exploration. It can be speculated that different inducers of the RNA-binding form of IRP might act upon different subsets of IRE-containing mRNAs. For example, H_2O_2 as a toxic byproduct of the respiratory chain might be able to act in a negative feedback control loop on mitochondrial aconitase, an enzyme of the citric acid cycle that "feeds" reducing equivalents into the respiratory chain. To investigate such questions, it will be interesting to examine the effects of endogenously produced H_2O_2 on the expression of IRE-containing mRNAs. In addition, experiments designed to identify further IRE-containing mRNAs, as well as new signaling molecules that can influence the IRE/IRP interaction might shed some more light on cellular signaling cascades and their linkage to biochemical pathways.

References

Aharon T, Schneider RJ (1993) Selective destabilization of short-lived mRNAs with the granulocyte-macrophage colony-stimulating factor AU-rich 3′ noncoding region is mediated by a cotranslational mechanism. Mol Cell Biol 13:1971–1980

Aziz N, Munro HN (1986) Both subunits of rat liver ferritin are regulated at a translational level by iron induction. Nucl Acids Res 14:915–927

Aziz N, Munro HN (1987) Iron regulates ferritin mRNA translation through a segment of its 5′ untranslated region. Proc Natl Acad Sci USA 84:8478–8482

Baggiolini M, Thelan M (1991) The phagocytes and the respiratory burst. In: Sies H(ed) Oxidative stress: oxidants and antioxidants. Harcourt Brace Jovanovich, London

Bakker GR, Boyer RF (1986) Iron incorporation into apoferritin. J Biol Chem 261:13182-13185

Barton HA, Eisenstein RS, Bomford A, Munro HN (1990) Determinants of the interaction between the iron-responsive element-binding protein and its binding site in rat L-ferritin mRNA. J Biol Chem 265:7000–7008

Bhasker CR, Burgiel G, Neupert B, Emery-Goodman A, Kühn LC, May BK (1993) The putative iron-responsive element in the human erythroid 5-aminolevulinate synthase mRNA mediates translational control. J Biol Chem 268:12699–12705

Binder R, Horowitz JA, Basilion JP, Koeller DM, Klausner RD, Harford JB (1994) Evidence that the pathway of transferrin receptor mRNA degradation involves an endonucleolytic cleavage within the 3′ UTR and does not involve poly(A) tail shortening. EMBO J 13:1969–1980

Boelens W, Jansen EJR, van Venrooij WJ, Stripecke R, Mattaj IW, Gunderson SI (1993) The human U1snRNP-specific protein U1A inhibits polyadenylation of its own pre-mRNA. Cell 72:881–892

Bothwell TH, Charlton RW, Motulski AG (1983) Idiopathic hemochromatosis. In: Stanbury SB, Wyngaarden JB, Fredrickson DS, Goldstein JL, Brown MS (eds.) The metabolic basis of inherited disease McGraw-Hill, New york

Bourgeade M-F, Silbermann F, Kühn L, Testa U, Peschle C, Memet S, Thang M, Besancon F (1992) Post-transcriptional regulation of transferrin receptor m-RNA by IFNγ. Nucl Acids Res 20:2997–3003

Braun V, Günter K, Hantke K (1991) Transport of iron across the outer membrane. Biol Met 4:14–22

Britton RS, Tavill AS, Bacon BR, Bacon BR (1994) Mechanisms of iron toxicity. In: Broch JH, Halliday JW, Pippard MJ, Powell LW (eds.) Iron metabolism in health and disease. Saunders, London, pp 311–351

Brown PH, Daniels-McQueen S, Walden WE, Patino MM, Gaffield L, Bielser D, Thach RE (1989) Requirements for the translational repression of ferritin transcripts in wheat germ extracts by a 90-kDa protein from rabbit liver. J Biol Chem 264:13383–13386

Casey JL, Hentze MW, Koeller DM, Caughman SW, Rouault TA, Klausner RD, Harford JB (1988) Iron-responsive elements: regulatory RNA sequences that control mRNA levels and translation. Science 240:924–928

Casey JL, Koeller DM, Ramin VC, Klausner RD, Harford JB (1989) Iron regulation of transferrin receptor mRNA levels requires iron-responsive elements and a rapid turnover determinant in the 3′ untranslated region of the mRNA. EMBO J 8:3693–3699

Castro L, Rodriguez M, Radi R (1994) Aconitase is readily inactivated by peroxynitrite, but not by its precursor, nitric oxide. J Biol Chem 269:29409–29415

Caughmann SW, Hentze MW, Rouault TA, Harford JB, Klausner RD (1988) The iron-responsive element is the single element responsible for iron-dependent translational regulation of ferritin biosynthesis. Evidence for function as the binding site for a translational repressor. J Biol Chem 263:19048–19052

Chu E, Koeller DM, Casey JL, Drake JC, Chabner BA, Elwood PC, Zinn S, Allegra CJ (1991) Autoregulation of human thymidylate synthase messenger RNA translation by thymidylate synthase. Proc Natl Acad Sci USA 88:8977–8981

Chu E, Takimoto CH, Voeller D, Grem JL, Allegra CJ (1993a) Specific binding of human dihydrofolate reductase protein to dihydrofolate reductase messenger RNA in vitro. Biochemistry 32:4756–4760

Chu E, Voeller D, Koeller DM, Drake JC, Takimoto CH, Maley GF, Maley F, Allegra CJ (1993b) Identification of an RNA-binding site for human thymidylate synthase. Proc Natl Acad Sci USA 90:517–521

Constable A, Quick S, Gray NK, Hentze MW (1992) Modulation of the RNA-binding activity of a regulatory protein by iron in vitro: switching between enzymatic and genetic function? Proc Natl Acad Sci USA 89:4554–4558

Coulson RMR, Cleveland DW (1993) Ferritin synthesis is controlled by iron-dependent translational derepression and by changes in synthesis/transport of nuclear ferritin RNAs. Proc Natl Acad Sci USA 90:7613–7617

Cox TC, Bawden MJ, Martin A, May BK (1991) Human erythroid 5-aminolevulinate synthase: promoter analysis and identification of an iron-responsive element in the mRNA. EMBO J 10:1891–1902

Dancis A, Roman DG, Anderson GJ, Hinnebusch AG, Klausner RD (1992) Ferric reductase of *Saccharomyces cerevisiae:* molecular characterization, role in iron uptake, and transcriptional control by iron. Proc Natl Acad Sci USA 89:3869–3873

Dandekar T, Stripecke R, Gray NK, Goossen B, Constable A, Johansson HE, Hentze MW (1991) Identification of a novel iron-responsive element in murine and human erythroid δ-aminolevulinic acid synthase mRNA. EMBO J 10:1903–1909

Dickey LF, Wang Y-H, Shull GE, Wortman III IA, Theil EC (1988) The importance of the 3′-untranslated region in the translational control of ferritin mRNA. J Biol Chem 263:3071-3074

Dierks P (1990) Molecular biology of eukaryotic 5-aminolevulinate synthase. In: Dailey HA (ed.) Biosynthesis of heme and chlorophylls. McGraw-Hill, New York

Drapier JC, Pellat C, Henry Y (1991) Generation of EPR-detectable nitrosyl-iron complexes in tumor target cells cocultured with activated macrophages. J Biol Chem 266:10162–10167

Drapier JC, Hirling H, Wietzerbin J, Kaldy P, Kühn LC (1993) Biosynthesis of nitric oxide activates iron regulatory factor in macrophages. EMBO J 12:3643–3649

Drysdale JW, Munro HN (1965) Regulation of synthesis and turnover of ferritin in rat liver. J Biol Chem 241:3630–3637

Eide D, Davis-Kaplan S, Jordan I, Sipe D, Kaplan J (1992) Regulation of iron uptake in *Saccharomyces cerevisiae.* J Biol Chem 267:20774–20781

Emery-Goodman A, Hirling H, Scarpellino L, Henderson B, Kühn LC (1993) Iron regulatory factor expressed from recombinant baculovirus: conversion between the RNA-binding apoprotein and Fe-S cluster containing aconitase. Nucl Acids Res 21:1457–1461

Goossen B, Hentze MW (1992) Position is the critical determinant for function of iron-responsive elements as translational regulators. Mol Cell Biol 12:1959–1966

Goossen B, Caughman SW, Harford JB, Klausner RD, Hentze MW (1990) Translational repression by a complex between the iron-responsive element of ferritin mRNA and its specific cytoplasmic-binding protein is position-dependent in vivo. EMBO J 9:4127–4133

Gray N (1994) The mechanism of translational repression by iron regulatory protein. Ph D Thesis Univ Glasgow, Glasgow

Gray NK, Hentze MW (1994) Iron regulatory protein prevents binding of the 43S translation preinitiation complex to ferritin and eALAS mRNAs. EMBO J 13:3882–3891

Gray NK, Quick S, Goossen B, Constable A, Hirling H, Kühn LC, Hentze MW (1993) Recombinant iron regulatory factor functions as an iron-responsive element-binding protein, a translational repressor and an aconitase. A functional assay for translational repression and direct demonstration of the iron switch. Eur J Biochem 218:657–667

Gray NK, Pantopoulos K, Dandekar T, Ackrell BAC, Hentze MW (1996) Translational regulation of mammalian and Drosophila citric acid cycle enzymes via iron-responsive elements. Proc Natl Acad Sci USA 93:4925–4930

Guo B, Yu Y, Leibold EA (1994) Iron regulates cytoplasmic levels of a novel iron-responsive element-binding protein without aconitase activity. J Biol Chem 269:24252–24260

Gutteridge JMC (1989) Iron and oxygen: a biologically damaging mixture. Acta Pediatr Scan Suppl 361:78–85

Haile DJ, Rouault TA, Tang CK, Chin J, Harford JB, Klausner RD (1992a) Reciprocal control of RNA-binding and aconitase activity in the regulation of the iron-responsive element binding protein: role of the iron-sulfur cluster. Proc Natl Acad Sci USA 89:7536–7540

Haile DJ, Rouault TA, Harford JB, Kennedy MC, Blodin GA, Beinert H, Klausner RD (1992b) Cellular regulation of the iron-responsive element binding protein: disassembly of the cubane iron-sulfur cluster results in high-affinity RNA binding. Proc Natl Acad Sci USA 89:11735–11739

Halliday JW, Ramm GA, Powell LW (1994) Cellular iron processing and storage: the role of ferritin. In: Brock JH, Halliday JW, Pippard MJ, Powell LW (eds.) Iron metabolism in health and disease. Saunders, London, pp 97–112

Harrell CM, McKenzie AR, Patino MM, Walden WE, Theil EC (1991) Ferritin mRNA: interactions of iron regulatory element with translational regulator protein P-90 and the effect on base-paired flanking regions. Proc Natl Acad Sci USA 88:4166–4170

Hausladen A, Fridovich I (1994) Superoxide and peroxynitrite inactivate aconitases, but nitric oxide does not. J Biol Chem 269:29405–29408

Henderson BR, Seiser C, Kühn LC (1993) Characterization of a second RNA-binding protein in rodents with specificity for iron-responsive elements. J Biol Chem 268:27327–27334

Hentze MW, Argos P (1991) Homology between IRE-BP, a regulatory RNA-binding protein, aconitase and isopropylmalate isomerase. Nucl Acids Res 19:1739–1740

Hentze MW, Caughman SW, Rouault TA, Barriocanal JG, Dancis A, Harford JB, Klausner RD (1987a) Identification of the iron-responsive element for the translational regulation of human ferritin mRNA. Science 238:1570–1573

Hentze MW, Rouault TA, Caughman SW, Dancis A, Harford JB, Klausner RD (1987b) A *cis*-acting element is necessary and sufficient for translational regulation of human ferritin expression in response to iron. Proc Natl Acad Sci USA 84:6730–6734

Hentze MW, Caughman SW, Casey JL, Koeller DM, Rouault TA, Harford JB, Klausner RD (1988) A model for the structure and function of iron-responsive elements. Gene 72:201-208

Hershey JWB (1991) Translational control in mammalian cells. Annu Rev Biochem 60:717–755

Hirling H, Henderson BR, Kühn LC (1994) Mutational analysis of the [4Fe-4S]-cluster converting iron regulatory factor from its RNA-binding form to cytoplasmic aconitase. EMBO J 13:453–461

Jaffrey SR, Haile DJ, Klausner RD, Harford JB (1993) The interaction between the iron-responsive element binding protein and its cognate RNA is highly dependent upon both RNA sequence and structure. Nucl Acids Res 21:4627–4631

Kennedy MC, Mende-Mueller L, Blondin GA, Beinert H (1992) Purification and characterization of cytosolic aconitase from beef liver and its relationship to the iron-responsive element binding protein. Proc Natl Acad Sci USA 89:11730–11734

Koeller DM, Casey JL, Hentze MW, Gerhardt EM, Chan L-NL, Klausner RD, Harford JB (1989) A cytosolic protein binds to structural elements within the iron regulatory region of the transferrin receptor mRNA. Proc Natl Acad Sci USA 86:3574–3578

Koeller DM, Horowitz JA, Casey JL, Klausner RD, Harford JB (1991) Translation and the stability of mRNAs encoding the transferrin receptor and *c-fos*. Proc Natl Acad Sci USA 88:7778–7782

Leibold EA, Munro HN (1988) Cytoplasmic protein binds in vitro to a highly conserved sequence in the 5′ untranslated regions of ferritin heavy- and light-subunit mRNAs. Proc Natl Acad Sci USA 85:2171-2175

Leibold EA, Laudano A, Yu Y (1990) Structural requirements of iron-responsive elements for binding of the protein involved in both transferrin receptor and ferritin mRNA post-transcriptional regulation. Nucl Acids Res 18:1819-1824

Lowenstein CJ, Snyder SH (1992) Nitric oxide, a novel biologic messenger. Cell 70:705–707

Mattia E, Rao K, Shapiro DS, Sussman HH, Klausner RD (1984) Biosynthetic regulation of the human transferrin receptor by desferrioxamine in K562 cells. J Biol Chem 259:2689–2692

McClelland A, Kühn LC, Ruddle FH (1984) The human transferrin receptor gene: genomic organization, and the complete primary structure of the receptor deduced from a cDNA sequence. Cell 39:267–274
Melefors Ö, Goossen B, Johansson HE, Stripecke R, Gray NK, Hentze MW (1993) Translational control of 5-aminolevulinate synthase mRNA by iron-responsive elements in erythroid cells. J Biol Chem 268:5974–5978
Müllner EW, Kühn LC (1988) A stem-loop in the 3′ untranslated region mediates iron-dependent regulation of transferrin receptor mRNA stability in the cytoplasm. Cell 53:815–825
Müllner EW, Neupert B, Kühn LC (1989) A specific mRNA binding factor regulates the iron-dependent stability of cytoplasmic transferrin receptor mRNA. Cell 58:373–382
Nathan C (1992) Nitric oxide as a secretory product of mammalian cells. FASEB J 6:3051–3164
Neilands JB, Nakamura K (1985) Regulation of iron assimilation in microorganisms. Nutr Rev 43:193–203
Neupert B, Thompson NA, Meyer C, Kühn LC (1990) A high yield affinity purification method for specific RNA-binding proteins: isolation of the iron regulatory factor from human placenta. Nucl Acids Res 18:51–55
Oliveira C, Goosen B, Zanchin N, McCarthy J, Hentze MW, Stripecke R (1993) Translational repression by the human iron-regulatory factor (IRF) in *Saccharomyces cerevisiae*. Nucl Acids Res 21:5316–5322
Ostareck-Lederer A, Ostareck DH, Standart N, Thiele B (1994) Translation of 15-lipoxygenase mRNA is inhibited by a protein that binds to a repeated sequence in the 3′ untranslated region. EMBO J 13:1476–1481
Owen D, Kühn LC (1987) Noncoding 3′ sequences of the transferrin receptor gene are required for mRNA regulation by iron. EMBO J 6:1287–1293
Pantopoulos K, Hentze MW (1995a) Nitric oxide signaling to iron-regulatory protein (IRP): direct control of ferritin mRNA translation and transferrin receptor mRNA stability in transfected firbroblasts. PNAS 92:1267–1271
Pantopoulos K, Hentze MW (1995b) Rapid responses to oxidative stress mediated by iron regulatory protein. EMBO J 14:2917–2924
Pantopoulos K, Gray NK, Hentze MW (1995) Differential regulation of two related RNA binding proteins, iron regulatory protein (IRP) and IRP-B. RNA 1:155–163
Patino MM, Walden WE (1992) Cloning of a functional cDNA for the rabbit ferritin mRNA repressor protein. J Biol Chem 267:19011–19016
Pelicci PG, Tabillio A, Thomopoulos P, Titeux M, Vainchenker W, Rochant H, Testa U (1982) Hemin regulates the expression of transferrin receptors in human hepatopoietic cell lines. FEBS Lett 145:350–354
Philpott CC, Rouault TA, Klausner RD (1991) Sequence and expression of the murine iron-responsive element-binding protein. Nucl Acids Res 19:6333
Philpott CC, Haile D, Rouault TA, Klausner RD (1993) Modification of a free Fe-S cluster cysteine residue in the active iron-responsive element-binding protein prevents RNA binding. J Biol Chem 268:17655–17658
Rao KK, Shapiro D, Mattia E, Bridges K, Klausner RD (1985) Effects of alterations in cellular iron on biosynthesis of the transferrin receptor in K562 cells. Mol Cell Biol 5:595–600
Rao K, Harford JB, Rouault T, McClelland A, Ruddle FH, Klausner RD (1986) Transcriptional regulation by iron of the gene for the transferrin receptor. Mol Cell Biol 6:236–240
Roberts KP, Griswold MD (1990) Characterization of rat transferrin receptor cDNA: the regulation of transferrin receptor mRNA in testes and in sertoli cells in culture. Mol Endocrinol 4:531–542
Rogers J, Munro H (1987) Translation of ferritin light and heavy subunit mRNAs is regulated by intracellular chelatable iron levels in rat hepatoma cells. Proc Natl Acad Sci USA 84:2277–2281
Rothenberger S, Müllner EW, Kühn LC (1990) The mRNA-binding protein which controls ferritin and transferrin receptor expression is conserved during evolution. Nucl Acids Res 18:1175-1179
Rouault TA, Hentze MW, Caughman SW, Harford JB, Klausner RD (1988) Binding of a cytosolic protein to the iron-responsive element of human ferritin messenger RNA. Science 241:1207–1210

Rouault TA, Hentze MW, Haile DJ, Harford JB, Klausner RD (1989) The iron-responsive element binding protein: a method for the affinity purification of a regulatory RNA-binding protein. Proc Natl Acad Sci USA 86:5768–5772

Rouault TA, Tang CK, Kaptain S, Burgess WH, Haile DJ, Samaniego G, McBride OW, Harford JB, Klausner RD (1990) Cloning of the cDNA encoding an RNA regulatory protein – the human iron-responsive element-binding protein. Proc Natl Acad Sci USA 87:7958–7962

Rouault TA, Stout CD, Kaptain S, Harford JB, Klausner RD (1991) Structural relationship between an iron-regulated RNA-binding protein (IRE-BP) and aconitase: functional implications. Cell 64:881-883

Samaniego F, Chin J, Kazuhiro I, Rouault TA, Klausner RD (1994) Molecular characterization of a second iron-responsive element binding protein, iron regulatory protein 2. J Biol Chem 269:30904–30910

Schmidt HHW, Walter U (1994) NO at work. Cell 78:919–925

Skikne B, Baynes RD (1994) Iron absorption. In: Broch JH, Halliday JW, Pippard MJ, Powell LW (eds.) Iron metabolism in health and disease. Saunders, London, pp 151–187

Stamler JS (1994) Redox signaling: nitrosylation and related target interactions of nitric oxide. Cell 78:931–936

Stripecke R (1993) m-RNA specific translational control in eucaryotic cells by proteins binding to the 5′ untranslated region. Ph D Thesis Univ Heidelberg, Heidelberg

Stripecke R, Hentze MW (1992) Bacteriophage and spliceosomal proteins function as position-dependent *cis/trans* repressors of mRNA translation in vitro. Nucl Acids Res 20:5555–5564

Stripecke R, Oliveira CC, McCarthy JEG, Hentze MW (1994) Proteins binding to 5′ untranslated region sites: a general mechanism for translational regulation of mRNAs in human and yeast cells. Mol Cell Biol 14:5898–5909

Veyrune JL, Carillo S, Vié A, Blanchard JM (1995) c-fos mRNA instability determinants present within both the coding and the 3′ non-coding region link the degradation of this mRNA to its translation. Oncogene 11:2127–2134

Walden WE, Daniels-McQueen S, Brown PH, Gaffield L, Russel DA, Bielser D, Bailey LC, Thach RE (1988) Translational repression in eukaryotes: partial purification and characterization of a repressor of ferritin mRNA translation. Proc Natl Acad Sci USA 85:9503–9507

Walden WE, Patino MM, Gaffield L (1989) Purification of a specific repressor of ferritin mRNA translation from rabbit liver. J Biol Chem 264:13765–13769

Ward JH, Kushner JP, Kaplan J (1982) Regulation of HeLa cell transferrin receptors. J Biol Chem 257:10317–10323

Weiss G, Goossen B, Doppler W, Fuchs D, Pantopoulos K, Werner–Felmayer G, Wachter H, Hentze MW (1993) Translational regulation via iron-responsive elements by the nitric oxide/NO-syntase pathway. EMBO J 12:3651-3657

Weiss G, Werner-Felmayer G, Werner ER, Grünewald K, Wachter H, Hentze MW (1994) Iron regulates nitric oxide synthase activity by controlling nuclear transcription. J Exp Med 180:969–976

White K, Munro HN (1988) Induction of ferritin subunit synthesis by iron is regulated at both the transcriptional and translational levels. J Biol Chem 263:8938–8942

Winstall E, Gamache M, Raymond V (1995) Rapid mRNA degradation mediated by the cfos AU-rich element and that mediated by the granulocyte-macrophage colony-stimulating factor 3′AU-rich element occur through similar polysome associated mechanisms. Mol Cell Biol 15:3796–3804

Yu Y, Radisky E, Leibold, EA (1992) The iron-responsive element binding protein. Purification, cloning and regulation in rat liver. J Biol Chem 267:19005–19010

Zähringer J, Baliga BS, Munro HN (1976) Novel mechanism for translational control in regulation of ferritin synthesis by iron. Proc Natl Acad Sci USA 73:857–861

Zheng L, Kennedy MC, Blondin GA, Beinert H, Zalkin H (1992) Binding of cytosolic aconitase to the iron responsive element of porcine mitochondrial aconitase mRNA. Arch Biochem Biophys 299:356–360

Interaction Between Iron-Regulatory Proteins and Their RNA Target Sequences, Iron-Responsive Elements

B.R. Henderson[1] and L.C. Kühn[2]

1 Introduction

Iron is perhaps the most essential inorganic element necessary for cell growth, and, acting as both donor and acceptor of electrons, has evolved as an integral cofactor and component of many proteins (Crichton and Ward 1992; Aisen 1994). Indeed, variability in the protein ligands coordinating iron can evoke a wide-ranging flexibility in redox potentials. Regulation of the amount of free intracellular iron is critical, and in mammals is mediated at several steps, including the uptake of iron by endocytosis of receptor-bound transferrin, and the storage of iron by ferritin within the cell (Baker and Morgan 1994). Insufficient iron will prevent cell division, whereas iron overload is toxic, due in large part to the potential for oxidative damage (Crichton 1991).

The maintenance of cellular iron homeostasis represents a unique model system of posttranscriptional gene regulation. The mRNAs encoding proteins involved in haem precursor biosynthesis (erythroid 5-aminolevulinic acid synthase, eALAS), iron uptake (transferrin receptor) and iron storage (ferritin) each contain, in their 5′ or 3′ untranslated region, conserved stem-loop structures known as iron-responsive elements (IREs). The IRE hairpin binds specifically a ~98-kDa cytoplasmic protein named iron-regulatory protein (IRP-1), and previously referred to as IRE-binding protein (Rouault et al. 1988), iron-regulatory factor (Müllner et al. 1989), p90 (Harrell et al. 1991) or ferritin-repressor protein (Walden et al. 1988). A second IRE-binding protein, IRP-2, was recently identified (Henderson et al. 1993; Guo et al. 1994; Samaniego et al. 1994). Under low iron conditions, the IRPs bind actively to IREs and thereby mediate two separate gene-regulatory events. Binding the single IRE in the 5′UTR of ferritin or eALAS mRNAs results in repressed translation of these proteins, and interaction with several of the five IREs clustered together in the transferrin receptor mRNA 3′UTR leads to mRNA stabilization. These separate actions increase both iron uptake and the availability of free iron within the cell. Conversely, as the iron level rises too high, IRP-1 and IRP-2 bind poorly to the IRE and their effect on mRNA translation and stability are

[1]MRC Laboratory of Molecular Biology, Hills Road, Cambridge, CB2 2QH, UK
[2]Genetics Unit, Swiss Institute for Experimental Cancer Research (ISREC), Chemin des Boveresses 155, 1066 Epalinges s/Lausanne, Switzerland

Progress in Molecular and Subcellular Biology, Vol. 18
Ph. Jeanteur (Ed.)

reversed. A more detailed description including the likely mechanisms by which the IRPs mediate these posttranscriptional events is presented in Muckenthaler and Hentze (this Vol.).

This chapter will focus on the RNA-protein interaction itself. The expression of IRP RNA-binding activity and the IRE structure both demonstrate a strong evolutionary conservation (Rothenberger et al. 1990; Kühn 1994; Theil 1994). Recent studies identified IRP-1 as the cytoplasmic form of the Krebs cycle enzyme, aconitase (Haile et al. 1992b; Kennedy et al. 1992), an intriguing finding, given that several other different metabolic enzymes have recently been found to display RNA-binding activity (summarised in Hentze 1994). One might envisage that the evolution of certain metabolic enzymes may perhaps have maintained conservation of both enzymatic and RNA-binding functions, endowing the cell with a range of dual-function cytoplasmic RNA regulators. This speculation no longer seems so implausible, and elucidation of the RNA-binding domain(s) and RNA-binding specificity of the IRPs, together with the structure of their RNA target(s), should prove important to our understanding of such RNA:protein interactions. The following perspective accounts for our current knowledge of the IRP:IRE interaction, and of its modulation both in vitro and in vivo.

2 Biochemistry of IRP-1

IRP-1 was first identified via its ability to form complexes with ^{32}P-labelled IRE probes on RNA band-shift gels (Leibold and Munro 1988; Rouault et al. 1988; Müllner et al. 1989). The protein binds in vitro to IREs from different genes including ferritin light-chain (Leibold and Munro 1988) and heavy-chain (Leibold and Munro 1988; Rouault et al. 1988), transferrin receptor (Koeller et al. 1989; Müllner et al. 1989), erythroid 5-aminolevulinic acid synthase (eALAS; Cox et al. 1991; Dandekar et al. 1991), porcine mitochondrial aconitase (Dandekar et al. 1991; Zheng et al. 1992), and *Drosophila melanogaster* succinate dehydrogenase subunit b (Kohler et al. 1995). IRP-1 binds a minimal 28-base ferritin IRE stem-loop (Aziz and Munro 1987; Hentze et al. 1987a,b) with high affinity (K_d, 10–90 pM; Haile et al. 1989; Barton et al. 1990), and to the aconitase IRE with slightly lower affinity (K_d, 250 pM; Zheng et al. 1992). UV-cross-linking first identified IRP-1 as a protein of 87–100 kDa in rodents and man (Leibold and Munro 1988; Müllner et al. 1989). The subsequent cloning of the protein predicted an 889-amino-acid peptide of 98 kDa mass in different species (Rouault et al. 1990; Philpott et al. 1991; Hirling et al. 1992; Patino and Walden 1992; Yu et al. 1992).

Despite the initial detection of IRP-1 through its RNA-binding activity, the protein has since been discovered to also possess enzymatic activity. Computer analyses revealed a 30% amino acid identity between IRP-1 and the mitochondrial enzyme aconitase (Hentze and Argos 1991; Rouault et al. 1991), and, more recently, a surprisingly better match was found with the bacterial aconitase (53% identity; Prodromou et al. 1992). Within a short time after the initial observation, IRP-1 was shown to possess aconitase activity (Kaptain et al. 1991; Haile et al. 1992a). This subsequently led to its identification as the cytosolic aconitase (Kennedy et al. 1992), an isozyme first observed about 30 years ago (see Beinert and Kennedy 1993 and references therein). Enzymatic activity of mitochondrial aconitase, which converts citrate to isocitrate, requires the presence of a [4Fe-4S] cluster embedded within the active site of the protein (reviewed in Beinert and Kennedy 1993). The conservation of active site residues in IRP-1, which by computer predictions could fold into a structure similar to that of mitochondrial aconitase (Rouault et al. 1991), indicated that IRP-1 might also coordinate insertion of a [4Fe-4S] cluster. Indeed, the purification and in vitro reconstitution of the [4Fe-4S] cluster-containing form of IRP has been demonstrated (Constable et al. 1992; Haile et al. 1992b, Kennedy et al. 1992; Emery-Goodman et al. 1993; Basilion et al. 1994b).

IRP-1 is known to switch between high-affinity (low iron) and low-affinity (high iron) RNA-binding forms (Haile et al. 1989). The confounding observation that the RNA-binding activity of IRP-1 from iron-treated cells could be induced in vitro by reduction with 2% 2-mercaptoethanol, first led to the proposal of a "sulphydryl switch" that modulated IRP activity in response to cellular redox potential (Hentze et al. 1989). The identification of an [Fe-S] cluster provides now a more reasonable explanation, and an intriguing model system. The IRP-1 is envisaged as a molecule with two independent and mutually exclusive functions: the holo-protein (with a [4Fe-4S] cluster) has enzyme activity but does not bind RNA (Constable et al. 1992; Haile et al. 1992b; Emery-Goodman et al. 1993), and the apo-protein (no cluster) can bind IREs but lacks aconitase activity (Haile et al. 1992b; Emery-Goodman et al. 1993). Thus, the IRP-1 [Fe-S] cluster would "sense" intracellular iron levels and thereby modulate RNA-binding and aconitase activities (see Fig. 1; Klausner et al. 1993; Kühn 1994). Support for this model comes from studies in which the cysteine residues predicted to ligate the [Fe-S] cluster were mutagenised; such mutants, unlike the wild-type IRP, were refractory to cluster insertion and enzymatically inactivate (Philpott et al. 1993; Hirling et al. 1994). Notably, these mutant IRPs did not respond to iron levels in vivo, but were constitutively active in IRE-binding following transfection into living cells (Hirling et al. 1994; Philpott et al. 1994). Such unregulated expression of IRP-1 was recently found to result in the iron-independent overexpression of cellular transferrin receptor, and the repression of ferritin synthesis (DeRusso et al. 1995).

The mutagenesis experiments explained other puzzling characteristics of IRP-1. For instance, in vitro binding is inhibited following oxidation of IRP-1 by diamide or alkylation by N-ethylmaleimide (Hentze et al. 1989; Neupert et al. 1990). These effects are now easily reconciled. Diamide appears to inactivate IRP-1 by inducing disulphide bonds between cysteine 437 and the other cluster-ligating residues, cysteines 503 and 506 (Philpott et al. 1993;

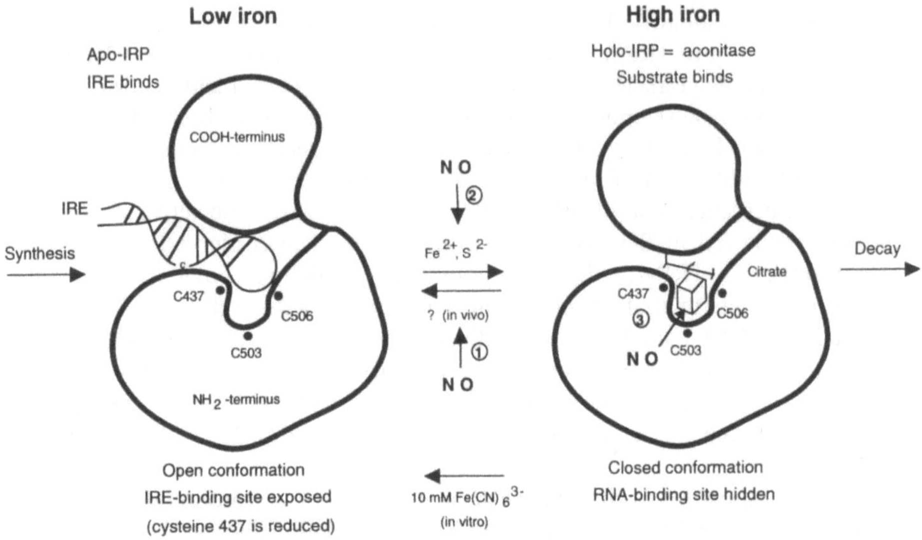

Fig. 1. IRP Switching between RNA-binding and enzymatic forms. IRP-1 is the cytosolic aconitase (Kennedy et al. 1992) and can be folded in a conformation similar to that of pig heart mitochondrial aconitase (Robbins and Stout 1989). Under low iron, apo-IRP-1 is newly synthesised and active in RNA binding. The IRE is thought to bind within the cleft between the upper carboxy-terminal lobe (domain 4) and the lower amino-terminal lobe (domains 1-3), as discussed in Section 2.2. IRP binding of IREs in ferritin and transferrin receptor mRNAs results in a posttranscriptionally mediated increase in free intracellular iron (see Muckenthaler and Hentze, this Vol.). When iron levels are high, IRP-1 is receptive to insertion of a [4Fe-4S] cluster, ligated by three essential cysteines (C437, C503, C506). The holo-form of IRP-1 is enzymatically active, and its inability to bind RNA is most likely due to a change in conformation. It is possible that upon iron starvation, the holo-protein is posttranslationally converted to the apo-form via cluster removal, or may be degraded, thus allowing the apo-form to accumulate by renewed synthesis. Small molecules such as nitric oxide (NO) may activate RNA binding of IRP-1 (Drapier et al. 1993; Weiss et al. 1993) in three possible ways as indicated by *arrows*: (1) cluster removal, (2) chelation of free iron or (3) by directly interacting with the [Fe-S] cluster

Hirling et al. 1994). Cysteine 437 is also the target for alkylation by N-ethylmaleimide. Curiously, alkylation of IRP-1 by the smaller iodoacetamide group was less inhibitory toward RNA-binding, suggesting that N-ethylmaleimide may repress RNA-binding by steric interference (Hirling et al. 1994). Mutagenesis studies of this kind may also prove fruitful when applied to other cytoplasmic RNA-binding proteins sensitive to in vitro oxidation or alkylation, such as AU-binding protein(s) (Malter and Hong 1991) and the enzyme thymidylate synthase (Chu et al. 1994).

The ability of 2% 2-mercaptoethanol to activate in vitro IRE-binding by the holoform of IRP-1 is less well explained, as the reductant at this high concentration exerts only a transitory effect, and the cluster is not removed (Haile et al. 1992b; Hirling et al. 1994). It is assumed that 2-mercaptoethanol can modulate one of the cysteine-Fe interactions to favour RNA-binding (Hirling et al. 1994). It was concluded that IRP-1 must be in the apo-form and that cysteine 437 should be fully reduced to ensure RNA binding activity in vivo (Hirling et al. 1994). This situation may arise when IRP-1 is newly translated in the cell and maintained in the reduced apo-form by cellular glutathione, and explains the previous finding that IRP-1 was transiently but reversibly inactivated by direct treatment of cells with diamide (Müllner et al. 1992).

2.1 Modulation of IRP-1 Binding Activity

The conversion from apo-IRP-1 to the [4Fe-4S] cluster-containing form is likely a reversible post-translational event (see Fig. 1; Klausner et al. 1993; Kühn 1994); however, it is not clear whether the [Fe-S] cluster inserts spontaneously, or whether its introduction is chaperoned by another protein or accessory factor. Upon considering the reverse situation, removal of the [4Fe-4S] cluster from holo-IRP, it must be borne in mind that maximal induction of IRP-1 binding activity by iron-chelator treatment of cells is not rapid, but takes about 20 h. Complete cluster removal is thus normally a slow process and may require the action of other proteins. It was demonstrated that removal of a single iron atom generates a [3Fe-4S]-intermediate form of IRP, which is neither enzymatically active nor able to bind RNA (Haile et al. 1992b; Kennedy et al. 1992; Basilion et al. 1994b). The mechanism by which the entire cluster is removed in vivo to generate the RNA-binding form is unclear. It is possible that in vivo, the natural degradation of the [4Fe-4S] or [3Fe-4S] cluster forms of IRP-1 may partially account for changes in IRE-binding activity, with the apo-form generated not by conversion, but via new synthesis (see Fig. 1; Henderson and Kühn 1995).

Some workers propose that IRP-1 inactivation by chelatable iron and haemin follows different pathways, and that haemin directly affects IRP-1 turnover in a growth rate dependent manner (Goessling et al. 1994). Haemin certainly appears to cross-link to IRP-1 in vitro (Lin et al. 1991); however, its role in IRP degradation remains controversial (Haile et al. 1990; Goessling et al. 1992; Müllner et al. 1992; Tang et al. 1992; Goessling et al. 1994). A correlation was made between the effect of haemin on IRP-1 turnover and ferritin synthesis (Goessling et al. 1994), yet haemin is also known to induce ferritin mRNA in certain cells (Liau et al. 1991), and has the potential to act indirectly via the induction of alternate messengers such as cAMP (Lander et al. 1992).

The potential for cross-talk between metabolic pathways and signal-transduction or growth-regulatory pathways is very exciting, particularly in the light of recent reports describing growth-regulated metabolic enzymes (e.g. dihydrofolate reductase, thymidylate synthase) capable of binding and regulating the translation of mRNAs (Chu et al. 1993, 1995). In particular, thymidylate synthase autoregulates its own translation in addition to that of c-myc in vitro (Chu et al. 1995), thus providing a possible link between normal metabolism and oncogenesis. Interestingly, IRP-1 is also activated following stimulation of growth in certain cell types, such as human T lymphocytes (Testa et al. 1991), mouse spleen cells (Teixeira and Kühn 1991) and a mouse T cell line (Seiser et al. 1993). While some of these situations reflect an alteration in the intracellular iron pool (see Kühn 1994), there is most likely an additional iron-independent event that contributes to IRP-1 activation by interleukin-2 (Seiser et al. 1993). In this context, Eisenstein et al. (1993) recently demonstrated the in vitro phosphorylation of IRP-1 by protein kinase C. Moreover, treatment of human HL60 cells with phorbol 12-myristate 13-acetate, an activator of protein kinase C, appeared to induce IRP-1 activity by up to threefold (Eisenstein et al. 1993), suggesting that IRP-1 may also be responsive to signalling pathways in certain cell types.

Two independent studies have observed that IRP-1 RNA-binding activity was also induced by the small signalling molecule, nitric oxide (NO; Drapier et al. 1993; Weiss et al. 1993). Activation in cells required a time of treatment similar to desferrioxamine, and while it remains possible that NO indirectly induces IRP-1 via modulation of intracellular iron levels, the in vitro analysis of recombinant IRP-1 suggested that NO can directly affect the [Fe-S] cluster (see Fig. 1 and Drapier et al. 1993). Indeed, this is a distinct possibility, given that NO has previously been demonstrated to interact directly with the [4Fe-4S] cluster of mitochondrial aconitase (see Henry et al. 1993 and references therein). Furthermore, exposure of cells to hydrogen peroxide (H_2O_2) was recently observed to lead to the rapid induction of IRP-1 activity (Pantopoulos and Hentze 1995). The effect apparently involved disassembly of the [4Fe-4S] cluster and was independent of new protein synthesis. Interestingly, the protein phosphatase inhibitor, okadaic acid, blocked induction of IRP-1 by H_2O_2 (Pantopoulos and Hentze 1995), suggesting a link between oxidative stress and kinase/phosphatase pathways in the iron-independent induction of IRP-1.

2.2 *RNA-Binding Domain of IRP-1*

Initial attempts to localise the RNA-binding domain of IRP-1 by deletion mapping suggested involvement of both the amino-terminal and carboxy-terminal domains (Hirling et al. 1992). IRP-1 does not comprise any conserved RNA-binding motifs or structural domains typical of other RNA-binding proteins (Burd and Dreyfuss 1994). Rather, it is expected that folding of IRP-1 is critical, and that adoption of the correct structural conformation favours RNA binding. Mutagenesis approaches based on the structure of mitochondrial aconitase (Fig. 1) have proven useful in this regard, and, as mentioned above, mutation of the IRP-1 cluster-ligating cysteine 437 residue did not abolish binding, but prevented the effects of alkylation, implying that the RNA binds somewhere proximal to this active site residue (Philpott et al. 1993; Hirling et al. 1994). Basilion et al. (1994a) adopted a strategy based on sequencing IRP peptides that became covalently bound to the IRE following

UV-crosslinking, and thereby identified a ten-amino-acid sequence which comprised two active-site residues and a modified serine 127, proposed to contact the RNA. A peptide overlapping this region was mapped independently using a novel approach based on insertion of hydroxylamine chemical cleavage sites and UV-cross-linking of IRP-1 (Neupert et al. 1995).

The IRE is generally thought bind IRP-1 within the active site cleft, as outlined in Fig. 1. Normally masked, the binding domain might become accessible between the two IRP-1 globular NH_2- and COOH-terminal domains (connected by a hinge/linker) due to the conformational change accompanying loss of the [4Fe-4S] cluster (Kühn and Hentze 1992; Klausner et al. 1993; Basilion et al. 1994a; Hirling et al. 1994). The RNA orientation and its precise positioning within the protein are not yet known; however, it is expected that specificity of the interaction is, in part, determined by multiple sites that contact the RNA. The single human IRP-1 peptide that cross-links to the IRE is predicted to lie well inside the IRP active site cleft (Basilion et al. 1994a; Neupert et al. 1995). Studying rabbit IRP-1 by proteolytic cleavage, Swenson and Walden (1994) identified a different UV-cross-link region. The 143-amino-acid peptide they identify precedes the hinge and encompasses certain active-site residues in the lower globular IRP-1 domain. Indeed, site-directed mutation of two active-site arginine residues (536 and 541) in this region not only inactivates enzyme activity, but greatly decreases RNA binding (Philpott et al. 1994). Another arginine (780), involved in substrate recongnition and located in the upper globular domain above the predicted cleft, also influences IRE binding (Philpott et al. 1994). Collectively, these studies indicate that RNA contacts IRP-1 in regions proximal to the active site. The further identification of different charged amino acids in the IRP-1 cleft, that actually maintain enzymatic activity and thus protein conformation when mutated, will provide a more controlled delineation of the RNA-binding domain. Ultimately, as for other proteins (Rould et al. 1991; Cavarelli et al. 1993; Oubridge et al. 1994; Valegard et al. 1994), complete elucidation of the IRP-1 RNA-binding domain awaits crystal structure analyses.

3 Evidence for a Second IRE-Binding Protein

The regulation of cytoplasmic mRNA stability or translation can involve recognition of the same mRNA signal element by different proteins, thus facilitating flexibility of control such as cell type or tissue-specific expression, and cross-talk between alternate signalling pathways. AU-rich elements involved in cytokine and oncogene mRNA instability (Shaw and Kamen 1986) represent linear sequences that are recognised by multiple proteins (Levine et al. 1993; Zhang et al. 1993; Nakagawa et al. 1995). RNAs that fold as specific stem-loops, such as the HIV Tat-activation region, are also capable of interacting with different cellular proteins (reviewed in Gait and Karn 1993). In this context, the identification of new IRE stem-loop binding proteins might open the possibility for more complex control of mRNAs encoding proteins of iron metabolism. The characterisation of a second IRE-binding protein in mouse and rat (Henderson et al. 1993) was therefore highly relevant.

The existence of a second IRE-binding protein was suggested earlier by several key observations. First, an incomplete cDNA was cloned in parallel with IRP-1 from human T-cells, predicting a protein 57% identical with and 75% similar to IRP-1 (Rouault et al. 1990). The complete cDNA for this IRP-1 related protein was latter cloned and named IRP-2 (Rouault et al. 1992). Second, two band-shift complexes form between ^{32}P-labelled ferritin or transferrin receptor IREs and proteins from rodent cell extracts (Leibold and Munro 1988; Müllner et al. 1989), correlating with a doublet observed by UV-cross-linking (Müllner et al. 1989). The two complexes detected in rat extracts could be separated by ion-exchange chromatography (Barton et al. 1990; Leibold et al. 1990). While a doublet was also observed in SDS-polyacrylamide gels with affinity-purified preparations of placental IRP (Neupert et al. 1990), this most likely reflected differentially reduced forms of IRP-1 (Kennedy et al. 1992). The rodent form of IRP-2 was the first to be characterised. This is partly because rodent IRP-1 bears a different charge to human IRP-1, and so migrates in an IRE-bound complex separated from rodent IRP-2 in non-denaturing band-shift gels (Philpott et al. 1991; Yu et al. 1992). Conversely, the IRE-binding activity of human IRP-2 is normally masked by its comigration with IRP-1 in band-shift gels (Henderson et al. 1993, 1996), thus making it difficult to be detected in human cells.

The second IRE-binding protein in rodents was initially named IRP_B (Henderson et al. 1993). However, the recent purification of rat IRP_B (Guo et al. 1994) and sequencing of the cDNA (Guo et al. 1995a) revealed its amino acid sequence to be 93% identical to that of the human IRP-2 cDNA (Rouault et al. 1992). Moreover, antibodies raised against regions from mouse and human IRP-2 that are not conserved in IRP-1, cross-reacted specifically with IRP_B from rat (Guo et al. 1994) and mouse (Samaniego et al. 1994). Therefore, IRP_B is now referred to as rodent IRP-2.

4 Characterisation of IRP-2

Rodent IRP-2 has a mass of ~105 kDa and shares several characteristics in common with IRP-1 (Henderson et al. 1993). For instance, the IRE-binding activity of both proteins is coordinately regulated by intracellular iron levels in vivo, and inhibited by oxidation and alkylation in vitro (Henderson and Kühn 1995). Not surprising in light of the 79% similarity in amino acid sequence between full-length IRP-1 and IRP-2 (Rouault et al. 1992; Guo et al. 1995a), the two proteins display different but related proteolytic digestion patterns after UV-cross-linking, and also bind IREs from ferritin and other mRNAs with equally high affinity (Henderson et al. 1993; Guo et al. 1994). These features, in addition to the observation that both proteins appear to protect the same nucleotides of the rat ferritin IRE from RNase T1 digestion (Leibold and Munro 1988), suggest similarities in the RNA-binding domains of these proteins. Moreover, IRP-1 and IRP-2 are each able to repress in vitro translation in wheat germ extract (Guo et al. 1994) or rabbit reticulocyte lysate (Kim et al. 1995) systems, suggesting that both proteins may function in iron-regulated control of mRNA translation.

Despite these similarities, IRP-2 does not appear to possess aconitase activity (Guo et al. 1994), and therefore, unlike IRP-1, it is not a bifunctional molecule. This may relate to its conservation of all IRP-1 active site residues except for three, which are important for aconitase substrate recognition and catalysis (Rouault et al. 1992). Another factor distinguishing IRP-2 from IRP-1 is that its RNA-binding activity cannot be recovered by 2% 2-mercaptoethanol treatment of extracts from iron-treated cells (Henderson et al. 1993). The inability to reactivate IRP-2 in vitro by high reductant appears to reflect a diminution in the total amount of IRP-2 protein. In support of this Guo et al. (1994) showed by immunoblot analysis that IRP-2 had decreased to undetectable levels following 4-h iron treatment of rat FTO2B hepatoma cells. Similar findings were reported by Samaniego et al. (1994), who observed a concordance between IRP-2 protein levels and RNA-binding activity in response to changes in iron levels, both in mouse cells and in transfected human cells. In HeLa cells, however, Samaniego et al. (1994) reported that 16-h iron treatment reduced RNA-binding activity of a transfected c-myc-tagged human IRP-2, but did not affect endogenous IRP-2 protein levels. Other studies show that shorter periods of iron treatment decrease endogenous IRP-1 and IRP-2 in HeLa cells (Guo et al. 1995b; Henderson et al. 1996).

4.1 Differential Iron Regulation of IRP-1 and IRP-2

The RNA-binding activities of both IRPs are modulated in response to iron, but by different mechanisms. While IRP-1 is posttranslationally converted between active and inactive RNA-binding forms (see Sect. 2.1), IRP-2 activity is increased by iron starvation through the renewed synthesis of stable IRP-2 protein (Henderson and Kühn 1995; Pantopoulos et al. 1995). The inactivation of IRP-2 by iron reflects degradation of the IRP-2 protein (Guo et al. 1994; Samaniego et al. 1994), and inhibitor studies indicate that this is mediated through the proteasome (Guo et al. 1995b; Iwai et al. 1995). IRP-2 contains a unique 73-amino-acid exon insertion not present in IRP-1 (Rouault et al. 1992; Guo et al. 1995a; Iwai et al. 1995), and this region is required for iron-mediated protein degradation (Iwai et al. 1995). In a convincing set of experiments, Iwai et al. (1995) demonstrated that deletion of this 73-amino-acid sequence abolished rapid turnover of IRP-2, and that transfer of this sequence to IRP-1 conferred iron-dependent degradation. Mutation analysis identified specific cysteine residues within this IRP-2-specific sequence that are required for iron-mediated degradation (Iwai et al. 1995). The three IRP-2 cysteines corresponding to those that coordinate [4Fe-4S] cluster ligation in IRP-1 are not involved in iron-dependent IRP-2 turnover (Iwai et al. 1995). It is not yet clear if IRP-2 can ligate an Fe-S cluster, and if it does, whether or not iron-binding is necessary for its degradation. Furthermore, the additional cysteines within the IRP-2 turnover region may contribute to the susceptibility of this protein to in vitro inactivation by oxidation and modification by alkylating agents (Henderson and Kühn 1995).

Surprisingly, when the degradation of IRP-2 was prevented by the addition of proteasome inhibitors, IRP-2 RNA-binding activity was no longer responsive to iron (Guo et al. 1995b; Iwai et al. 1995). Thus, IRP-2 degradation is essential for modulation of IRP-2 activity. This finding further explains why IRP-2 activity, which can be modulated by oxidation/reduction in vitro (Henderson and Kühn 1995), is not recovered in extracts from iron-treated cells by 2-mercaptoethanol treatment (Henderson et al. 1993). The precise mechanism by which the 73-amino-acid sequence targets IRP-2 for degradation by the proteasome is not yet clear, but may involve an independent labile protein, as IRP-2 degradation by iron is a translation-dependent process (Guo et al. 1995b; Henderson and Kühn 1995).

IRP-2 displays a different tissue distribution of activity to IRP-1. Whereas IRP-1 activity and protein levels are highest in liver, intestine and kidney, IRP-2 was most active in intestine and brain, two organs active in the transport of iron across cell layers (Henderson et al. 1993). IRP-2 mRNA levels were also reported to be highest in brain (Samaniego et al. 1994). The expression of IRP-2 displays a more pronounced variation in different rodent and human cell lines (Henderson and Kühn 1995; Henderson et al. 1996).

It is not yet clear if these differences reflect alternate iron-independent mechanisms that modulate IRP-2 activity, synthesis or turnover. However, it is noteworthy that both IRP-1 and IRP-2 can be activated in response to nitric oxide production in macrophages (Weiss et al. 1993) or growth stimulation of a T-cell line by interleukin-2 (Seiser et al. 1993). Furthermore, a recent study correlated the increased binding of IRP-2, but not IRP-1, with transferrin receptor mRNA stabilisation during rat liver regeneration (Cairo and Pietrangelo 1994). IRP-2 may thus be activated in rapidly dividing cells in vivo. Conversely, IRP-2 activity appeared to be preferentially decreased in animals in response to oxidative stress (Cairo et al. 1995). These animal studies are of potential interest; however, caution should be exercised in their interpretation, as the changes they report in IRP-2 activity were rather modest.

5 Sequence and Structure of Iron-Responsive Elements

Iron-responsive elements are specific RNA stem-loop structures that bind with high affinity to IRP-1 or IRP-2. IREs were first identified in ferritin mRNA 5′ untranslated regions as sequences conserved between the light (L) and heavy (H) chain ferritin genes of rat (Aziz and Munro 1987) and human (Hentze et al.

1987a). The minimal human ferritin H-chain IRE sequence necessary to confer iron-dependent translation control on a heterologous reporter gene is 32 bases in length (Hentze et al. 1987b). Similar-sized elements were subsequently identified in the transferrin receptor 3′ untranslated region, and these IREs were also found capable of conferring iron-dependent translation control in vivo (Casey et al. 1988). As for the single IRE in the ferritin 5′ untranslated regions, transferrin receptor mRNAs from different species maintain striking conservation of five IRE stem-loops, predicted to cluster in spatial proximity (Koeller et al. 1989; Müllner et al. 1989; Horowitz and Harford 1992). Phylogenetic comparison of IREs from different species identifies strong conservation of an unpaired (bulge) cytosine 5′ of the 5-bp upper stem and of a highly conserved loop sequence (5′ CAGUGN 3′), as illustrated in Fig. 2 (reviewed in Kühn and Hentze 1992; Klausner et al. 1993; Theil 1994). The primary sequence of the upper and lower stem appears to be less important than is the ability to form Watson-Crick base pairing within these regions, as evident by the difference in stem sequence between ferritin and transferrin receptor IREs (Fig. 2). It should, in addition, be noted that the context in which the minimal IRE is placed is also important, and that flanking sequences can influence the translation or stability of the endogenous ferritin (Goossen et al. 1990; Goossen and Hentze 1992; Dix et al. 1993; Rogers et al. 1994) and transferrin receptor (Müllner and Kühn 1988; Casey et al. 1989) mRNAs, respectively.

The functionality of IREs was found to be mediated via binding to proteins now recognised as IRP-1 and IRP-2 (Leibold and Munro 1988; Rouault et al. 1988; Müllner et al. 1989). Structural analyses of the ferritin and transferrin receptor IREs have defined a minimal 28-base "core" sequence necessary for IRP binding (Leibold et al. 1990; Harrell et al. 1991). Base-pairing of the upper and lower stems was confirmed by mutagenesis experiments in which IREs with non-pairing stems were found to bind IRP-1 with greatly reduced affinity (Leibold et al. 1990; Bettany et al. 1992). Moreover, increasing or decreasing the size of the six-base loop or the size of the upper stem between the loop and the 5′ bulge, also decreased binding to IRP-1 (Bettany et al. 1992; Jaffrey et al. 1993). Probing IRE secondary structure by chemical and enzyme cleavage confirmed the double-stranded nature of the stems, and revealed both the bulge and loop to be single-stranded (Wang et al. 1990; Harrell et al. 1991; Bettany et al. 1992; Horowitz and Harford 1992). As shown in Fig. 2A, different ferritin H-chain or L-chain IREs contain a three-base UGC bulge located five bases 5′ of the loop, as revealed by computer folding predictions, cleavage analysis, and most recently an NMR spectroscopy study (Wang et al. 1990; Bettany et al. 1992; Sierzputowska-Gracz et al. 1995). IREs from other known genes including transferrin receptor, erythroid 5-aminolevulinic acid synthase, mitochondrial aconitase and succinate dehydrogenase b comprise a single cytosine bulge (Fig. 2B; Koeller et al. 1989; Dandekar et al. 1991; Bettany et al. 1992). The bulge itself may bend or "kink" the IRE, and its conformation is clearly important; a three-base UGC bulge is optimal and promotes a three-

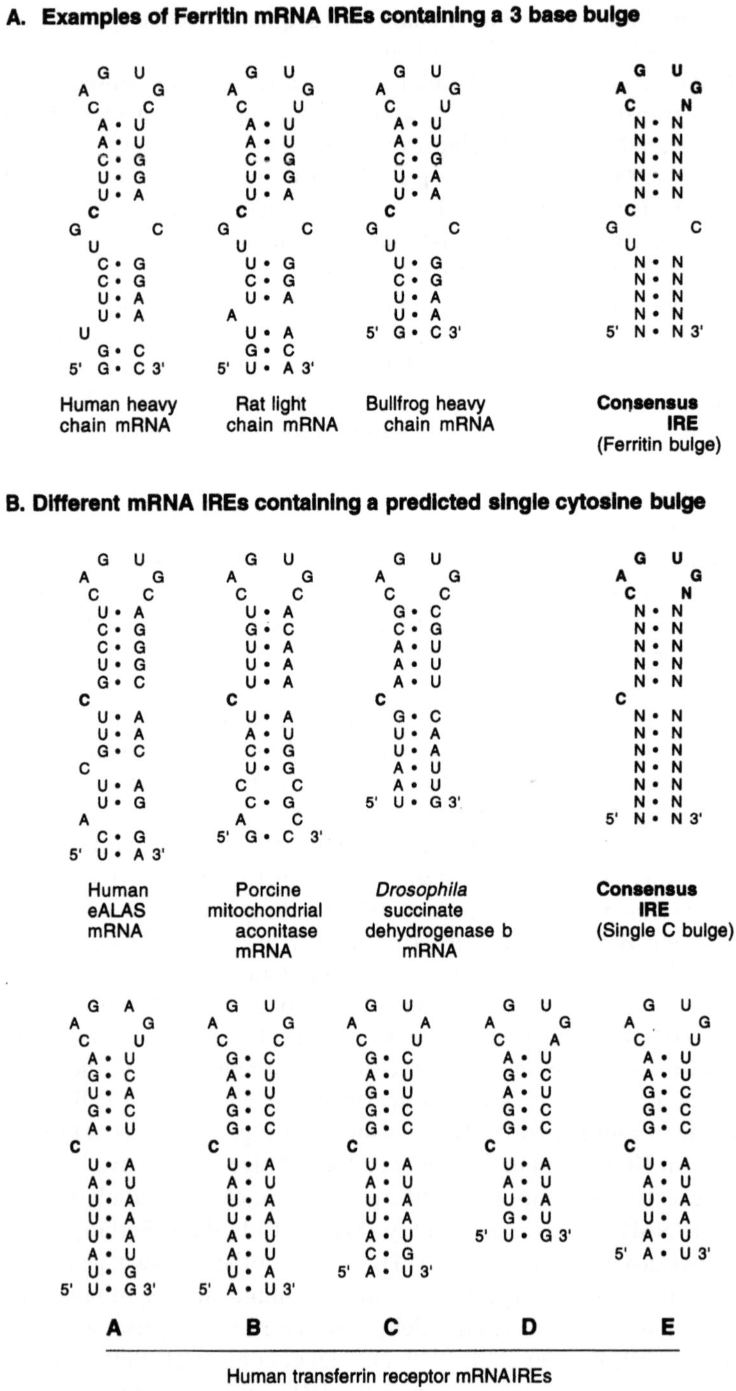
A. Examples of Ferritin mRNA IREs containing a 3 base bulge
Human heavy chain mRNA
Rat light chain mRNA
Bullfrog heavy chain mRNA
Consensus IRE (Ferritin bulge)
B. Different mRNA IREs containing a predicted single cytosine bulge
Human eALAS mRNA
Porcine mitochondrial aconitase mRNA
Drosophila succinate dehydrogenase b mRNA
Consensus IRE (Single C bulge)
A
B
C
D
E
Human transferrin receptor mRNAIREs

fold stronger binding to recombinant IRP-1 than a single cytosine (Henderson et al. 1994).

Phylogenetic comparisons predict binding specificity to be determined by the seven highly conserved nucleotides present in the loop and bulge cytosine (Kühn and Hentze 1992; Theil 1994). The loop sequence (5′ CAGUGN 3′) is well conserved, the main exception being transferrin receptor IRE stem-loop A, where the loop base U^4 is replaced by an A in humans and rat, or a C in chicken (Kühn and Hentze 1992). The first studies to probe these regions by introducing single base substitutions into the loop or bulge, observed little effect or at most a sevenfold decrease in binding to IRP-1 in vitro (Barton et al. 1990; Leibold et al. 1990). Deletion of nucleotides at these sites reduced binding up to 50-fold (Hentze et al. 1988; Haile et al. 1989). A latter study analysing IRE mutants observed more drastic effects on binding; all IRE base changes or deletions generated a 6- to 654-fold decrease in binding to IRP-1 (Jaffrey et al. 1993). A single IRE base change converting the loop nucleotide A^2 to G^2 has been detected in the ferritin light-chain gene of patients with inherited hyperferritinaemia and cataract (Beaumont et al. 1995). This study is the first to correlate an IRE mutation that prevents binding by the IRPs, with a specific iron-related disorder.

6 Role of IRE Loop Structure in Protein Recognition

A recent analysis of IRE sequence and structure applied an in vitro genetic selection strategy to the human ferritin H-chain IRE, in which the seven nucleotides of the loop and bulge were randomised, and selected RNA sequences were then tested for binding to purified recombinant human IRP-1 (Henderson et al. 1994). The selection identified two major classes of high affinity ligands, the optimal loop sequences of these were 5′ CAGUGN 3′ (wild-type) and

Fig. 2A, B. Sequence and secondary structure of iron responsive elements. IREs from different mRNAs are shown divided into two groups. Panel **A** illustrates the strong sequence conservation of IREs derived from the 5′ untranslated region (UTR) of ferritin heavy (H) or light (L) chain genes from different species (see Aziz and Munro 1987; Hentze et al. 1987a; Wang et al. 1990). The predicted secondary structure, including the three-base UGC bulge specific to the ferritin IREs, is supported by enzymatic, chemical cleavage and NMR studies (Wang et al. 1990; Harrell et al. 1991; Bettany et al. 1992; Sierzputowska-Gracz et al. 1995). Panel **B** depicts IREs from other genes, which fold as stem-loops containing a single cytosine residue. Comparison of IREs from the 5′ UTR of mRNAs encoding erythroid ALAS (Cox et al. 1991; Dandekar et al. 1991), mitochondrial aconitase (Dandekar et al. 1991) and succinate dehydrogenase subunit b (Kohler et al. 1995), and the five IREs present in the 3′ UTR of transferrin receptor mRNA (Casey et al. 1988; Koeller et al. 1989; Müllner et al. 1989), demonstrates the diversity of primary sequence in the upper and lower IRE stems. A consensus IRE is shown to illustrate the seven conserved bases (*bold type*) of the loop and the cytosine bulge located five nucleotides 5′ of the loop. The conformation of the ferritin IRE three-base bulge confers a threefold stronger binding to IRP-1 than the single cytosine bulge. (Henderson et al. 1994)

5′ UAGUAN 3′ (Henderson et al. 1994). The novel prediction from this work was that the six-base IRE loop is not single-stranded, but forms an apical structure in which bases at loop positions 1 and 5 are paired. The formation of such a structure would make nucleotides at loop positions 2, 3 and 4 (AGU) accessible for interaction with IRP-1. Indeed, substitution of other bases at these positions reduced binding by 36 to 99% (Henderson et al. 1994). IRP binding invariably favoured a cytosine at the unpaired RNA bulge. Further mutagenesis studies of the ferritin IRE loop have more recently revealed that RNA-binding by IRP-2 also requires an intraloop 1:5 base pair (Henderson et al. 1996). This may suggest similarity in the RNA-binding domains of these proteins, although the type of intraloop base pair can influence binding by either IRP (summarised in Fig. 3). A model for IRE secondary structure based on these mutagenesis experiments is shown in Fig. 3, where the bulge cytosine and the loop bases 2 to 4 are proposed to make direct contact with the IRPs.

Previous investigators likewise inferred base pairing within the six-nucleotide HIV TAR hairpin loop based on secondary structure analysis (Colvin and Garcia-Blanco 1992), although a subsequent NMR spectroscopy study of a

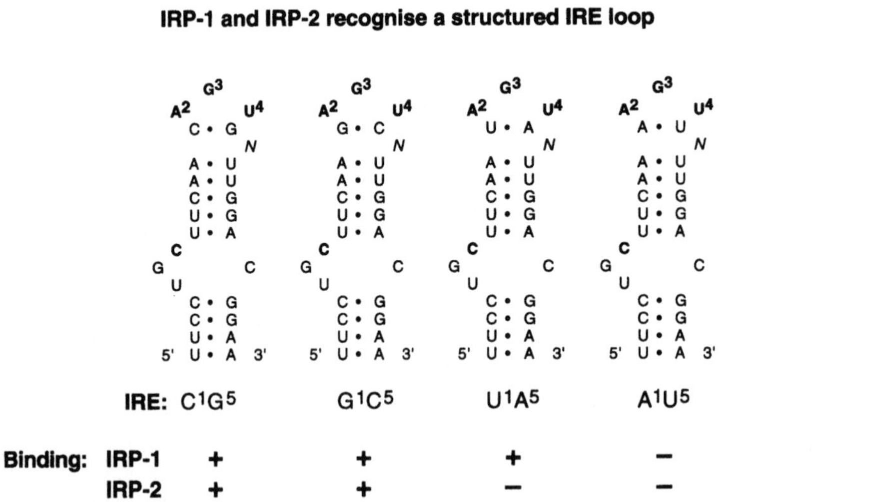

Fig. 3. Secondary structure of the IRE loop. Extensive mutagenesis studies of the human ferritin H-chain IRE reveal that optimal RNA binding by IRP-1 and IRP-2 require internal base pairing between IRE loop positions 1 and 5 (Henderson et al. 1994, 1996). We propose that all known naturally occurring IREs adopt a loop structure as shown (C^1G^5-loop), and that unpaired loop nucleotides 2, 3 and 4 and the bulge cytosine (*bold type*) are made accessible for direct protein contact (Henderson et al. 1994). In the context of the ferritin IRE, the type of 1:5 intraloop base pair can influence binding, and only C^1G^5 and G^1C^5 -type IREs bind to both IRP-1 and IRP-2 (Henderson et al. 1996). This may, in part, reflect variation in the bend angle of the first loop nucleotides as they stack on bases of the IRE upper stem. Optimal binding by both IRPs occurs when the sixth base (*N*) of the IRE loop does not pair with the first (Henderson et al. 1994, 1996). Our model explains why the transferrin receptor IRE stem-loop C (see Fig. 2B), which contains a single loop base change (G^5 to A^5) predicted to disrupt 1:5 pairing, binds poorly to the IRPs. (Koeller et al. 1989)

different but related HIV RNA stem loop did not detect such intraloop pairing (Colvin et al. 1993). The proposed IRE loop structure does explain some previous findings. Jaffrey et al. (1993) found that out of 15 different single base changes in the IRE loop, the most detrimental effect on binding resulted from mutation of the loop nucleotide G^5. The same team earlier showed that a G^5 to A^5 substitution in the transferrin receptor IRE stem-loop C caused a 100-fold decrease in IRP binding (Koeller et al. 1989). These observations are easily reconciled in that they would disrupt predicted 1:5 base pairing in the IRE loop.

The prediction of intraloop pairing within the IRE is further supported by a recent examination of ferritin IRE structure. Sierzputowska-Gracz et al. (1995) applied NMR spectroscopy to a wild-type bull-frog ferritin H-chain IRE and to an IRE mutant in which the loop base G^5 was changed to A^5, and observed differences in the spectra for these RNAs at different temperatures. They concluded that, in solution, the IRE loop is structured due to formation of a C^1:G^5 base pair (Sierzputowska-Gracz et al. 1995). Intraloop base pairing of the IRE may therefore be thermodynamically favourable in solution, a conclusion supported by earlier secondary structure analyses which suggested that the loop base G^5 is less susceptible to modification or enzyme attack than the middle nucleotide G^3 (Wang et al. 1990; Bettany et al. 1992). Does the IRE therefore adopt its optimal folding prior to interaction with IRP-1 or IRP-2? This issue remains unresolved, as the above-mentioned NMR analysis was rather indirect, and performed in the absence of $MgCl_2$ and KCl (Sierzputowska-Gracz et al. 1995), which are normally required for optimal interaction between the IRE and the IRPs in vitro (Leibold and Munro 1988). Further studies are necessary to define the solution and crystal structures of the IRE, and to determine whether intraloop pairing is a prerequisite for binding, or whether it is induced by a conformational change in the IRE following "docking" with the IRP proteins. Such alterations in RNA stem-loop conformation have been reported to be induced by different RNA-binding proteins such as HIV Rev (Kjems et al. 1991; Battiste et al. 1994) and the U1A spliceosome factor (Oubridge et al. 1994), upon binding their target RNAs.

In retrospect, the finding of a structured loop in the IRE hairpin is not so surprising in light of previous NMR studies on the sarcin/ricin loop from 28S rRNA (Szewczak et al. 1993), which was found to adopt a very compact structure displaying unusual base pairing and cross-strand stacking interactions. Considering the IRE loop secondary structure model (Fig. 3), how might specific nucleotides make protein contact? The middle IRE loop bases might splay outwards and directly contact the IRPs, as was previously shown for other RNA:protein interactions including that between the $tRNA^{Gln}$ and $tRNA^{Asp}$ anticodon loops and their respective glutaminyl- and aspartyl-tRNA synthetases (Rould et al. 1991; Cavarelli et al. 1993), and, more recently, for the U1 snRNA stem-loop and U1A spliceosomal protein (Oubridge et al. 1994). Resolution of the IRE:IRP interaction ultimately awaits more refined structural examination.

7 IRP-1 and IRP-2 Bind Distinct Sets of RNA Targets

Several human IRP-1 binding sequences were tested for binding to rat IRP-1 and IRP-2, and while the binding specificities of rat and human IRP-1 were similar, IRP-2 differed markedly in its RNA-binding preference (Henderson et al. 1994). IRP-1 and IRP-2 bind the wild-type ferritin IRE with equal avidity (Henderson et al. 1993; Guo et al. 1994); however, the introduction of changes in the type of intraloop base pair in IREs selected to bind IRP-1 dramatically decreased recognition by IRP-2 (see Fig. 4; Henderson et al. 1994). For in-

IRP-1 specific binding

(Change in 1:5 intraloop base pairing)

IRE	Loop position 1	2	3	4	5	6	Binding Ratio (%) IRP-1	IRP-2
CG42 (wild-type)	**C**	A	G	U	**G**	C	100	100
CG34	**U**	A	G	U	**A**	C	92	6
GG17	**G**	A	G	U	**G**	U	54	2
AG25	**A**	A	G	U	**G**	C	18	1
GA39	**G**	A	G	U	**A**	C	19	1

IRP-2 specific binding

(Double mutation of unpaired loop bases 2 and 4)

IRE	Loop position 1	2	3	4	5	6	Binding Ratio (%) IRP-1	IRP-2
CG42 (wild-type)	C	A	G	U	G	C	100	100
CG125	C	**C**	G	**A**	G	C	0.1	14
CG284	C	**U**	G	**G**	G	C	1	15
CG147	C	**C**	G	**C**	G	C	1	12
GC157	G	**C**	G	**C**	C	G	0	10

Fig. 4. IRE structural features that determine specific binding by IRP-1 or IRP-2. Specific IRE binding by IRP-1 can be achieved by altering the type of intraloop base pair, or by changing the "bulge" cytosine to guanine (Henderson et al. 1994). The *top panel* compares the ability of IREs with different intraloop base pairs (*bold type*) to form IRE:IRP complexes (expressed as the % binding ratio, relative to a wild-type ferritin IRE). IRP-2 bound poorly to IREs containing a U^1A^5 -type loop and different non-Watson-Crick base pairs found in other RNAs (data taken from Henderson et al. 1994). By contrast, preferential binding by IRP-2 was conferred by double mutation of the two unpaired bases (*bold type* and *underlined*) at loop positions *2* and *4* (*lower panel*). These findings (taken from Henderson et al. 1996) are not restricted to C^1G^5 -type IREs. Two IREs containing the same altered apical sequence ($C^2G^3C^4$) but different 1:5 base pairs (C^1G^5, mutant CG147 and G^1C^5, mutant GC157) each bound with similar preference to IRP-2. The price of achieving IRP-2 specificity, however, is a six- to tenfold reduction in binding affinity. Binding data were obtained by bandshift assays using labelled RNAs transcribed from the plasmids indicated

stance, only IRP-1 recognised IRE loops containing a U^1A^5 pair, and some less stable non-Watson-Crick base pairs (Henderson et al. 1994, 1996). Both IRPs recognise IREs comprising C^1G^5 or G^1C^5 intraloop base pairs (Fig. 3; Henderson et al. 1996), but binding is reduced if the RNAs fold as a tetraloop (Henderson et al. 1994). A single selected RNA (clone 55; loop sequence, 5′ CCGUGC 3′) bound more strongly to IRP-2 than to IRP-1 (Henderson et al. 1994), suggesting that each protein may target additional RNAs to which it can bind exclusively. Indeed, IRE-like hairpins with novel loop sequences were subsequently selected which bind preferentially to IRP-2 (Henderson et al. 1996). Whereas IRP-1 specificity is gained by altering the type of 1:5 loop base pair, specific double mutation of the two unpaired bases at loop positions 2 and 4 conferred preferential interaction with IRP-2 (outlined in Fig. 4; Henderson et al. 1996). These findings reveal that the IRPs bind overlapping, but quite distinct, sets of RNA targets.

This scenario is highly reminiscent of the RNA bacteriophage coat proteins MS2 and GA: related translational regulators which recognise similar RNA stem-loop structures differing primarily in the loop sequence (see Lim et al. 1994). The MS2 and GA protein amino acid sequences are highly homologous, and discernment of the different RNA loop sequences could be attributed to a single amino acid (Lim et al. 1994). The relatedness of human IRP-1 and IRP-2 cDNAs (Rouault et al. 1992; Guo et al. 1995a) may account for their shared capacity to bind a wild-type IRE, but equally as interesting will be the amino acid differences which decide the individual specificities of these two proteins. Furthermore, since the distinct IRP RNA-binding specificities are at least partly conserved between different species (Henderson et al. 1996), we propose that IRP-1 and IRP-2 each regulate unique mRNA targets in vivo. The functional importance of any in vitro-selected IRE variant must ultimately be assessed by in vivo assays, and transfection experiments will prove important in dissecting further this posttranscriptional system.

8 Summary

In this chapter, we have focused on the biochemistry of IRP-1 and the features which distinguish it from the related RNA-binding protein, IRP-2. IRP-1 is the cytoplasmic isoform of the enzyme aconitase, and, depending on iron status, may switch between enzymatic and RNA-binding activities. IRP-1 and IRP-2 are *trans*-acting regulators of mRNAs involved in iron uptake, storage and utilisation. The finding of an IRE in the citric acid cycle enzymes, mitochondrial aconitase and succinate dehydrogenase, suggests that the IRPs may also influence cellular energy production. These two proteins appear to bind RNAs with different but overlapping specificity, suggesting that they may regulate the stability or translation of as yet undefined mRNA targets, possibly extending their regulatory function beyond that of iron homeostasis. The interaction between the IRPs and the IRE represents one of the best characterised model systems for posttranscriptional gene control, and given that each IRP can also recognise its own unique set of RNAs, the search for new in vivo mRNA targets is expected to provide yet more surprises and insights into the fate of cytoplasmic mRNAs.

References

Aisen P (1994) Iron metabolism: an evolutionary perspective. In: Brock JH, Halliday JW, Pippard MJ, Powell LW (ed) Iron metabolism in health and disease. Saunders, London, p 1

Aziz N, Munro HN (1987) Iron regulates ferritin mRNA translation through a segment of its 5′ untranslated region. Proc Natl Acad Sci USA 89:8478–8482

Baker E, Morgan EH (1994) Iron transport. In: Brock JH, Halliday JW, Pippard MJ, Powell LW (ed) Iron metabolism in health and disease. Saunders, London, p 63

Barton HA, Eisenstein RS, Bomford A, Munro HN (1990) Determinants of the interaction between the iron-responsive element-binding protein and its binding site in rat L-ferritin mRNA. J Biol Chem 265:7000–7008

Basilion JP, Rouault TA, Massinople CM, Klausner RD, Burgess WH (1994a) The iron-responsive element-binding protein: localization of the RNA-binding site to the aconitase active-site cleft. Proc Natl Acad Sci USA 91:574–578

Basilion JP, Kennedy MC, Beinert H, Massinople CM, Klausner RD, Rouault TA (1994b) Overexpression of iron-responsive element-binding protein and its analytical characterization as the RNA-binding form, devoid of an iron-sulfur cluster. Arch Biochem Biophys 311:517–522

Battiste JL, Tan R, Frankel AD, Williamson JR (1994) Binding of an HIV rev peptide to rev responsive element RNA induces formation of purine-purine base pairs. Biochemistry 33:2741–2747

Beaumont C, Leneuve P, Devaux I, Scoazec J-Y, Berthier M, Loiseau M-N, Grandchamp B, Bonneau D (1995) Mutation in the iron responsive element of the L Ferritin mRNA in a family with dominant hyperferritinaemia and cataract. Nature Genet 11:444–446

Beinert H, Kennedy MC (1993) Aconitase, a two-faced protein: enzyme and iron regulatory factor. FASEB J 7:1442–1449

Bettany AJE, Eisenstein RS, Munro HN (1992) Mutagenesis of the iron-regulatory element further defines a role for RNA secondary structure in the regulation of ferritin and transferrin receptor expression. J Biol Chem 267:16531–16537

Burd CG, Dreyfuss G (1994) Conserved structures and diversity of functions of RNA-binding proteins. Science 265:615–621

Cairo G, Pietrangelo A (1994) Transferrin receptor gene expression during rat liver regeneration: evidence for post-transcriptional regulation by iron regulatory factor$_B$, a second iron-responsive element binding protein. J Biol Chem 269:6405–6409

Cairo G, Tacchini L, Pogliaghi G, Anzon E, Tomasi A, Bernelli-Zazzera A (1995) Induction of ferritin synthesis by oxidative stress: transcriptional and post-transcriptional regulation by expansion of the free iron pool. J Biol Chem 270:700–703

Casey JL, Hentze MW, Koeller DM, Caughman SW, Rouault TA, Klausner RD, Harford JB (1988) Iron-responsive elements: regulatory RNA sequences that control mRNA levels and translation. Science 240:924–928

Casey JL, Koeller DM, Ramin VC, Klausner RD, Harford JB (1989) Iron regulation of transferrin receptor mRNA levels requires iron-responsive elements and a rapid turnover determinant in the 3′ untranslated region of the mRNA. EMBO J 8:3693–3699

Cavarelli J, Rees B, Ruff M, Thierry J-C, Moras D (1993) Yeast $tRNA^{Asp}$ recognition by its cognate class II aminoacyl-tRNA synthetase. Nature 362:181–184

Chu E, Koeller DM, Casey JL, Drake JC, Chabner BA, Elwood PC, Zinn S, Allegra CJ (1991) Autoregulation of human thymidylate synthase messenger RNA translation by thymidylate synthase. Proc Natl Acad Sci USA 88:8977–8981

Chu E, Takimoto CH, Voeller D, Grem JL, Allegra CJ (1993) Specific binding of human dihydofolate reductase protein to dihydrofolate reductase messenger RNA in vitro. Biochemistry 32:4756–4760

Chu E, Voeller DM, Morrison PF, Jones KL, Takechi T, Maley GF, Maley F, Allegra CJ (1994) The effect of reducing reagents on binding of thymidylate synthase protein to thymidylate synthase messenger RNA. J Biol Chem 269:20289–20293

Chu E, Takechi T, Jones KL, Voeller DM, Copur SM, Maley GF, Maley FM, Segal S, Allegra CJ (1995) Thymidylate synthase binds to c-myc RNA in human colon cancer cells and in vitro. Mol Cell Biol 15:179–185

Colvin RA, Garcia-Blanco MA (1992) Unusual structure of the human immunodeficiency virus type 1 *trans* -activation response element. J Virol 66:930–935

Colvin RA, White SW, Garcia-Blanco MA, Hoffman DW (1993) Structural features of an RNA containing the CUGGGA loop of the human immunodeficiency virus type 1 *trans*-activation response element. Biochemistry 32:1105–1112

Constable A, Quick S, Gray NK, Hentze MW (1992) Modulation of the RNA-binding activity of a regulatory protein by iron in vitro: switching between enzymatic and genetic function? Proc Natl Acad Sci USA 89:4554–4558

Cox TC, Bawden MJ, Martin A, May BK (1991) Human erythroid 5-aminolevulinate synthase: promoter analysis and identification of an iron-responsive element in the mRNA. EMBO J 10:1891–1902

Crichton RR (1991) Iron homeostasis and cellular iron release. In: Crichton RR (ed) Inorganic biochemistry of iron metabolism. Horwood, New York, p 163

Crichton RR, Ward RJ (1992) Iron metabolism – new perspectives in view. Biochemistry 31:11255–11264

Dandekar T, Stripecke R, Gray NK, Goossen B, Constable A, Johansson HE, Hentze MW (1991) Identification of a novel iron-responsive element in murine and human erythroid 5-aminolevulinic acid synthase mRNA. EMBO J 10:1903–1909

DeRusso P, Philpott CC, Iwai K, Mostowski HS, Klausner RD, Rouault TA (1995) Expression of a constitutive mutant of iron regulatory protein 1 abolishes iron homeostasis in mammalian cells. J Biol Chem 270:15451–15454

Dix DJ, Lin PN, McKenzie AR, Walden WE, Theil EC (1993) The influence of the base-paired flanking region on structure and function of the ferritin mRNA iron regulatory element. J Mol Biol 231:230–240

Drapier J-C, Hirling H, Wietzerbin J, Kaldy P, Kühn LC (1993) Biosynthesis of nitric oxide activates iron regulatory factor in macrophages. EMBO J 12:3643–3649

Eisenstein RS, Tuazon PT, Schalinske KL, Anderson SA, Traugh JA (1993) Iron-responsive element binding protein: phosphorylation by protein kinase C.J. Biol Chem 268:27363–27370

Emery-Goodman A, Hirling H, Scarpellino L, Henderson B, Kühn LC (1993) Iron regulatory factor expressed from recombinant baculovirus: conversion between the RNA-binding apoprotein and Fe-S cluster containing aconitase. Nucl Acids Res 21:1457–1461

Gait MJ, Karn J (1993) RNA recognition by the human immunodeficiency virus Tat and Rev proteins. Trends Biochem Sci 18:255–259

Goessling LS, Daniels-McQueen S, Bhattacharyya-Pakrasi M, Lin JJ, Thach RE (1992) Enhanced degradation of the ferritin repressor protein during induction of ferritin messenger RNA translation. Science 256:670–673

Goessling LS, Mascotti DP, Bhattacharyya-Pakrasi M, Gang H, Thach RE (1994) Irreversible steps in the ferritin synthesis induction pathway. J Biol Chem 269:4343–4348

Goossen B, Hentze MW (1992) Position is the critical determinant for function of iron-responsive elements as translational regulators. Mol Cell Biol 12:1959–1966

Goossen B, Caughman SW, Harford JB, Klausner RD, Hentze MW (1990) Translational repression by a complex between the iron-responsive element of ferritin mRNA and its specific cytoplasmic binding protein is position-dependent in vivo. EMBO J 9:4127–4133

Guo B, Yu Y, Leibold EA (1994) Iron regulates cytoplasmic levels of a novel iron-responsive element-binding protein without aconitase activity. J Biol Chem 269:24252–24260

Guo B, Brown FM, Phillips JD, Yu Y, Leibold EA (1995a) Characterization and expression of iron regulatory protein 2 (IRP2): presence of multiple IRP2 transcripts regulated by intracellular iron levels. J Biol Chem 270:16529–16535

Guo B, Phillips JD, Yu Y, Leibold EA (1995b) Iron regulates the intracellular degradation of iron regulatory protein 2 by the proteasome. J Biol Chem 270:21645–21651

Haile DJ, Hentze MW, Rouault TA, Harford JB, Klausner RD (1989) Regulation of interaction of the iron-responsive element binding protein with iron-responsive elements. Mol Cell Biol 9:5055–5061

Haile DJ, Rouault TA, Harford JB, Klausner RD (1990) The inhibition of the iron responsive element RNA-protein interaction by heme does not mimic in vivo iron regulation. J Biol Chem 265:12786–12789

Haile DJ, Rouault TA, Tang CK, Chin J, Harford JB, Klausner RD (1992a) Reciprocal control of RNA-binding and aconitase activity in the regulation of the iron-responsive element binding protein: role of the iron-sulfur cluster. Proc Natl Acad Sci USA 89:7536–7540

Haile DJ, Rouault TA, Harford JB, Kennedy MC, Blondin GA, Beinert H, Klausner RD (1992b) Cellular regulation of the iron-responsive element binding protein: disassembly of the cubane iron-sulfur cluster results in high-affinity RNA binding. Proc Natl Acad Sci USA 89:11735–11739

Harrell CM, McKenzie AR, Patino MM, Walden WE, Theil EC (1991) Ferritin mRNA: Interactions of iron regulatory element with translational regulatory protein P-90 and the effect on base-paired flanking regions. Proc Natl Acad Sci USA 88:4166–4170

Henderson BR, Kühn LC (1995) Differential modulation of the RNA-binding proteins IRP-1 and IRP-2 in response to iron: IRP-2 inactivation requires translation of another protein. J Biol Chem 270:20509–20515

Henderson BR, Seiser C, Kühn LC (1993) Characterisation of a second RNA-binding protein in rodents with specificity for iron-responsive elements. J Biol Chem 268:27327–27334

Henderson BR, Menotti E, Bonnard C, Kühn LC (1994) Optimal sequence and structure of iron responsive elements: selection of RNA stem-loops with high affinity for iron regulatory factor. J Biol Chem 269:17481–17489

Henderson BR, Menotti E, Kühn LC (1996) Iron regulatory proteins −1 and −2 bind distinct sets of RNA target sequences. J Biol Chem 271:4900–4908

Henry Y, Lepoivre M, Drapier J-C, Ducrocq C, Boucher J-L, Guissani A (1993) EPR characterization of molecular targets for NO in mammalian cells and organelles. FASEB J 7:1124–1134

Hentze MW (1994) Enzymes as RNA-binding proteins: a role for (di)nucleotide-binding domains? Trends Biochem Sci 19:101–103

Hentze MW, Argos P (1991) Homology between IRE-BP, a regulatory RNA-binding protein, aconitase, and isopropylmalate isomerase. Nucl Acids Res 19:1739–1740

Hentze MW, Rouault TA, Caughman SW, Dancis A, Harford JB, Klausner RD (1987a) A *cis*-acting element is necessary for translational regulation of human ferritin expression in response to iron. Proc Natl Acad Sci USA 84:6730–6734

Hentze MW, Caughman SW, Rouault TA, Barriocanal JG, Dancis A, Harford JB, Klausner RD (1987b) Identification of the iron-responsive element for the translational regulation of human ferritin mRNA. Science 238:1570–1573

Hentze MW, Caughman SW, Casey JL, Koeller DM, Rouault TA, Harford JB, Klausner RD (1988) A model for the structure and functions of iron-responsive elements. Gene 72:201–208

Hentze MW, Rouault TA, Harford JB, Klausner RD (1989) Oxidation-reduction as a molecular mechanism of a regulatory RNA-protein interaction. Science 244:357–359

Hirling H, Emery-Goodman A, Thompson N, Neupert B, Seiser C, Kühn LC (1992) Expression of active iron regulatory factor from a full-length human cDNA by in vitro transcription/translation. Nucl Acids Res 20:33–39

Hirling H, Henderson BR, Kühn LC (1994) Mutational analysis of the [4Fe-4S] cluster converting iron regulatory factor from its RNA-binding form to cytoplasmic aconitase. EMBO J 13:453–461

Horowitz JA, Harford JB (1992) The secondary structure of the regulatory region of the transferrin receptor mRNA deduced by enzymatic cleavage. New Biol 4:330-338

Iwai K, Klausner R, Rouault TA (1995) Requirements for iron-regulated degradation of the RNA binding protein, iron regulatory protein 2. EMBO J 14:5350–5357

Jaffrey SR, Haile DJ, Klausner RD, Harford JB (1993) The interaction between the iron-responsive element binding protein and its cognate RNA is highly dependent upon both RNA sequence and structure. Nucl Acids Res 24:4627–4631

Kaptain S, Downey WE, Tang C, Philpott C, Haile D, Orloff DG, Harford JB, Rouault TA, Klausner RD (1991) A regulated RNA-binding protein also possesses aconitase activity. Proc Natl Acad Sci USA 88:10109–10113

Kennedy MC, Mende-Mueller L, Blondin GA, Beinert H (1992) Purification and characterization of cytosolic aconitase from beef liver and its relationship to the iron-responsive element binding protein. Proc Natl Acad Sci USA 89:11730–11734

Kim H-Y, Klausner RD, Rouault TA (1995) Translational repressor activity is equivalent and is quantitatively predicted by in vitro RNA binding for two iron-responsive element-binding proteins, IRP1 and IRP2. J Biol Chem 270:4983–4986

Kjems J, Brown M, Chang DD, Sharp PA (1991) Structural analysis of the interaction between the human immunodeficiency virus rev protein and the rev response element. Proc Natl Acad Sci USA 88:683–687

Klausner RD, Rouault TA, Harford JB (1993) Regulating the fate of mRNA: the control of cellular iron metabolism. Cell 72:19–26

Koeller DM, Casey JL, Gerhardt EM, Chan LN, Klausner RD, Harford JB (1989) A cytosolic protein binds to the structural elements within the iron regulatory region of the transferrin receptor mRNA. Proc Natl Acad Sci USA 86:3574–3578

Kohler SA, Henderson BR, Kühn LC (1995) Succinate dehydrogenase b mRNA of Drosophila melanogaster has a functional iron-responsive element in its 5′-untranslated region. J Biol Chem 270:30781–30786

Kühn LC (1994) Molecular regulation of iron proteins. In: Hershko C (ed) Baillière's Clinical Haematology, vol 7. Baillière Tindall, London, pp 763–785

Kühn LC, Hentze MW (1992) Coordination of cellular iron metabolism by post-transcriptional gene regulation. J Inorg Biochem 47:183–195

Lander HM, Levine DM, Novogrodsky A (1992) Hemin stimulation of cAMP production in human lymphocytes. FEBS Lett 303:242–246

Leibold EA, Munro HN (1988) Cytoplasmic protein binds in vitro to a highly conserved sequence in the 5′ untranslated region of ferritin heavy and light subunit mRNAs. Proc Natl Acad Sci USA 85:2171–2175

Leibold EA, Laudano A, Yu Y (1990) Structural requirements of iron-responsive elements for binding the protein involved in both transferrin receptor and ferritin mRNA post-transcriptional regulation. Nucl Acids Res 18:1819–1824

Levine TD, Gao F-B, King PH, Andrews LG, Keene JD (1993) Hel-N1: an autoimmune RNA-binding protein with specificity for 3′ uridylate-rich untranslated regions of growth factor mRNAs. Mol Cell Biol 13:3494–3504

Liau G, Chan LM, Feng P (1991) Increased ferritin gene expression is both promoted by cAMP and a marker of growth arrest in rabbit vascular smooth muscle cells. J Biol Chem 266:18819–18826

Lim F, Spingola M, Peabody DS (1994) Altering the RNA binding specificity of a translational repressor. J Biol Chem 269:9006–9010

Lin JJ, Patino MM, Gaffield L, Walden WE, Smith A, Thach RE (1991) Crosslinking of hemin to a specific site on the 90-kDa ferritin repressor protein. Proc Natl Acad Sci USA 88:6068–6071

Malter JS, Hong Y (1991) A redox switch and phosphorylation are involved in the post-transcriptional up-regulation of the adenosine-uridine binding factor by phorbol ester and ionophore. J Biol Chem 266:3167–3171

Müllner EW, Kühn LC (1988) A stem-loop in the 3′ untranslated region mediates iron-dependent regulation of transferrin receptor mRNA stability in the cytoplasm. Cell 58:373–382

Müllner EW, Neupert B, Kühn LC (1989) A specific mRNA binding factor regulates the iron-dependent stability of cytoplasmic transferrin receptor mRNA. Cell 58:373–382

Müllner EW, Rothenberger S, Müller AM, Kühn LC (1992) In vivo and in vitro modulation of the mRNA-binding activity of iron-regulatory factor: tissue distribution and effects of cell proliferation, iron levels and redox state. Eur J Biochem 208:597–605

Nakagawa J, Waldner H, Meyer-Monard S, Hofsteenge J, Jeno P, Moroni C (1995) AUH, a gene encoding an AU-specific RNA binding protein with intrinsic enoyl-CoA hydratase activity. Proc Natl Acad Sci USA 92:2051–2055

Neupert B, Thompson NA, Meyer C, Kühn LC (1990) A high affinity purification method for specific RNA-binding proteins: isolation of the iron regulatory factor from human placenta. Nucl Acids Res 18:51–55

Neupert B, Menotti E, Kühn LC (1995) A novel method to identify nucleic acid-binding sites in proteins by scanning mutagenesis: application to iron regulatory protein. Nucl Acids Res 23:2579–2583

Oubridge C, Ito N, Evans PR, Teo C-H, Nagai K (1994) Crystal structure at 1.92 A resolution of the RNA-binding domain of the U1A spliceosomal protein complexed with an RNA hairpin. Nature 372:432–438

Pantopoulos K, Hentze MW (1995) Rapid responses to oxidative stress mediated by iron regulatory protein. EMBO J 14:2917–2924

Pantopoulos K, Gray NK, Hentze MW (1995) Differential regulation of two related RNA-binding proteins, iron regulatory protein (IRP) and IRP_B. RNA 1:155–163

Patino MM, Walden WE (1992) Cloning of a functional cDNA for the rabbit ferritin mRNA repressor protein: demonstration of a tissue-specific pattern of expression. J Biol Chem 267:19011–19016

Philpott CC, Rouault TA, Klausner RD (1991) Sequence and expression of the murine iron-responsive element binding protein. Nucl Acids Res 19:6333

Philpott CC, Haile D, Rouault TA, Klausner RD (1993) Modification of a free Fe-S cluster cysteine residue in the active iron-responsive element binding protein prevents RNA binding. J Biol Chem 268:17655–17658

Philpott CC, Klausner RD, Rouault TA (1994) The bifunctional iron-responsive element binding protein/cytosolic aconitase: the role of active site residues in ligand binding and regulation. Proc Natl Acad Sci USA 91:7321–7325

Prodromou C, Artymiuk PJ, Guest JR (1992) The aconitase of *Escherichia coli*: nucleotide sequence of the aconitase gene and amino acid sequence similarity with mitochondrial aconitases, the iron-responsive element-binding protein and isopropylmalate isomerases. Eur J Biochem 204:599–609

Robbins AH, Stout CD (1989) The structure of aconitase. Proteins 5:289–312

Rogers JT, Andriotakis JL, Lacroix L, Durmowicz GP, Kasschau KD, Bridges KR (1994) Translational enhancement of H-ferritin mRNA by interleukin-1B acts through 5′ leader sequences distinct from the iron responsive element. Nucl Acids Res 22:2678–2686

Rothenberger S, Müllner EW, Kühn LC (1990) The mRNA-binding protein which controls ferritin and transferrin receptor expression is conserved during evolution. Nucl Acids Res 18:1175–1179

Rouault TA, Hentze MW, Caughman SW, Harford JB, Klausner RD (1988) Binding of a cytosolic protein to the iron-responsive element of human ferritin messenger RNA. Science 241:1207–1210

Rouault TA, Tang CK, Kaptain S, Burgess WH, Haile DJ, Samaniego G, McBride OW, Harford JB, Klausner RD (1990) Cloning of the cDNA encoding an RNA regulatory protein – the human iron-responsive element-binding protein. Proc Natl Acad Sci USA 87:7958–7962

Rouault TA, Stout CD, Kaptain S, Harford JB, Klausner RD (1991) Structural relationship between an iron-regulated RNA-binding protein (IRE-BP) and aconitase: functional implications. Cell 64:881–883

Rouault TA, Haile DJ, Downey WE, Philpott CC, Tang C, Samaniego F, Chin J, Orloff D, Harford JB, Klausner RD (1992) An iron-sulfur cluster plays a novel regulatory role in the iron-responsive element binding protein. Biometals 5:131–140

Rould MA, Perona JJ, Steitz TA (1991) Structural basis of anticodon loop recognition by glutaminyl-tRNA synthetase. Nature 352:213–218

Samaniego F, Chin J, Iwai K, Rouault T, Klausner RD (1994) Molecular characterization of a second iron-responsive element binding protein, iron regulatory protein 2: structure, function, and post-translational regulation. J Biol Chem 269:30904–30910

Seiser C, Teixeira S, Kühn LC (1993) Interleukin-2-dependent transcriptional and post-transcriptional regulation of transferrin receptor mRNA. J Biol Chem 268:13074–13080

Shaw G, Kamen R (1986) A conserved AU sequence from the 3′ untranslated region of GM-CSF mRNA mediates selective mRNA degradation. Cell 46:659–667

Sierzputowska-Gracz H, McKenzie RA, Theil EC (1995) The importance of a single G in the hairpin loop of the iron responsive element (IRE) in ferritin mRNA for structure: an NMR spectroscopy study. Nucl Acids Res 23:146–153

Swenson GR, Walden WE (1994) Localization of an RNA binding element of the iron responsive element binding protein within a proteolytic fragment containing iron coordination ligands. Nucl Acids Res 22:2627–2633

Szewczak AA, Moore PB, Chan Y-L, Wool IG (1993) The conformation of the sarcin/ricin loop from 28S ribosomal RNA. Proc Natl Acad Sci USA 90:9581–9585

Tang CK, Chin J, Harford JB, Klausner RD, Rouault TA (1992) Iron regulates the activity of the iron-responsive element binding protein without changing its rate of synthesis or degradation. J Biol Chem 267:24466–24470

Teixeira S, Kühn LC (1991) Post-transcriptional regulation of the transferrin receptor and 4F2 antigen heavy chain mRNA during growth activation of spleen cells. Eur J Biochem 202:819–826

Testa U, Kühn L, Petrin M, Quaranta MT, Pelosi E, Peschle C (1991) Differential regulation of iron regulatory element-binding protein(s) in cell extracts of activated lymphocytes versus monocytes-macrophages. J Biol Chem 266:13925–13930

Theil EC (1994) Iron regulatory elements (IREs): a family of mRNA non-coding sequences. Biochem J 304:1–11

Valegard K, Murray JB, Stockley PG, Stonehouse NJ, Liljas L (1994) Crystal structure of an RNA bacteriophage coat protein-operator complex. Nature 371:623–626

Walden WE, Daniels-McQueen S, Brown PH, Gaffield L, Russell DA, Bielser D, Bailey LC, Thach RE (1988) Translational repression in eukaryotes: partial purification and characterisation of a repressor of ferritin mRNA translation. Proc Natl Acad Sci USA 85:9503–9507

Wang Y-H, Sczekan SR, Theil EC (1990) Structure of the 5′ untranslated regulatory region of ferritin mRNA studied in solution. Nucl Acids Res 18:4463–4468

Weiss G, Goossen B, Doppler W, Fuchs D, Pantopoulos K, Werner-Felmayer G, Wachter H, Hentze MW (1993) Translation regulation via iron-responsive elements by the nitric oxide/NO-synthase pathway. EMBO J 12:3651–3657

Yu Y, Radisky E, Leibold EA (1992) The iron-responsive element binding protein: purification, cloning and regulation in rat liver. J Biol Chem 267:19005–19010

Zhang W, Wagner BJ, Ehrenman K, Schaeffer AW, DeMaria CT, Crater D, DeHaven K, Long L, Brewer G (1993) Purification, characterization, and cDNA cloning of an AU-rich element RNA-binding protein, AUF1. Mol Cell Biol 13:7652–7665

Zheng L, Kennedy MC, Blondin GA, Beinert H, Zalkin H (1992) Binding of cytosolic aconitase to the iron responsive element of porcine mitochondrial aconitase mRNA. Arch Biochem Biophys 299:356–360

Cytoplasmic Fate of Eukaryotic mRNA: Identification and Characterization of AU-Binding Proteins

J.A. JARZEMBOWSKI and J.S. MALTER

1 Introduction

Messenger RNA (mRNA) turnover has only recently been appreciated as an important control point in the regulation of gene expression. Initially thought to be a static process, mRNA decay is now known to be a highly regulated pathway capable of temporally, qualitatively, and quantitatively altering the cytoplasmic levels of specific mRNAs. The heterogeneity of different mRNA half-lives has a direct impact on gene expression by varying the levels of protein production (for reviews, see Raghow 1987; Hargrove and Schmidt 1989; Carter and Malter 1991; Peltz et al. 1991). In addition, the stability of a single mRNA can be differentially regulated by the cell in response to various stimuli. Alterations in mRNA decay rates are critical for both normal development and adult homeostasis (Kafatos 1972; Atwater et al. 1990). Dysregulated mRNA degradation can contribute to the development of a variety of human diseases, including cancer (Eick et al. 1985; Piechaczyk et al. 1985; Bauer et al. 1989; Ross et al. 1991), autoimmune syndromes (Lerner and Steitz 1979; Levine et al. 1993), and possibly Alzheimer's disease (Zaidi and Malter 1995).

Control of gene expression occurs at several points during the information flow from DNA to RNA to protein. Transcriptional rates and the production of mRNA are modulated by DNA-binding proteins through interactions with the target gene sequence and other components of the transcriptional complex. Each transcription factor may affect the expression of an individual gene or an entire class of genes, based on the genomic prevalence of its target DNA sequence. In addition, the activity of these proteins may be controlled by a host of factors including environmental stimuli, position in the cell cycle, and various signal transduction pathways.

After mRNA has been spliced, processed, and transported into the cytoplasm, its translational efficiency is a function of many variables, such as the composition and length of 5′ leader sequences (Kozak 1991), overall nucleotide context of the start codon (Kozak 1987), and the presence of a ^{7m}G cap (Both et al. 1975). The absolute amount of protein produced ultimately depends upon the steady-state level of cognate mRNA. The steady-state mRNA level is, in

Department of Pathology and Laboratory Medicine, University of Wisconsin Hospital and Clinics, 600 Highland Avenue, Madison, WI 53792-2472, USA

Progress in Molecular and Subcellular Biology, Vol. 18
Ph. Jeanteur (Ed.)

turn, determined by the net sum of transcription and degradation; typically, an increase in the cytoplasmic level of an mRNA will lead to increased protein production. Therefore, alterations in mRNA turnover rates can have profound consequences for gene expression.

Stable mRNAs can accumulate to high steady-state levels in the cytoplasm, providing a large number of translational targets and, in the absence of repressing *cis*-acting sequences or *trans*-acting factors, resulting in high levels of protein expression. It is bioenergetically frugal for a single mRNA molecule to serve reiteratively as a template for multiple rounds of translation. Moreover, a stable message can buffer the effects of transcriptional shutdown and facilitate the constant production of protein. Such regulation is especially appropriate for "housekeeping" genes, whose protein products are needed throughout the cell cycle. In eukaryotes, typical members of this group include β-globin (Aviv et al. 1976; Volloch and Housman 1981) and actin (Wreschner and Rechavi 1988), whose mRNAs have half-lives greater than 20 h.

At the other extreme, rapidly degraded mRNAs may be destroyed just as quickly as they can be produced. Under such conditions, mRNA accumulation is negligible, as is ensuing translation. Thus, protein production is severely restricted and dependent on alterations in the rates of transcription or degradation to replenish the mRNA population. Proteins which have potent growth or developmental effects, or whose function is temporally restricted, for example, during a particular phase of the cell cycle, are often encoded by unstable mRNAs. Members of this class in eukaryotes include proto-oncogenes such as *c-myc* (Dani et al. 1984; Herrick and Ross 1994) and *c-fos* (Greenberg and Ziff 1984), cytokines such as GM-CSF (Shaw and Kamen 1986), IL-2 (Kaempfer and Efrat 1985; Lindsten et al. 1989), and IL-3 (Wodnar-Filipowicz and Moroni 1990), and cell cycle-regulated genes such as histone H3 (Alterman et al. 1984) and p53 mRNA (Voelkerding et al. 1995), all of which have half-lives shorter than 1 h.

Between these two extremes of stability are thousands of mRNAs with independent decay rates (Singer and Penman 1973; Berger and Cooper 1975); Exactly how the cellular machinery is able to distinguish between specific targets within the vast cytoplasmic mRNA pool remains largely unknown. Individual mRNAs appear to be identified by information they carry in the form of *cis*-acting sequences. These may be recognized as primary sequence elements, secondary structures such as stem-loops or hairpins, or, in certain cases, a combination of both features.

The cellular factors able to recognize particular *cis*-elements are most likely cytoplasmic mRNA-binding proteins. These activities may stabilize cognate mRNAs by preventing recognition or destruction by ribonucleases (RNases) or conversely, may facilitate degradation by altering mRNA secondary structure or physically recruiting RNases to particular mRNAs. In most cases, the *cis-trans* interactions are modified by ongoing cell processes such as position in the cell cycle, intracellular redox state, kinase and phosphatase activity, overall nutritional status, engagement of cell-surface receptors, and degree of translation. Thus, the degradation of individual or classes of mRNAs and their relevant regulatory factors function at multiple, interdigitating levels.

2 Stable mRNAs

The structural features which define a stable mRNA remain unclear. Stability might be the result of specific, *cis*-acting elements which prevent turnover or, alternatively, the absence of destabilizing elements may, by default, create a stable mRNA. The former model is supported by the identification of a protein complex associated with the 3′ UTR of α-globin mRNA (Wang et al. 1995). Mutations in this region which disrupted complex formation were associated with decreased mRNA stability in erythroid cells. Although the constituent proteins of the complex were present in a variety of cell types, both wild-type and mutant mRNAs were stable in nonerythroid cells (Weiss and Liebhaber 1994). The authors propose the existence of an erythroid cell-specific ribonuclease responsible for α-globin turnover. Together, these data suggest a model where α-globin is generally stable in nonerythroid cells, but its turnover in erythroid cells is regulated by a combination of *cis*-acting sequences, *trans*-acting protein factors, and ribonucleases.

3 Unstable mRNAs

Far more infomation is available about mRNAs which rapidly decay in eukaryotic cells. In mammalian cells, mRNAs encoding growth factors (e.g., cytokines) or proteins responsible for regulating intracellular processes (e.g., transcription factors) are typically unstable (Raghow 1987; Brawerman 1989). This group of mRNAs is under quantitative, qualitative, and temporal control at both the transcriptional and posttranscriptional levels. Because the products of these genes are usually needed in significant quantities for a limited period of time, transient gene expression is achieved by a short burst of transcription concomitant with stabilization of the mRNA. Transcriptional shutdown and mRNA destabilization then effectively quench gene expression at the appropriate time.

The best-studied group of unstable mRNAs includes those coding for cytokines and proto-oncogenes. These proteins mediate cell cycle progression, intercellular communication, and various responses to environmental stimuli. The mRNAs encoding these proteins are usually highly labile in resting cells, but transiently stabilized preceding their expression at particular points during the cell cycle (Saini and Summerhayes 1991). A prototypical member of the cytokine group is granulocyte-macrophage colony-stimulating factor (GM-CSF), a hematopoietic growth factor necessary for the development and differentiation of granulocytes and macrophages (Metcalf 1985).

Shaw and Kamen (1986) and Caput et al. (1986) simultaneously demonstrated that GM-CSF mRNA was unstable, decaying in lymphoid cell lines with a half-life of less than 30 min. Treatment of these cells with mitogenic doses of phorbol ester (TPA) and phytohemagglutinin (PHA) stabilized the mRNA indefinitely to a half-life in excess of 4 h. These data demonstrated that signal transduction cascades, presumably through TPA/PHA-induced protein kinase C (PKC) activation, had dramatic effects on mRNA decay. Since a variety of signaling pathways converge on PKC, these data implied that multiple cell surface triggers could have comparable effects, and that signals which affect transcription also influence cytoplasmic mRNA turnover. As will be discussed later, phosphorylation cascades involving PKC, as well as other kinases, appear to regulate cytokine and other mRNA turnover rates (Stephens et al. 1992; Aksamit et al. 1993; Gorospe et al. 1993).

4 *Cis*-Acting Elements

In an effort to decipher which signals were responsible for regulating GM-CSF mRNA turnover in T-lymphocytes, Shaw and Kamen aligned the primary sequences of the murine and human GM-CSF cDNAs. Both the 5′- and 3′-untranslated regions (UTRs) demonstrated extremely high homology, which was unanticipated for noncoding regions. A distinguishing feature of the 3′ UTR of GM-CSF mRNA was the presence of multiple AUUUA sequence motifs, or AU-rich elements (AREs), which were >90% conserved between the murine and human versions of the gene. Both the number of elements and their location were conserved, suggesting a potential regulatory function.

In two corollary experiments, the investigators evaluated the role of the AREs in GM-CSF mRNA metabolism. First, after mutating the AUUUA motifs, GM-CSF cDNA was transfected into lymphoid cell lines. The resulting mRNA was stable, with a half-life greater that 3 h (Shaw and Kamen 1986). Second, to assess whether the ARE could function in a chimeric context, it was fused to the coding region of β-globin mRNA. This message was transiently expressed in T-cells, where it decayed with a half-life of about 45 min, or >20-fold faster than wild-type β-globin mRNA (Shaw and Kamen 1986). These data directly demonstrated that the ARE of GM-CSF mRNA was necessary and sufficient to confer instability. As addition of the ARE could destabilize β-globin mRNA *in trans*, it must contain the necessary sequence information in an appropriate context to be recognized by ribonucleases responsible for accelerated decay. Presumably, these are the same activities which degrade wild-type GM-CSF mRNA. The most likely candidate for this functional *cis* element was the reiterated AUUUA motifs, which composed 50% of the region.

Database searches identified similar reiterated AUUUA motifs in the 3′-UTRs of a variety of proto-oncogene and lymphokine mRNAs (Caput et al. 1986; Shaw and Kamen 1986). The elements were restricted to the 3′ UTRs of messages encoding growth-promoting or regulatory proteins. With rare exceptions, mRNAs encoding "housekeeping" genes or constituitively expressed proteins, such as actin and β-globin, lacked multiple AREs (Chen and Shyu 1994). This lent further support to the hypothesis that the AREs are destabilizing elements, functional in the 3′ UTR of highly labile, post transcriptionally regulated mRNAs.

A striking feature of the AREs is their diversity (see Fig. 1). Human GM-CSF mRNA contains eight AUUUA motifs, five of which are in tandem (i.e., ...AUUUAUUUA...) and share interior adenosines (Wong et al. 1985), whereas murine GM-CSF has six such motifs, four of which are tandem (Gough et al. 1984). The AREs of interferon gamma (IFN-γ) mRNA, on the other hand, consist of three AUUUA motifs separated by five nucleotides, within an AU-rich region (Gray et al. 1982). The AREs of *c-fos* and *c-myc* mRNAs contain single copies of the AUUUA motifs within much longer U-rich regions (Battey et al. 1983; van Straaten et al. 1983). All known AREs are

Human GM-CSF
UAUUUAUAUAUUUAUAUUUUUAAAAUAUUUAUUUAUUUA
UUUAUUUAAGUUCAUAUUCCAUAUUUA

Murine GM-CSF
AUAUUUAUUUAUAUAUUUAUAUUUUUUAAAUAUUUAUUU
AUUUAUUUAUUUAAUUUUGCACUCUAUUUAUUGA

Human *c-myc*
GAAUUAUUUAUGAUAAGAAUAAUUUAAUU

Human *c-fos*
UUAAUUUAUUUAUUAAGAUGGAUUCUCAGAUAUUUAUUUU
UAUUUUAUUUUUUUCUA

Murine *c-fos*
CAAUUUAUUUAUUAAGAUGGAUUCUCAGAUAUUUAUAUUU

IFN-γ
AUCUAUUUAUUAAUAUUUAACAUUAUUUAUAUGGGGAAUA
UAUUUUUAG..... GUAUUUAUAAUAG

IL-2
CAUUUACUGCUGGAUUUACAGAUGAUUUUGAAUGGAAUUAA
UGUAAGUAUAUUUCCUUUCUUACUAAAAUUAUUACAUUUA

IL-3
UAUUUAUUUAUGUAUGUAUGUAUUUAUUUAUUUAUUGCCU
GGAGUGUGAACUGUAUUUAUUUUAGC

HSUR 1 (5′ UTR)
^{7m}GACACUACAUAUUUAUUUAUUUAUUUCUUA

Rat TNF-β
GUUAAGUUAUUUAUUUAUAGGGAA

Fig. 1. The 3′ UTR ARE sequences of several ARE mRNAs. (Refer to text for further discussion and appropriate references)

located within the 3′ UTR of unstable mRNAs, at different, yet conserved positions relative to the coding body and poly(A) tail (Shaw and Kamen 1986).

Two dominant approaches have been used to demonstrate the function of the ARE. These include constructing chimeric mRNAs composed of a stable mRNA body and an ARE-containing 3′ UTR ("gain-of-function" or destabilization), or deleting the AREs from a typically unstable wild-type mRNA ("loss-of-function" or stabilization). Despite the ARE's competence to confer instability, chimeric mRNAs rarely demonstrate normal behavior. To the extent that mRNA context influences or synergizes with the ARE, chimeras may not be regulated by the full functional spectrum of this element. For example, when three consecutive AUUUA motifs are fused to the coding body of β-globin or chloramphenicol acetyltransferase (CAT) reporter gene constructs, the chimeric mRNA decay is slower than when the same sequence is fused to a wild-type GM-CSF coding body (Akashi et al. 1994). Nonetheless, in all cases examined so far, chimeric mRNAs composed of a previously stable mRNA body and a 3′ UTR derived from an ARE-containing cytokine or proto-oncogene are destabilized. Rarely do these contructs decay as rapidly as the comparable wild-type ARE mRNA (Bonnieu et al. 1989; Shyu et al. 1989), implying context effects or the existence of ancillary or independent domains. Therefore, while all AREs are capable of destabilizing an mRNA in *cis*, differences in their arrangement or surrounding context most likely confer different decay rates.

4.1 ARE Sequence Requirements

Recent work from several laboratories has begun to elucidate the minimal ARE sequence necessary for destabilization. Koeffler and coworkers have sequentially mutated the eight AUUUA motifs of human GM-CSF mRNA, and observed the decay rates of the resulting mRNAs in NIH/3T3 fibroblasts (Akashi et al. 1994). When three or more of the original eight AUUUA motifs remain intact, the mRNAs are labile and have half-lives of approximately 2 h. "Extra" motifs have no additional effects in this context; nonetheless, other sequences must also be important, because the wild-type GM-CSF mRNA was even more unstable ($t_{1/2} \approx 30$ min). However, when two or fewer motifs were present, the mRNAs were substantially stabilized, with half-lives greater than 20 h. These data suggest that, when presented in the context of GM-CSF, a minimum of three consecutive AUUUA elements is necessary for recognition by the *trans*-acting factors of the rapid degradation pathway.

Data from the laboratories of Moroni and Goodall suggest that the absolute number of AUUUA elements is less important than their arrangement. Moroni's group has demonstrated that IgE-receptor activation stabilizes IL-3 mRNA in PB-3c cells (Wodnar-Filipowicz and Moroni 1990; Lagnado et al. 1994). By mutational analysis of the eight AUUUA repeats in the 3′ UTR of IL-3 mRNA, they showed that six of these were critical, and when mutated, had a stabilizing effect comparable to that obtained by deleting the entire 3′ UTR (Stoecklin et al. 1994). Mutations which selectively disrupted the three adjacent motifs within the critical six elements also interfered with destabilization. These experiments suggested that three consecutive AUUUA elements comprise the essential component of the ARE.

Such data are consistent with our observation that the decay of amyloid precursor protein (APP) message is moderately slow ($t_{1/2} \approx 4$ h) in G_0/G_1 lymphocytes, despite the presence of four 3′ UTR AUUUA motifs (Zaidi and Malter 1994). These motifs are spaced at approximately 50 nucleotide intervals without a poly(U) context. Similar data have been observed by Gorospe and Baglioni (1994), who showed that interleukin-1α (IL-1α) mRNAs containing three clustered AUUUA motifs were unstable in rabbit reticulocyte lysates (RRLs), whereas mutant mRNAs with four scattered AUUUA elements were stable. Therefore, it is probable that both the spacing and context of the ARE affect its function.

Originally, based on sequence homology, the sequence UUAUUUAU was proposed as the consensus ARE (Caput et al. 1986). More recently, Goodall's group has proposed that the minimal functional sequence element may be UAUUUA(U/A)(U/A) which, when present in a single copy, is sufficient to destabilize a chimeric mRNA (Lagnado et al. 1994). Neither single nor multiple AUUUA elements (e.g., AUUUAAUUUA), nor poly(U) stretches conferred instability. Likewise, another group has proposed the consensus ARE sequence UUAUUUAUU, based on the ability of mutant elements to destabilize chimeric *c-fos* constructs (Zubiaga et al. 1995). The reiterated, tandem AUUU repeats (AUUUAUUUAUUU), as well as isolated motifs in a U-rich context, form a destabilizing octamer commonly found in proto-oncogene mRNAs.

4.2 Genomic Rearrangements of the ARE

Many tumor cell lines overproduce transcription factors or cytokines which may provide a growth advantage. An identical phenotype can arise from either transcriptional upregulation or mRNA stabilization. Analysis of one tumor cell line showed that genetic recombination had deleted 67 nucleotides of the 3′ UTR ARE of *c-fos*. The mutant *fos* mRNA was stable, accumulated, and was well translated into protein (Meijlink et al. 1985). Likewise, the viral oncogene *v-fos* lacks a portion of this same ARE region, and its mRNA accumulates to two- to threefold higher levels than its cellular counterpart, suggesting an identical mechanism of oncogenicity (Miller et al. 1984). Thus, increased stability of *fos* mRNA can increase its steady-state level, leading to overproduction of this transforming gene product.

Cell lines with retroviral insertions into the 3′ UTRs of interleukin-2 (IL-2) and interleukin-3 (IL-3) genes have been identified. A gibbon leukemia cell line, MLA-144, maintains its proliferative state by constitutively producing IL-2

(Chen SJ et al. 1985). The increased IL-2 expression results from enhanced mRNA stability caused by insertion of the gibbon leukemia virus into the 3′ UTR of the gene. Likewise, a cell line that gained exogenous IL-3 independence was found to constitutively express its own IL-3, resulting in autocrine transformation (Algate and McCubrey 1993). Transposition of a provirus into the 3′ UTR of the gene had disrupted the ARE, stabilized the mRNA, and led to IL-3 overexpression. PB-3c mast cell tumors have a similar retroviral insertion in the 3′ UTR of the IL-3 gene, and secrete abnormally high levels of IL-3, allowing them to grow in an autocrine fashion (Hirsch et al. 1993).

In addition to transformed cell lines, cytokine mRNAs are stabilized in human tumor xenografts grown in mice (Ross et al. 1991). Induction of mRNA stabilization after mitogen treatment persists for an abnormally long time in these cells, and chimeric constructs with 3′ UTR AREs from unstable mRNAs accumulate to significant steady-state levels. Thus, dysregulated mRNA turnover can enhance or cause overexpression of a transforming gene product, potentially leading to oncogenesis or cellular transformation.

4.3 Other Sequences

The subtle differences between cytokine ARE sequences and the contributory effects of flanking nucleotide sequences may explain the selective stabilization of some normally labile messages in response to particular signals. For example, in T-cells stimulated by mitogenic antibodies (α-CD3 and α-CD28), the ARE-containing mRNAs encoding TNF-α, GM-CSF, IFN-γ, and IL-2 were all stabilized (Lindsten et al. 1989). However, *c-myc, c-fos,* and IL-2R mRNAs, which also contain AU-rich elements in their 3′ UTRs, were not affected. Whether this phenomenon is due to minor ARE sequence differences or the existence of additional elements is currently unclear; in both *c-fos* and *c-myc* mRNAs, additional destabilizing elements have been identified in the coding region (Kabnick and Housman 1988; Wisdom and Lee 1991; Schiavi et al. 1994). Irrespective of mechanism, cells can respond to specific environmental stimuli by stabilizing subclasses of ARE-containing mRNAs.

ARE-containing mRNAs can be rapidly stabilized by exposing the cell to phorbol esters (Taniguchi 1988; Lindsen et al. 1989; Wodnar-Filipowicz and Moroni 1990), ionophore (Wodnar-Filipowicz and Moroni 1990; Iwai et al. 1993), cAMP, okadaic acid (Stephens et al. 1992), or mitogenic anti-cell surface antibodies (Lindsten et al. 1989). However, chimeric constructs composed of a stable mRNA body fused to multiple AREs were not comparably stabilized by these stimuli (Akashi et al. 1994), suggesting that chimeric ARE mRNAs lack some essential element(s) necessary for stimulus-induced stabilization. This could be additional primary sequence(s) adjacent to or distant from the ARE. Cohen's laboratory has tentatively identified a TPA-response element located approximately 150 bases upstream of the AREs in GM-CSF mRNA (Iwai et al.

1991). Unexpectedly, this region shows no homology with 3′ UTRs derived from other TPA-responsive mRNAs. This suggests that higher-order structure, rather than primary sequence alone, may be critical for its function.

5 Translational Dependence of ARE-mediated mRNA Turnover

Cycloheximide (CHX) treatment of cells efficiently blocks translation, and also inhibits the rapid degradation of many ARE- and non-ARE-containing mRNAs (Kelly et al. 1983; Dani et al. 1984; Shaw and Kamen 1986; Fort et al. 1987; Wilson and Treisman 1988; Altus and Nagamine 1991). This has been interpreted as either (1) extremely labile protein factors are necessary for rapid mRNA decay, (2) CHX directly interferes with mRNA decay, or (3) degradation requires ongoing translocation of polyribosomes along the mRNA template. The stabilization effect is very rapid (Wilson and Treisman 1988), suggesting a direct effect of CHX on mRNA decay or a requirement for polyribosome transiting. As a wide range of translational inhibitors affect ARE mRNA decay, it is unlikely to be a specific effect of CHX.

Direct experiments to clarify the translational dependence of ARE mRNA decay have been performed. Mutant GM-CSF mRNAs with stable 5′ UTR secondary structures inhibitory to translational initiation (Aharon and Schneider 1993; Curatola et al. 1995), or with mutated start codons (Savant-Bhonsale and Cleveland 1992), were stable compared to wild-type GM-CSF mRNA. These data suggest that ARE-containing mRNAs must be mobilized onto ribosomes and actively translated to be rapidly degraded.

Normal regulation is lost, despite ongoing translation, if the ARE sequences themselves are translated (Savant-Bhonsale and Cleveland 1992). GM-CSF mRNAs without stop codons were unusually stable. Translation through the ARE may inhibit degradation by altering the conformation or accessibility of the 3′ UTR, thereby preventing recognition by *trans*-acting factors, including ribonucleases. Alternatively, transiting ribosomes may displace protein factors or other regulatory molecules from the ARE, thereby interfering with mRNA turnover. These data imply that the ARE's relative position to the stop codon is important. Indeed, the location of an ARE, and the number of AUUUA repeats, are typically conserved across species (Shaw and Kamen 1986).

The ARE not only serves to target mRNAs for rapid degradation, but may inhibit their translation. The minimal sequence UUAUUUAU, when present in the 3′ UTR, reduced the translation of IFN, GM-CSF, and *c-fos* mRNAs, possibly due to an interaction between the AU-rich sequences and the poly(A) tail (Grafi et al. 1993). This is consistent with observations that the length of the poly(A) tail regulates translational efficiency in maturing *Xenopus* oocytes (Salles et al. 1992; Sheets et al. 1994). Thus, the ARE in the 3′ UTR of proto-oncogene and growth factor mRNAs may regulate both mRNA stability and translation.

6 Poly(A) Shortening in ARE mRNA Decay

While the process of ARE mRNA degradation is not fully understood, it most likely requires both the *cis*-acting ARE sequence element(s) and ongoing translation. Once turnover has been initiated, the first degradative step in the pathway appears to be deadenylation.

Kinetic analysis of GM-CSF, *c-fos,* and *c-myc* mRNA decay have provided evidence for the accumulation of shortened, deadenylated intermediates (Brewer and Ross 1988; Wilson and Treisman 1988; Schiavi et al. 1994; Chen et al. 1995; Rajagopalan et al. 1995), similar to those seen in non-ARE-mediated mRNA turnover in yeast (Decker and Parker 1993). This process could occur through a 3′→5′ processive ribonucleolytic activity, or by a single endonucleolytic snip within the 3′ UTR of the mRNA. The product of a cleavage reaction would appear as a single species on a Northern blot (synchronous shortening), whereas the products of processive degradation would appear as a smear of many different size mRNAs (asynchronous shortening).

Poly(A) tail removal might serve (1) to reduce interactions between the ARE and the poly(A) tail, (2) to displace poly(A) binding proteins from the end of the mRNA, or (3) to unmask protected 3′ UTR domains for a 3′→5′ processive ribonucleolytic activity. Clearly, poly(A) mRNAs will be translationally silenced, further inhibiting gene expression. Since not all ARE mRNAs have been shown to decay through poly(A) shortening, it remains possible that this phenomenon represents only one of several possible ARE mRNA decay pathways.

7 Subcellular Localization of mRNA Decay

The sequestration of particular mRNAs to distinct cytoplasmic compartments (such as the cytoskeleton, polyribosomes, or mitochondria) may provide yet an additional level of regulation and complexity. For example, *hsp83* mRNA is initially present throughout the cytoplasm of *Drosophila* embryos, but during maturation is degraded everywhere except at the pole plasm (Ding et al. 1993). While the mechanism underlying this process is unknown, the localization of a stabilizing factor could preserve and concentrate *hsp83* mRNA at the pole plasm. Similarly, *nanos* protein represses the translation of *hunchback,* concomitantly destabilizing the mRNA as well (Wharton and Struhl 1991). As a general phenomenon, the sequestration of inactive maternal mRNAs in unfertilized *Xenopus* oocytes is most likely mediated by interacting ribonucleoproteins (for review, see St Johnston 1995).

Both the efficacy of polysome-based in vitro decay systems, and the translational dependence of ARE RNA turnover suggest that the decay machinery is localized to polyribosomes, and that free cytoplasmic mRNA may be stable or degraded by another mechanism. The loading of mRNA onto polysomes may in itself be regulated, thereby controlling the rapidity of mRNA turnover.

8 How Might the ARE Be Recognized?

The ability of mRNA to fold and adopt complex structures greatly increases the possible conformations of the ARE. Primary sequence, secondary structure, and higher-order folding may, individually or together, present the ARE to the decay machinery.

8.1 Primary Structure

Both the number of AUUUA repeats and constituent nucleotides are important in ARE-mediated decay. Belasco's and Goodall's groups have both shown that substitution of the center U with a G inhibited ARE function (Shyu et al. 1989; Lagnado et al. 1994). Similar mutations inhibited the binding of the stabilizing protein AUBF (Gillis and Malter 1991), suggesting that ARE-binding proteins and ribonucleases recognize similar or identical primary sequences.

8.2 Secondary Structure

The specific sequence requirements of the ARE most likely promote the formation of a secondary structure recognized by components of the degradation pathway. A precedent for this exists in the regulation of translation and degradation of ferritin and transferrin mRNAs by the iron-response element (IRE; for review, see Klausner and Harford 1989). The IRE is a 35-base sequence which folds into a stable stem-loop structure capable of interacting with the IRE-binding protein (IRE-BP). The IRE-BP binds to the IRE in the 5′-UTR of ferritin mRNA, inhibiting translation by sterically hindering ribosomal complex assembly (Walden et al. 1988; Goosen and Hentze 1992). In addition, the IRE-BP can interact with any of the five IREs in the 3′ UTR of transferrin mRNA, inhibiting mRNA degradation either by obscuring a destabilizing signal or blocking the action of a specific endoribonuclease (Mullner and Kuhn 1988; Binder et al. 1994). However, neither computer nor experimental modeling of the ARE has predicted any thermodynamically significant folding patterns.

ARE recognition could occur after long-range intramolecular interactions with other regions of the 3′ UTR, coding region, or 5′ UTR. Long-range interactions are required for the degradation of insulin-like growth factor II (IGF-II) mRNA (Scheper et al. 1995). Two 30-base 3′ UTR elements separated by about 1800 nucleotides interact to form a stable stem-loop structure (~100 kcal/mol). This double-stranded RNA region is endonucleolytically cleaved to create a 5′ fragment which contains the entire coding region, and a 3′ fragment which is polyadenylated. The deadenylated coding region is then rapidly degraded. Such a mechanism is unlikely for ARE-containing mRNAs, however, since chimeras with minimal AU-rich sequences are still destabilized. Nonetheless, for certain mRNAs such as *c-fos* (Roy et al. 1992; Chen CA et al. 1994; Schiavi et al. 1994) and *c-myc* (Bonnieu et al. 1988; Herrick and Ross 1994), sequences in the 5′ UTR and coding region have been shown to be necessary for rapid decay, and may exert their effects through such long-range intramolecular interactions.

9 Trans-Acting Factors

Recent evidence supports the existence of sequence-specific, protein-AUUUA interactions which participate in the mRNA degradation process. ARE-binding proteins (AUBPs or ARE BPs) recognize the ARE and most likely mediate selective mRNA stabilization and/or destabilization. AUBPs presumably accomplish this regulation by directing or inhibiting the activity of ribonucleases. As the ARE-binding proteins appear to differ in their precise affinities for specific RNA sequence (see Fig. 2), they may preferentially interact with different mRNAs in vivo (Malter and Hong 1991; Bohjanen et al. 1992). While the general ARE structure might confer instability upon all proto-oncogene and growth factor mRNAs, the subtle sequence differences between them may allow for selective stabilization by discrimination in binding of various AUBPs.

9.1 AUBF

The first ARE-specific RNA-binding protein to be identified was the adenosine-uridine binding factor (AUBF; Malter 1989). Crude cytoplasmic lysates from Jurkat cells, a human T-cell leukemia, were incubated with radiolabeled $(AUUUA)_4$ oligoribonucleotide probe. After binding, the reactions were

Protein	Location	Size (kDa)	poly(U)	*c-myc*	RRMs
AUBF	Cytoplasm	32	-	+	ND
p32	Nucleus	32	+	+	ND
hnRNP A0	Nucleus	32	+	+	Yes (2)
AUF-1	Cytoplasm	37, 40, 45	+	+	Yes (2)
AU-A	Shuttles	34	+	+	ND
AU-B	Cytoplasm	30	-	-	ND
AU-C	Cytoplasm	43	-	-	ND
hnRNP A1	Shuttles	35	ND	+	Yes (2)
hnRNP C	Shuttles	42	+	+	Yes (1)
Hel-N1	Shuttles	38	ND	+	Yes (3)
AU-H	Cytoplasm	32	+/-	+	No

Fig. 2. The various AUBPs, their properties, and their binding specificities. See the text for appropriate references; ND, not determined

quenched with either RNase A or RNase T1 to digest free probe and unprotected portions of the bound probe, and then electrophoresed on native gels or SDS-PAGE.

A band-shifted, radiolabeled complex was observed on native gels, which was sensitive to proteinase K digestion or pretreatment of the lysate with heat. UV-cross-linking followed by SDS-PAGE demonstrated a 42-kDa RNA-protein complex. The binding was specific, because the $(AUUUA)_4$ probe could only be effectively competed for by unlabeled $(AUUUA)_4$ RNA, but not $(AUGUA)_4$ or poly(U) RNAs. At least three contiguous AUUUA boxes, or a single motif in a poly(U)-rich context, were required for AUBF recognition (Gillis and Malter 1991; Malter, unpubl. data). When the lysate-probe mixture was electrophoresed on reducing SDS-PAGE after UV-cross-linking, three RNA-protein complexes of roughly 15, 17, and 19 kDa were observed (Malter 1989). This suggested that AUBF was a multimeric protein composed of disulfide-linked subunits, each in contact with the AUUUA RNA.

These data strongly suggest, but do not prove, that AUBF binds the AUUUA motifs. For example, AUUUA-containing RNAs might fold to present a distinct secondary structure which serves as the binding site for AUBF. To address this, $(AUUUA)_4$ probes were radiolabeled with either $[^{32}P]$rUTP or $[^{32}P]$rGTP and used in mobility shifts. If AUBF interacted and protected the AUUUA motifs, which lack guanosine residues, $[^{32}P]$rUTP-labeled AUUUA-RNA protein complexes should have greater radioactivity than $[^{32}P]$rGTP-labeled AUUUA RNA-protein complexes. Conversely, if AUBF protected a non-AUUUA region from RNase cleavage, equal amounts of radioactivity should be incorporated into complexes containing $[^{32}P]$rUTP or $[^{32}P]$rGTP. Fiftyfold more label was incorporated into the AUBF- mRNA complex when $[^{32}P]$rUTP-, as compared to $[^{32}P]$rGTP-labeled probe, was used (Malter, unpubl. data). This demonstrates that AUBF recognizes and protects the AUUUA portion of the probe, rather than guanosine-containing, peripheral regions. In addition, reducing the length of the target mRNA below 30 bases abolished binding, providing evidence that some secondary structure is most likely involved in AUBF-AUUA RNA recognition.

In addition to short, ARE-containing RNA ligands, AUBF also binds to the AREs of full-length GM-CSF, IFNγ, IL-3, *c-fos*, and *v-myc* mRNAs, suggesting a general role for AUBF in the stabilization of proto-oncogene and growth factor mRNAs (Gillis and Malter 1991). Although AUBF is inactive in resting peripheral blood mononuclear cells, it is activated by treatment with mitogens, such as TPA (Malter and Hong 1991). As the same stimuli both upregulate AUBF and stabilize ARE mRNAs, it is probable that AUBF activation is involved in regulating ARE-containing mRNA turnover. The constituitive levels of active AUBF in some tumor cell lines may also be responsible for the stabilization of proto-oncogene mRNAs, and the subsequent overexpression of these transforming gene products (Malter, unpubl. data).

9.2 p32 and hnRNP A0

Shenk and coworkers have described a 32-kDa protein (p32) which binds the 3′ UTR of GM-CSF, *c-fos,* and *c-myc* mRNAs (Vakalopoulou et al. 1991). This activity was present in both nuclear and cytoplasmic HeLa extracts, and required only a single AUUUA motif in a uridine-rich context for binding. Unlike AUBF, p32 binding was inhibited by unlabeled poly(U) RNA, and the resulting complex could not be resolved into three smaller peptides by reduction. Therefore, AUBF and p32 are believed to be distinct activities.

More recently, work by Steitz's laboratory has shown that a protein similar in molecular mass to p32 also binds to the AUUUA motifs of *Herpesvirus saimiri* U RNAs (HSURs; Myer et al. 1992; Geck et al. 1994). These RNAs are required for cellular transformation by the virus. When nuclear extracts were incubated with radiolabeled HSURs and subjected to UV-cross-linking, a 32-kDa complex was detected. This complex had both similar proteolytic digestion patterns and sensitivity to poly(U) RNA as p32 (Myer et al. 1992). The activity of p32 correlated with rapid mRNA decay, leading the authors to propose that it functions as a nuclear destabilizer of ARE mRNAs. They also hypothesized that the abundance of HSURs in the nucleus of infected cells titrates AUBPs, such as p32, away from proto-oncogene and growth factor mRNAs, thereby stabilizing them. The resulting increase in protein expression may assist in the subsequent cellular transformation. Recently, a 32-kDa protein, hnRNP A0 (Myer and Steitz 1995), which may represent p32, was purified and cloned. hnRNP A0 does not react with anti-hnRNP A1 antibodies, and displays a unique proteolytic digestion pattern.

9.3 hnRNPs

Heterogeneous nuclear ribonucleoproteins (hnRNPs) were first identified as factors involved in pre-mRNA processing (Choi et al. 1986). These proteins were originally thought to be localized solely to the nucleus, but have recently been identified in the cytoplasm (Hamilton et al. 1993), as well as on polysomes (Zaidi and Malter 1995). This suggests that hnRNP A1, which contains a nuclear localization sequence, may shuttle between the nuclear and cytoplasmic compartments (Siomi and Dreyfuss 1995).

Both hnRNP A and C have consensus RNA-binding motifs (RBDs or RRMs), and C-terminal glycine-rich domains. The optimal binding sequence for hnRNP C protein was ascertained by SELEX, and determined to be a 14-nucleotide poly(U) RNA (Gorlach et al. 1994). As only short RNA ligands were screened by this method, it remains possible that additional sequences of greater length may also specifically bind hnRNP C protein. Experiments by Rigby and coworkers showed that both hnRNP A1 and C also bind the AREs found in GM-CSF and IL-2 mRNAs (Hamilton et al. 1993). In addition, antibodies immunoprecipitated hnRNPs A1 and C, along with radiolabeled

$(AUUUA)_4$ probes, from lysate-RNA mixtures. Based on these data, hnRNP A1 and C proteins may play a role in labile mRNA turnover, potentially acting in both the nucleus and cytoplasm.

9.4 AUF-1

Brewer and Ross (1988) have shown that the rapid in vitro decay of *c-myc* mRNA on polyribosomes required the addition of the 130 000 *g* postribosomal supernatant (S130). In a search for the factor(s) mediating this decay, Brewer (1991) assayed purified sucrose gradient fractions from the S130 of cycling human erythroleukemia K562 cells for both ARE-binding activity and the ability to stimulate labile mRNA degradation in an in vitro, polysome-based system. Using the 3′ UTR of the *c-myc* proto-oncogene mRNA as a probe, Brewer demonstrated that the presence of two protein-RNA complexes of 37 and 40 kDa was correlated with accelerated mRNA degradation. These interactions could be competed for by both the 3′ UTR of GM-CSF and poly(U) RNA, but by neither poly(A) nor poly(G) RNAs. While the 37- and 40-kDa proteins were found in both the nucleus and cytoplasm, an immunologically related peptide of 45 kDa was exclusively localized to the nucleus (Zhang et al. 1993). This family of ARE-BPs has been cloned and denoted AUF-1. Binding assays with recombinant protein show ARE specificity, most likely mediated through the two RNA recognition moftis (RRMs) present in the N-terminus of AUF-1. A direct demonstration of the function of AUF-1 has yet to be published.

9.5 AU-A, AU-B, and AU-C

Thompson's group has identified three ARE-binding proteins, designated AU-A, AU-B, and AU-C. AU-A was a constitutively expressed 34-kDa nuclear activity which bound the AU-rich sequences in GM-CSF, IL-2, TNF-α, and *c-myc* mRNAs (Bohjanen et al. 1992). Recently, this protein was shown to shuttle between the nucleus and cytoplasm much like hnRNP A1, although proteolytic digestion demonstrated that these two activities are distinct (Katz et al. 1994). Based on localization, size, and binding specificity, AU-A and p32 may be the same factor.

AU-B specifically bound to the 3′ UTRs of IL-2 and TNF-α mRNAs (Bohjanen et al. 1991). This activity was localized to the cytoplasm, and was induced only in T-cells stimulated by antibodies to the TCR/CD3 complex. AU-B binding was competed by the 3′ UTR of GM-CSF and $(AUUUA)_4$ oligoriboprobes, but not by the 3′ UTR of *c-myc* mRNA. In addition, like AUBF, AU-B is incapable of binding monomeric AUUUA motifs, instead requiring tandem repeats of this sequence element. Following cell treatment with phorbol ester, increased steady-state GM-CSF levels were correlated with decreased AU-B activity, suggesting that AU-B binding serves to target labile mRNAs for rapid degradation. AU-C is a 43-kDa activity which is structurally related to AU-B, and shares its sequence specificity and induction pattern (Bohjanen et al. 1992).

9.6 *AU-H*

Recent work in Moroni's laboratory has identified AU-H, a 32-kDa protein which specifically binds to the 3′ UTR elements in IL-3, GM-CSF, *c-fos,* and *c-myc* mRNAs (Nakagawa et al. 1995). This protein has significant homology to enoyl-CoA hydratase, and exhibits weak hydratase activity. This duality in function is akin to the IRE-BP, an mRNA-binding protein originally identified as aconitase, an important metabolic enzyme in the Krebs cycle. Although a definite role for AU-H in regulating mRNA turnover has not yet been described, its binding specificity suggests it may participate in ARE mRNA turnover.

9.7 *Hel-N1*

Hel-N1 was identified as a neuronal-specific RNA-binding protein which binds the 3′ UTR of *c-myc, c-fos,* and GM-CSF mRNAs (Levine et al. 1993). This 38-kDa protein is homologous to the *Drosophila* elav protein, which participates in regulating neuronal development. Hel-N1 is present in neurons as well as T-cells. Antibodies from patients suffering from paraneoplastic encephalomyelitis cross-react with Hel-N1 (Szabo et al. 1991), suggesting this AUBP may be an autoimmune target, just as snRNPs are in systemic lupus erythematosis (SLE; Lerner and Steitz 1979).

10 Functions

The functions of ARE BPs have been inferred from their binding specificity, subcellular location, and regulation. Nuclear ARE BPs are unlikely to modulate cytoplasmic mRNA turnover, and are more likely to participate in the processing of hnRNA into mRNA, or transport of mRNA to the cytoplasm. Cytoplasmic localization on polysomes appears to be ideal for both stabilizing and destabilizing functions, as translation is an integral requirement for normal regulation. Proteins which are upregulated upon cell activation, cell cycling, or other physiological events associated with AUUUA mRNA stabilization most likely act as stabilizers, while ARE BPs downregulated by these same events are presumably destabilizers. Posttranslational modifications in response to such stimuli may alter the binding activity or subcellular location of the ARE BPs. Thus, a global assignment to stabilization or destabilization may be an oversimplification, as a single activity could serve different functions under different conditions. Clearly, direct demonstration of function is necessary.

Antisense RNAs, ribozymes, knock-out mutants, or dominant negative mutants (hnRNP C, for example, exists as a tetramer) might also be employed to block the function of individual AUBPs in vivo. Some of the ARE BPs have not yet been cloned, but concern that knocking out AUBP expression may be lethal has hindered in vivo work of this type, necessitating the exploitation of reasonable in vitro paradigms. Ideally, an in vitro system would accurately recapitulate intracellular events, be manipulatable, and include all relevant regulatory proteins. Unfortunately, such a system does not currently exist. The most widely employed approaches are polyribosomes (Brewer and Ross 1988) and rabbit reticulocyte lysates (Wreschner and Rechavi 1988; Gorospe and Baglioni 1994).

In either of these in vitro systems, by depleting or adding excess exogenous protein, and determining the subsequent effect on labile mRNA decay rates, the role of a specific AUBP in the degradation pathway can be determined. Proteins can be selectively removed by several methods, including immunoprecipitation or preincubation with competitor RNAs containing the target binding site. Alternatively, the protein(s) of interest can be added to the system. After the manipulations, the system is incubated under decay conditions, and RNA is collected at various timepoints and analyzed by Northern blotting, RNase protection assays, or quantitative RT-PCR.

Polysome-associated mRNAs are degraded when incubated at 20–37 °C under translation conditions (Brewer and Ross 1988). The rates of mRNA decay are variable and temperature-dependent but, in general, the relative rates of decay are maintained (e.g., *c-myc* < GAPDH < β-globin half-life). mRNAs in this system are not actively translated, which may be a significant deficiency. Accessory proteins may be lost during polysome preparation, with functional consequences. Finally, the disruption of the cytoplasm will eliminate any supramolecular organization which may influence mRNA decay. Despite these deficiencies, polysomes can be isolated from a variety of normal or malignant cell types, are easily manipulated, and can be stored at –80 °C for at least a year (Rajagopalan and Malter, unpubl. observ.).

Using the polysome system, Rajagopalan et al. (1995) indirectly assessed the function of AUBF in vitro. Polysomes were isolated from TPA/PHA-stimulated PBMCs, and shown to contain both GM-CSF mRNA and AUBF. After incubation under protein synthesis conditions and Northern blotting at various timepoints, GM-CSF decayed with a half-life of 90 min. AUBF was then removed from the system by titration with competitor $(AUUUA)_4$ RNA ligand, which accelerated GM-CSF mRNA decay by ~fivefold. A nonspecific competitor $(AUGUA)_4$ RNA had no effect on the decay rates, suggesting that the destabilization was caused by the removal of a specific ARE BP. In addition to demonstrating an effect of AUBF on ARE mRNA decay, these experiments also show that the polysome-associated components of the degradation apparatus appear to be intact, even in stimulated cells. This suggests that mRNA turnover is inhibited by mitogens, and not irreversibly turned off.

Raymond's laboratory used chimeric 5′ UTR IRE/β-globin coding region / 3′ UTR ARE-mRNAs to demonstrate the translational dependence of *c-fos* and GM-CSF mRNA decay (Winstall et al. 1995). A 5′ IRE was added to regulate the cytoplasmic/polysomal distribution of chimeric mRNAs under different iron concentrations. The chimeric mRNAs, containing the AREs from *fos* or GM-CSF, were stable when free in the cytoplasm, but when loaded onto polysomes were rapidly degraded. Similar results were obtained by others, who demonstrated the translational dependence of destabilization mediated by both the 3′ UTR and coding region elements (Veyrune et al. 1995). Thus, both *c-fos* and GM-CSF mRNAs are selectively degraded on polysomes concomitant with translation, and AREs from either are sufficient to direct the rapid decay of chimeric mRNAs.

Rabbit reticulocyte lysates (RRLs) retain translational competence and can be programmed with in vitro-transcribed mRNAs of interest. Whether RRLs and sucrose pad polysomes are biochemically identical is unknown. As the former is a lysate from a differentiated, nondividing, normal cell, whereas the latter is typically derived from a cycling tumor cell, these systems are probably different. The potential discrepancies between RRLs and polysomes hinder direct comparison of experiments performed in the two systems. However, work done in such a system has shown that unstable mRNAs such as *c-myc* (Wreschner and Rechavi 1988) and IL-1α (Gorospe and Baglioni 1994) demonstrate the expected behavior in vitro.

11 Posttranslational Modifications

11.1 Phosphorylation

Transient stabilization of normally labile mRNAs can be induced by treating cells with diverse agents, such as TPA/PHA, cAMP, calcium ionophore, okadaic acid, or CD3/CD28 (Lindsten et al. 1989). These stimuli are activators of multiple signal transduction pathways generally inducing posttranslational modifications including phosphorylation, methylation, and acylation. During growth transitions, the general population of mRNA-binding proteins, with the exception of a 32-kDa species, did not change significantly, suggesting that activation of some of these proteins was posttranscriptionally regulated (Berger and Sells 1992). The response by *trans*-acting factors is very rapid, and also consistent with posttranslational modification of existing proteins.

Many signal transduction pathways activate phosphorylation by protein kinases. Both AUF1 and Hel-N1 contain several consensus phosphorylation sites, suggesting that they may be regulated in this manner. hnRNP C proteins undergo cell cycle-regulated phosphorylation (Piñol-Roma and Dreyfuss 1993). Hypophosphorylated hnRNP C bound an adenovirus type 2 major late

RNA ligand, while hyperphosphorylated C protein did not. Thus, phosphorylation most likely modulates the RNA-binding activity of hnRNP C protein. In vitro AUBF binding activity can be abolished by pretreatment with phosphatases (Malter and Hong 1991).

The activities of the ribonucleotide reductase R1 and R2 mRNA binding proteins (R1-BP; R2-BP) were downregulated by cell treatment with TPA or okadaic acid, upregulated by protein kinase C inhibitors, and unaffected by protein kinase A inhibitors (Chen FY et al. 1994). R1 and R2 mRNA stability were inversely correlated with binding activity, suggesting that the proteins destabilize the mRNAs. This evidence supports a model in which the activation or inactivation of AUBPs occurs upon phosphorylation by protein kinases, through signal transduction pathways.

11.2 Metal Ions

Some nucleic acid-binding proteins require a metal ion for activity. This cation may serve to stabilize the quaternary structure of the protein, such as in zinc finger motifs (Freedman et al. 1988), or can be directly involved in ligand binding. AUBF binding was shown to require calcium or magnesium cations, and was sensitive to metal ion chelators such as EDTA or EGTA (Malter et al. 1990). The divalent metal ion requirement suggests a potential regulatory mechanism through intracellular calcium fluxes which often accompany protein kinase C (PKC) signalling pathways.

11.3 Methylation

Most of the hnRNPs are posttranslationally methylated on arginine residues (Liu and Dreyfuss 1995). While the importance of this process is not fully understood, the target arginines are located in the RGG box RNA-binding domain, suggesting that methylation may influence RNA-binding activity.

11.4 Redox State

AUBPs are sensitive to redox conditions. Reduced sulfhydryl groups are often essential for activity, and may function in chelating metal ions, maintaining the quaternary structure of the protein through disulfide bonds, or directly interacting with the RNA ligand. AUBF activity is reversibly inhibited by diamide treatment, and irreversibly blocked by n-ethylmaleimide (Malter and Hong 1991), suggesting that reduced sulfhydryls are essential for RNA binding, probably by maintaining higher-order structure of the protein subunits (Malter 1989).

The IRE-BP shows aconitase or RNA-binding activity, depending on the iron levels and redox state of the cell (for review, see Klausner and Harford 1989). Similar regulation likely pertains to a sarcotoxin IIA mRNA binding protein in *Sarcophaga peregina* (flesh fly), which is homologous to mitochondrial 3-oxoacyl-CoA thiolase (Nanbu et al. 1993), or AU-H, a hydratase with IL-3 mRNA-binding capability (Nakagawa et al 1995).

12 Purification

Although the purification of AUBPs has been slow, recent advances suggest that substantial progress will be made shortly. Classical methods such as affinity and size-exclusion chromatography, while still utilized, are often preludes to newer methods, which exploit a knowledge of the proteins' subcellular localization and RNA-binding affinity.

The choice of starting material is made based on its AUBP content and availability. While whole, normal tissue sources may be easily obtainable in large quantities, levels of AUBP activity are typically low; brain may be a notable exception. Large-scale culture of tumor cells can be used due to their high AUBP expression, especially after appropriate stimulation with mitogens. Based on the induction pattern of the protein's activity, different cell conditions may be preferred to specifically upregulate the activity of a particular AUBP.

Preliminary steps in AUBP purification are often designed to reduce the amount of contaminating protein while still maintaining the desired activity. Often, a small-scale purification is attempted, in an effort to identify steps that have higher yields, maintain more activity, or produce greater degrees of purification. These initial methods include chemical procedures such as ammonium sulfate precipitation or heat denaturation, and classical chromatographic techniques such as ion exchange, size exclusion, hydrophobic interaction, and HPLC chromatography. Once the bulk of contaminating protein has been removed, more specific methods, such as RNA affinity chromatography, may be applied.

Subcellular fractionation can eliminate large amounts of contaminating material. For example, AUBPs may display either cytoplasmic or nuclear localization. A single centrifugation step can separate these two fractions, and results in a quick purification with small losses of the desired activity. Cytoplasmic AUBPs are often localized to polysomes, suggesting preliminary steps to obtain high quality polysomes would merit time spent on fractionation procedures. While crude lysates can be made in kilogram quantities, it is difficult to obtain large amounts of subcellular fractions. In addition, the starting material must be very fresh and without appreciable degradation. Therefore, low-abundance proteins may be difficult to purify using this method.

Another purification strategy utilizes the RNA-binding specificity of AUBPs. The target RNA is coupled to a solid support, such as agarose beads, after in vitro transcription. Three dominant approaches, are (1) the inclusion of modest amounts of biotinylated nucleotide with binding to strepavidin-agarose beads (Rouault et al. 1989; Rajagopalan and Malter 1994), (2) chemically cross-linking the RNA to the Sepharose beads with cyanogen bromide (CNBr) (Gorospe and Baglioni 1994), or (3) transcribing and polyadenylating the target mRNA in vitro, and annealing it to commercially available oligo(dT)-agarose beads (Nakagawa et al. 1995).

In general, RNA affinity chromatography requires care to (1) avoid ribonuclease contamination which can destroy the ligand RNA, (2) make sufficient quantities of RNA in vitro, and (3) attach the RNA to the matrix while maintaining its ability to bind protein. RNase destruction can be reduced by keeping all reactions at 4 °C, and including commercially available ribonuclease inhibitors in each step. In addition, the use of traditional chromatographic steps prior to RNA-affinity work may decrease RNase contamination. The use of synthetic riboligomers, with chemically different, ribonuclease-resistant backbones, while expensive, eliminates this concern. Recent developments in biotechnology have significantly improved the yields from in vitro transcription kits, and allow for the production of sufficient RNA in several reaction batches. The greatest difficulty lies in attaching the RNA to the matrix. Annealing a polyadenylated target mRNA to an oligo(dT)-Sepharose matrix requires specific ionic conditions which may be incompatible with the protein's binding activity or assay conditions. The use of chemicals such as CNBr to couple the RNA is indiscriminate and results in a variety of linkage sites, often several per molecule, which may interfere with AUBP binding. The incorporation of biotinylated nucleotides (bio-NTPs) allows for some optimization by altering the NTP : bio-NTP nucleotide ratio, in the hopes of having one biotin per RNA, and avoiding the incorporation of a bio-NTP into the binding site. Some workers have added a poly(U) tail to the RNA ligand into which bio-UTP can be preferentially incorporated (Ruby and Abelson 1988).

Once the RNA has been coupled to the matrix, crude lysate (or a subcellular fraction) is incubated with the beads to allow protein binding to occur. The beads can then be collected (either on a magnet if paramagnetic beads are used, or in a column) and washed several times with buffer alone, then buffer containing competitor RNAs to remove nonspecifically bound proteins. The protein is then eluted off the beads with detergent or high salt washes. Unfortunately, many proteins share similar binding specificities (such as the entire family of AUBPs), and will therefore most likely copurify. However, this can be a successful part of a multistep purification strategy.

The purification of AUBPs is akin to that of an enzyme; the desired protein is followed at every step by assays of its activity. For example, fractions can be dialyzed against a binding buffer, and used in gel-mobility shift assays to measure activity. Unfortunately, the active portion may represent only a small percentage of the total protein, and thus much inactive protein may be lost at each step. All purification steps must be carried out under gentle conditions of temperature and ionic strength in order to maintain AUBP activity.

Several examples attest to the power of these methods. The IRE-BP was purified by using an IRE-containing, biotinylated RNA coupled to strepavidin agarose beads in a column (Rouault et al. 1989). AUF-1 was initially purified by taking a heparin-agarose purification of the S130 subcellular fraction from K562 cells, and passing this through a poly(U)-agarose column (Zhang et al. 1993). Currently, AUBF is being purified by first using several classical affinity and size-exclusion chromatographies, and then incubating this partially purified material with $(AUUUA)_4$-containing RNA coupled to streptavidin paramagnetic beads (Sebkova and Malter, unpubl. data).

AU-H was purified by passing human brain homogenate through a column consisting of a polyadenlyated $(AUUUA)_6$ RNA (based on the protein's IL-3 mRNA target) bound to an oligo(dT)-cellulose matrix (Nakagawa et al. 1995). Nonspecific proteins flowed through the column, whereas specific binding proteins were retained by the column and could later be removed with high salt washes. The eluate was then dialyzed, concentrated, and microsequenced to facilitate cloning efforts.

13 Cloning

Cloning efforts have been hampered by the lack of purified protein; however, both AUF-1 and Hel-N1 have been cloned by screening expression libraries with polyclonal antibodies against purified protein (Levine et al. 1993; Zhang et al. 1993). Due to the difficulty of purifying AUBPs, most proteins, such as AUBF and AU-H, are being purified in sufficient quantities for microsequencing. Peptide sequences then direct the synthesis of synthetic peptides or degenerate oligonucleotides. These, in turn, are used to produce monoclonal antibodies or cDNAs, which may be used to screen an expression library.

Once a putative cDNA clone, or highly purified fractions, of an AUBP has been obtained, the ultimate proof of identity lies in demonstrating activity. This can be shown through a binding assay, such as Northwestern blotting or gel shifts with recombinant protein. One laboratory has begun searching for RNA-binding proteins based on RNA target specificity (Qian and Wilusz 1993). They screened cDNA expression libraries by Northwestern blotting, and isolated DSEF-1, a protein with three RRMs, which bound to their probe, the DSEF-1 RNA recognition site. Since the RNA probe can easily be produced in vitro from any sequence, this method provides a rapid way of searching for *trans*-acting factors. One potential disadvantage is that some nucleic acid-binding proteins may lose binding specificity when immobilized on a membrane, thereby creating a false positive background (Zaidi and Malter 1995).

By screening a subtracted cDNA library with radiolabeled AUUUA RNA, one laboratory attempted to identify proteins whose mRNAs might be post-transcriptionally regulated through the ARE-mediated pathway (Gaugitsch et al. 1992). They cloned an integral membrane protein with a 3-kb 3′ UTR containing an ARE. Its mRNA was highly unstable in resting PBMC, myeloid, and lymphoid cell lines, but was rapidly induced upon cell stimulation. This procedure, in addition to PCR cloning with consensus ARE or RRM oligonucleotides, may allow the identification of other members of the ARE mRNA and binding protein family.

Another method of exploiting AUBP-binding specificity is to determine the consensus binding sequence once a protein has been obtained. Once Hel-N1 was purified, a random RNA library was screened to determine the optimal ligand. Briefly, the RNA pool was incubated with recombinant protein, and then immunoprecipitated using an epitope tag (Levine et al. 1993). The bound RNAs were amplified via reverse transcription and PCR, and again selected by the same method. Finally, the RNA pool was reverse transcribed, and the cDNAs cloned and sequenced. This procedure demonstrated that Hel-N1 had a high affinity for short U-rich sequences, such as those found in the 3′ UTRs of several ARE RNAs.

14 Sequence Information

Analysis of the amino acid sequences of cloned AUBPs reveals several similarities among this family of proteins. Most striking is the conservation of RNA recognition motifs (RRMs), which are glutamine-rich regions believed to be involved in RNA-protein interactions, and found in a number of diverse RNA-binding proteins, most notably the hnRNPs (A0, A1, C1, and C2) (Burd and Dreyfuss 1994), and AUF1 (Zhang et al. 1993). AU-H, however, lacks canonical RRMs, suggesting that other domains are also functional.

In general, the members of the ARE BP family share some common characteristics. They tend to be small (< 50 kDa), and are often posttranslationally regulated via phosphorylation (AUBF and AUF-1), methylation (hnRNPs), or redox conditions (AUBF). These proteins can usually be found distributed between the nuclear and cytoplasmic fractions, with the transport between compartments often correlated with activity. These proteins bind to similar RNA sequences, and often share protein homology in the form of RRMs. Some AUBPs (AU-H, and sarcotoxin IIA mRNA binding protein) have both RNA-binding and enzymatic functions.

15 ARE-Mediated Decay Model

Enough information about the AUBPs and their target mRNA sequences is now known to develop models of the ARE-mediated labile mRNA degradation pathway. In unstimulated and resting cells, basal transcription of protooncogene and growth factor sequences is low, and any mRNAs which are produced are degraded rapidly. These messages are transcribed, spliced, and transported out of the nucleus, possibly in association with hnRNP A0, A1, C1, and/or C2 proteins. Both spliced and unspliced transcripts are stabilized upon cell activation, suggesting that at least one of the stabilizing activities originates in the nucleus (Vakalopoulou et al. 1991). In the cytoplasm, the mRNAs are then loaded onto polysomes in anticipation of translation.

Recently, a 65-kDa, polysome-associated RNase E counterpart has been identified in mammalian cell lines, and shown to specifically cleave the sequence AUU/UA in vitro (Wennborg et al. 1995). The authors demonstrated a direct correlation between the number of AUUUA motifs in the substrate RNA and the rate of cleavage in vitro, suggesting an explanation for the short in vivo half-life of mRNAs containing multiple AUUUA repeats in their 3′ UTRs.

If such a model is correct, then ARE RNA decay rates are a function of both AUBP protection and RNase degradation. As mRNAs enter the cytoplasm, they are protected by nuclear AUBPs, such as the hnRNPs, and are loaded onto polysomes. Here, both AUBPs and ribonucleases are present, and the abundance or activity of each may be the control point for labile mRNA decay. These protein factors may compete for binding to similar recognition sequences, or the binding of an AUBP may alter a secondary recognition site. In either case, the presence of an AUBP may block ribonuclease cleavage or recognition. Unprotected ARE mRNAs on polysomes may be cleaved in the 3′ UTR by a single enzymatic step, or gradually deadenylated by a processive enzyme.

Translation may displace the AUBPs from the 3′ UTR long enough to allow an endoribonucleolytic snip by an AU-specific endonuclease (Bandyopadhyay et al. 1990; Wennborg et al. 1995; see Fig. 3), which would destroy the AUBP-binding site, and serve to separate the coding body from the poly(A) tail (Decker and Parker 1993). Some evidence exists to support the accumulation of such deadenylated transcripts in vitro (Brewer and Ross 1988) and in vivo (Stoeckle 1991). Such mRNAs are not translated, but are either decapped and undergo $5' \rightarrow 3'$ exonucleolytic digestion (Muhlrad et al. 1994), or, alternatively, are directly degraded $3' \rightarrow 5'$ by a processive exoribonuclease (Decker and Parker 1994).

Alternatively, the unbound ARE might signal a processive enzyme to begin poly(A) tail shortening. This deadenylation, while only seen in yeast, may be a common step in eukaryotic mRNA turnover. Without a poly(A) tail, the message is no longer translated, and probably dissociates from the complex once the transiting ribosomes finish. In the cytoplasm, the mRNA may be targeted for rapid decay by a missense turnover pathway.

Both of these models would predict biphasic decay kinetics, as the two degradation steps – on the polysome and in the cytoplasm – need not occur at equal rates. If ARE cleavage or deadenylation on polysomes was rate-limiting, decay intermediates would not be seen, as the cytoplasmic machinery would rapidly remove them. However, if the final cleavage steps along missense pathway were slower, the deadenylated intermediates formed on polysomes should accumulate in the cytoplasm. Because decay experiments have been performed in a variety of in vitro and in vivo systems, and in different cell and extract types, the exact mechanism of labile mRNA turnover remains unclear.

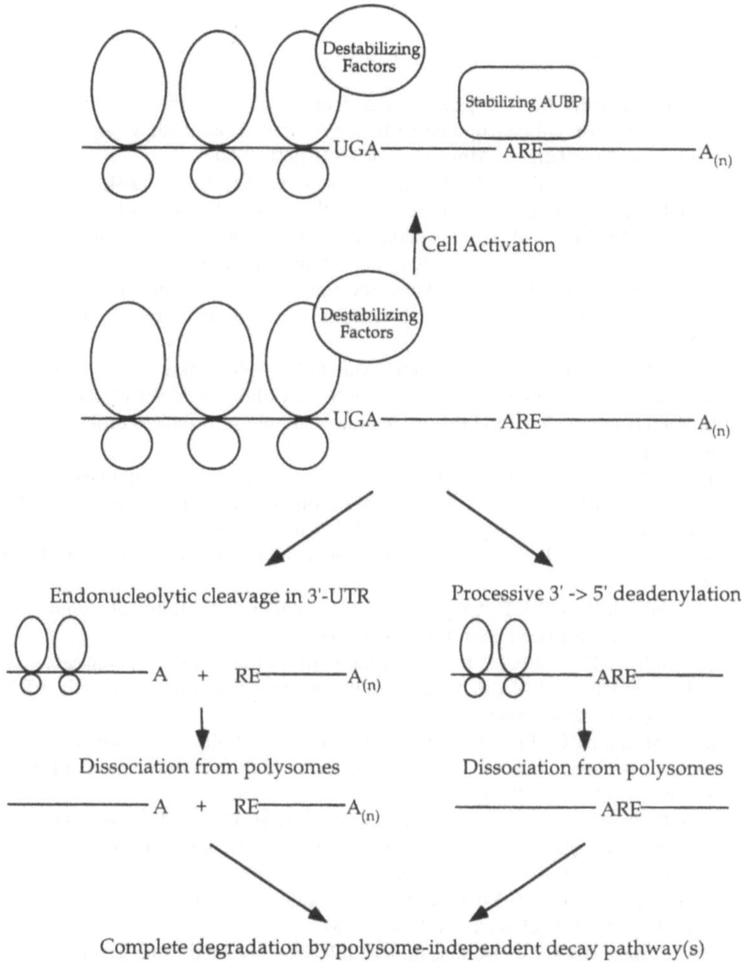

Fig. 3. Polysome-bound ARE mRNAs are targeted for rapid turnover in the absence of stabilizing AUBPs. The initial event in polysome-associated turnover is either a processive deadenylation or an endonucleolytic cleavage in the ARE. Poor translational reinitiation of these decay intermediates leads to their dissociation from polysomes, followed by complete degradation

Under specific cell conditions, the ARE may be occupied by stabilizing or destabilizing factors. Stabilizing factors, such as AUBF, most likely obscure the AUUUA motifs from both nucleases and destabilizing factors. Presumably, deadenylation is also slowed with ultimate inhibition of body cleavage. Destabilizing factors, such as AUF-1, presumably target the mRNA for destruction by directing the action of nucleases. Thus, the interplay at the same site between multiple factors may ultimately determine the rapidity of ARE-containing RNA decay.

References

Aharon T, Schneider RJ (1993) Selective destabilization of short-lived mRNAs with the granulocyte-macrophage colony-stimulating factor AU-rich 3′ noncoding region is mediated by a cotranslational mechanism. Mol Cell Biol 13(3):1971–1980

Akashi M, Shaw G, Hachiya M, Elstner E, Suzuki G, Koeffler P (1994) Number and location of AUUUA motifs: role in regulating transiently expressed RNAs. Blood 83(11):3182–3187

Aksamit TR, Monick MM, Hunninghake GW (1993) Protein kinase C modulates the amounts of IL-1 receptor mRNA in human lung fibroblasts. J Immunol 151(1):284–290

Algate PA, McCubrey JA (1993) Autocrine transformation of hemopoietic cells resulting from cytokine message stabilization after intracisternal A particle transposition. Oncogene 8:1221–1232

Alterman RBM, Ganguly S, Schulze DH, Marzluff WF, Schildkraut CL, Skoultchi AI (1984) Cell cycle regulation of mouse H3 histone mRNA metabolism. Mol Cell Biol 4:4123–4132

Altus MS, Nagamine Y (1991) Protein synthesis inhibition stabilizes urokinase-type plasminogen activator mRNA. J Biol Chem 266(31):21190–21196

Amara FM, Chen FY, Wright JA (1994) Phorbol ester modulation of a novel cytoplasmic protein binding activity at the 3′-untranslated region of mammalian ribonucleotide reductase R2 mRNA and role in message stability. J Biol Chem 269(9):6709–6715

Atwater JA, Wisdom R, Verma IM (1990) Regulated mRNA stability. Annu Rev Genet 24:519–541

Aviv H, Valloch V, Bastos R, Levy S (1976) Biosynthesis and stability of globin mRNA in cultured erythroleukemia Friend cells. Cell 8:495–503

Bandyopadhyay R, Coutts M, Krowczynska A, Brawerman G (1990) Nuclease activity associated with mammalian mRNA in its native state: possible basis for selectivity in mRNA decay. Mol Cell Biol 10(5):2060–2069

Battey J, Moudling C, Taub R, Murphy W, Stewart T, Potter H, Lenoir G, Leder P (1983) The human *c-myc* oncogene: structural consequences of translocation into the IgH locus in Burkitt lymphoma. Cell 34:779–787

Bauer S, Piechaczyk M, Nepveu A, Marcu K, Nordan R, Potter M, Mushinski F (1989) Mutations present in two plasmacytomas increase abundance of *c-myc* mRNA by increasing both transcription and mRNA stability. Oncogene 4:615–623

Berger LC, Cooper HL (1975) Very short-lived and stable mRNAs from resting human lymphocytes. Proc Natl Acad Sci USA 72:3873–3877

Berger LC, Sells BH (1992) Modification of mRNA-associated proteins during changes in growth conditions. Biochem Cell Biol 70:528–534

Bickel M, Cohen RB, Pluznik DH (1990) Post-transcriptional regulation of granulocyte-macrophage colony-stimulating factor synthesis in murine T cells. J Immunol 145(3):840–845

Binder R, Horowitz JA, Basilion JP, Koeller DM, Klausner RD, Harford JB (1994) Evidence that the pathway of transferrin receptor mRNA degradation involves an endonucleolytic cleavage within the 3′ UTR and does not involve poly(A) tail shortening. EMBO J 13(8):1969–1980

Bohjanen PR, Petryniak B, June CH, Thompson CB, Lindsten T (1991) An inducible cytoplasmic factor (AU-B) binds selectively to AUUUA multimers in the 3′ untranslated region of lymphokine mRNA. Mol Cell Biol 11(6):3288–3295

Bohjanen PR, Petryniak B, June CH, Thompson CB, Lindsten T (1992) AU RNA-binding factors differ in their binding specificities and affinities. J Biol Chem 267(9):6302–6309

Bonnieu A, Piechaczyk M, Marty L, Cuny M, Blanchard J, Fort P, Jeanteur P (1988) Sequence determinants of *c-myc* mRNA turnover: influence of 3′ and 5′ noncoding regions. Oncogene Res 3:155–166

Bonnieu A, Rech J, Jeanteur P, Fort P (1989) Requirements for *c-fos* mRNA down regulation in growth stimulated murine cells. Oncogene 4:881–888

Both GW, Furuichi Y, Muthukrishnan S, Shatkin AJ (1975) Ribosome binding to reovirus mRNA in protein synthesis requires 5′ terminal 7-methylguanosine. Cell 9:185–195

Brawerman G (1989) mRNA decay: finding the right targets. Cell 57:9–10

Brewer G (1991) An A+U rich element RNA-binding factor regulates *c-myc* mRNA stability in vitro. Mol Cell Biol 11(5):2460–2466

Brewer G, Ross J (1988) Poly(A) shortening and degradation of the 3′ A+U-rich sequences of human *c-myc* mRNA in a cell-free system. Mol Cell Biol 8(4):1697–1708

Burd CG, Dreyfuss G (1994) Conserved structures and diversity of functions of RNA-binding proteins. Science 265:615–621

Caput D, Beutler B, Hartog K, Thayer R, Brown-Shimer S, Cerami A (1986) Identification of a common nucleotide sequence in the 3′-untranslated region of mRNA molecules specifying inflammatory mediators. Proc Natl Acad Sci USA 83:1670–1674

Carter BZ, Malter JS, (1991) Regulation of mRNA stability and its relevance to disease. Lab Invest 65(6):610–621

Chen CA, Shyu A (1994) Selective degradation of early-response gene mRNAs: functional analyses of sequence features of the AU-rich elements. Mol Cell Biol 14(12):8471–8482

Chen CA, Chen T, Shyu A (1994) Interplay of two functionally and structurally distinct domains of the *c-fos* AU-rich element specifies its mRNA-destabilizing function. Mol Cell Biol 14(1):416–426

Chen CA, Xu N, Shyu A (1995) mRNA decay mediated by two distinct AU-rich elements from *c-fos* and granulocyte-macrophage colony-stimulating factor transcripts: different deadenylation kinetics and uncoupling from translation. Mol Cell Biol 15(10):5777–5788

Chen FY, Amara FM, Wright JA (1994) Regulation of mammalian ribonucleotide reductase R1 mRNA stability is mediated by a ribonucleotide reductase R1 mRNA 3′-untranslated region *cis-trans* interaction through a protein kinase C-controlled pathway. Biochem J 302:125–132

Chen SJ, Holbrook NJ, Mitchell KF, Vallone CA, Greengard JS, Crabtree GR, Lin Y (1985) A viral long terminal repeat in the interleukin 2 gene of a cell line that constituitively produces interleukin 2. Proc Natl Acad Sci USA 82:7284–7288

Choi YD, Grabowski PJ, Sharp PA, Dreyfuss G (1986) Heterogeneous nuclear ribonucleoproteins: role in RNA splicing. Science 231:1534–1539

Curatola AM, Nadal MS, Schneider RJ (1995) Rapid degradation of AU-rich element (ARE) mRNAs is activated by ribosome transit and blocked by secondary structure at any position 5′ to the ARE. Mol Cell Biol 15(11):6331–6340

Dani C, Blanchard JM, Piechaczyk M, El Sabrouty S, Marty L, Jeanteur P (1984) Extreme unstability of *c-myc* RNA in normal and tumor cells. Proc Natl Acad Sci USA 81:7046–7050

Decker CJ, Parker R (1993) A turnover pathway for both stable and unstable mRNAs in yeast: evidence for a requirement for deadenylation. Genes Dev 7:1632–1643

Decker CJ, Parker R (1994) Mechanisms of mRNA degradation in eukaryotes. TIBS 19:336–340

Ding D, Parkhurst SM, Halsell SR, Lipshitz HD (1993) Dynamic *hsp*83 RNA localization during *Drosophila* oogenesis and embryogenesis. Mol Cell Biol 13:3773–3781

Eick D, Piechaczyk M, Heinglein B, Blanchard JM, Traub B, Koffer E, Wiest S, Lenoir G, Bornkamm GW (1985) Aberrant *c-myc* RNAs of Burkitt's lymphoma cells have longer half-lives. EMBO J 4:3717–3725

Fort P, Rech J, Vie A, Piechaczyk M, Bonnieu A, Jeanteur P, Blanchard JM (1987) *c-fos* expression in serum-stimulated hamster fibroblasts: transcriptional elongation block and mRNA stability. Nucl Acids Res 15:5657–5667

Freedman LP, Luisi BF, Korszun ZR, Basavappa R, Sigler PB, Yamamoto KR (1988) The function and structure of the metal coordination sites within the glucocorticoid receptor DNA binding domain. Nature 334:543–546

Gaugitsch HW, Prieschl EE, Kalthoff F, Huber NE, Baumruker T (1992) A novel transiently expressed, integral membrane protein linked to cell activation. J Biol Chem 267(16):11267–11273

Geck P, Whitaker SA, Medveczky MM, Medveczky PG (1994) Small RNA expression from the oncogenic region of a highly oncogenic strain of herpesvirus saimiri. Virus Genes 8:25–34

Gillis P, Malter JS (1991) The adenosine-uridine binding factor recognizes the AU-rich elements of cytokine, lymphokine, and oncogene mRNAs. J Biol Chem 266(5):3172–3177

Goosen B, Hentze MW (1992) Position is the critical determinant for function of iron-responsive elements as translational regulators. Mol Cell Biol 12:1959–1966

Gorlach M, Burd CG, Dreyfuss G (1994) The determinants of RNA-binding specificity of the heterogeneous nuclear ribonucleoprotein C proteins. J Biol Chem 269(37):23074–23078

Gorospe M, Baglioni C (1994) Degradation of unstable interleukin-1α mRNA in a rabbit reticulocyte cell-free system. J Biol Chem 269(16):11845–11851

Gorospe M, Kumar S, Baglioni C (1993) Tumor necrosis factor increases stability of interleukin-1 mRNA by activating protein kinase C. J Biol Chem 268(9):6214–6220

Gough NM, Metcalf D, Gough J, Grail D, Dunn AR (1984) Molecular cloning of cDNA encoding a murine haematopoietic growth regulator granulocyte-macrophage colony stimulating factor. Nature 309:763–767

Grafi G, Sela I, Galili G (1993) Translational regulation of human beta interferon mRNA: association of the 3′ AU-rich sequence with the poly(A) tail reduces translation efficiency in vitro. Mol Cell Biol 13(6):3487–3493

Gray PW, Leung DW, Pennica D, Yelverton E, Najarian R, Simonsen CC, Derynck, Sherwood PJ, Wallace DM, Berger SL, Levinson AD, Goeddel DV (1982) Expression of human interferon cDNA in *E. coli* and monkey cells. Nature 295:503–508

Greenberg ME, Ziff EB (1984) Stimulation of 3T3 cells induces transcription of the *c-fos* proto-oncogene. Nature 311:433–438

Hamilton BJ, Nagy E, Malter JS, Arrick BA, Rigby WFC (1993) Association of heterogeneous ribonucleoprotein A1 and C proteins with reiterated AUUUA sequences. J Biol Chem 268(12):8881–8887

Hargrove JL, Schmidt FH (1989) The role of mRNA and protein stability in gene expression. FASEB J 3(12):2360–2370

Herrick DJ, Ross J (1994) The half-life of c-*myc* mRNA in growing and serum-stimulated cells: influence of the coding and 3′ untranslated regions and role of ribosome translocation. Mol Cell Biol 14(3):2119–2128

Hirsch HH, Nair APK, Moroni C (1993) Suppressible and nonsuppressible autocrine mast cell tumors are distinguished by insertion of an endogenous retroviral element (IAP) into the interleukin-3 gene. J Exp Med 178:403–411

Huang LY, Tholanikunnel BG, Vakalopoulou E, Malbon CC (1993) The M_r 35,000 β-adrenergic receptor mRNA-binding protein induced by agonists requires both an AUUUA pentamer and U-rich domains for RNA recognition. J Biol Chem 268(34):25769–25775

Iwai Y, Bickel M, Pluznik DH, Cohen RB (1991) Identification of sequences within the murine granulocyte-macrophage colony-stimulating factor mRNA 3′-untranslated region that mediate mRNA stabilization induced by mitogen treatment of EL-4 thymoma cells. J Biol Chem 266(27):17959–17965

Iwai Y, Akahane K, Pluznik DH, Cohen RB (1993) Ca^{2+} ionophore A23187-dependent stabilization of granulocyte-macrophage colony-stimulating messenger RNA in murine thymoma EL-4 cells is mediated through two distinct regions in the 3′-untranslated region. J Immunol 150(10):4386–4394

Jones TR, Cole MD (1987) Rapid cytoplasmic turnover of *c-myc* mRNA: requirement of the 3′-untranslated sequences. Mol Cell Biol 7(12):4513–4521

Kabnick KS, Housman DE (1988) Determinants that contribute to cytoplasmic stability of human *c-fos* and β-globin mRNAs are located at several sites in each mRNA. Mol Cell Biol 8:3244–3250

Kaempfer R, Efrat S (1985) Regulation of human IL-2 gene expression. In: Song C, Schimpl A (eds) Cellular and molecular biology of lymphokines. Academic Press, Orlando

Kafatos F (1972) mRNA stability and cellular differentiation. Acta Endocrinol 168:319–345

Katz DA, Theodorakis NG, Cleveland DW, Lindsten T, Thompson CB (1994) AU-A, an RNA-binding activity distinct from hnRNP A1, is selective for AUUUA repeats and shuttles between the nucleus and the cytoplasm. Nucl Acids Res 22(2):238-246

Kelly K, Cochran B, Stiles CD, Leder P (1983) Cell-specific regulation of the *c-myc* gene by lymphocyte mitogens and platelet-derived growth factor. Cell 35:603–610

Klausner RD, Harford JB (1989) *Cis-trans* models for post-transcriptional gene regulation. Science 246:870–872

Kozak M (1987) At least six nucleotides preceding the AUG initiator codon enhance translation in mammalian cells. J Mol Biol 196:947–950

Kozak M (1991) Structural features in eukaryotic mRNAs that modulate the initiation of translation. J Biol Chem 266(30):19867–19870

Kruys V, Marinx O, Shaw G, Deschamps J, Huez G (1989) Translational blockade imposed by cytokine-derived UA-rich sequences. Science 245 :852–855

Lagnado CA, Brown CY, Goodall GJ (1994) AUUUA is not sufficient to promote poly(A) shortening and degradation of an mRNA: the functional sequence within AU-rich elements may be UUAUUUA(U/A)(U/A). Mol Cell Biol 14(12):7984–7995

Lerner MR, Steitz JA (1979) Antibodies to small nuclear RNAs complexed with proteins are produced by patients with systemic lupus erythematosus. Proc Natl Acad Sci USA 76:5495–5499

Levine TD, Gao F, King PH, Andrews LG, Keene JD (1993) Hel-N1: an autoimmune RNA-binding protein with specificity for 3′ uridylate-rich untranslated regions of growth factor mRNAs. Mol Cell Biol 13(6):3494–3504

Lindsten T, June CH, Ledbetter JA, Stella G, Thompson CB (1989) Regulation of lymphokine messenger RNA stability by a surface-mediated T cell activation pathway. Science 244:339–345

Liu Q, Dreyfuss G (1995) In vivo and in vitro arginine methylation of RNA-binding proteins. Mol Cell Biol 15(5):2800–2808

Malter JS (1989) Identification of an AUUUA-specific messenger RNA binding protein. Science 246:664–666

Malter JS, Hong Y (1991) A redox switch and phosphorylation are involved in the post-translational up-regulation of the adenosine-uridine binding factor by phorbol ester and ionophore. J Biol Chem 266(5):3167–3171

Malter JS, McCrory WA, Wilson M, Gillis P (1990) Adenosine-uridine binding factor requires metals for binding factor requires metals for binding to granulocyte-macrophage colony-stimulating factor mRNA. Enzyme 44:203–213

Meijlink F, Curran T, Miller AD, Verma IM (1985) Removal of a 67 base-pair sequence in the nonconing, region of proto-oncogene *c-fos* converts it to a transforming gene. Proc Natal Acad Sci USA 82:4987–4991

Metcalf D (1985) The granulocyte-macrophage colony-stimulating factor. Nature 229:16–22

Miller AD, Curran T, Verma IM (1984) *c-fos* protein can induce cellular transformation: a novel mechanism of activation of a cellular oncogene, Cell 36:51–60

Muhlrad D, Decker CJ, Parker R (1994) Deadenylation of the unstable mRNA encoded by the yeast MFAZ gene leads to decapping followed by 5′ → 3′ digestion of the transcript. Genes Dev 8:855–866

Müller WEG, Slor H, Pfeifer K, Hühn P, Bek A, Orsulic S, Ushijima H, Schröder HC (1992) Association of AUUUA-binding protein with A+U-rich mRNA during nucleo-cytoplasmic transport. J Mol Biol 226:721–733

Mullner EW, Kuhn LC (1988) A stem-loop in the 3′ untranslated region mediates iron-denpendent regulation of transferrin receptor mRNA stability in the cytoplasm. Cell 53:815–825

Myer VE, Steitz, JA (1995) Isolation and characterization of a novel, low abundance hnRNP protein: A0. RNA 1(2):1995

Myer VE, Lee SI, Steitz JA (1992) Viral small nuclear ribonucleoproteins bind a protein implicated in messenger RNA destabilization. Proc Natl Acad Sci USA 89:1296–1300

Nakagawa J, Waldner H, Meyer-Monard S, Hofsteende, Jeno P, Moroni C (1995) AUH, a gene encoding an AU-specific RNA binding protein with intrinsic enoyl-CoA hydratase activity. Proc Natl Acad Sci USA 92:2051–2055

Nanbu R, Kubo T, Hashimoto T, Natori S (1993) Purification of an AU-rich RNA binding protein from *Sarcophaga peregrina* (flesh fly) and its identification as a thiolase. J Biochem 114:432–437

Peltz SW, Brewer G, Bernstein PJ, Hart PA, Ross J (1991) Regulation of mRNA turnover in eukaryotic cells. Crit Rev Euk Gene Exp 1:99–126

Piechaczyk M, Yang JQ, Blanchard JM, Jeanteur P, Marcu K (1985) Post-translational mechanisms are responsible for accumulation of truncated *c-myc* RNAs in murine plasma cell tumors. Cell 42:589–597

Piñol-Roma S, Dreyfuss G (1993) Cell cycle-regulated phosphorylation of the pre-mRNA-binding (heterogeneous nuclear ribonucleoprotein) C proteins. Mol Cell Biol 13(9):5762–5770

Qian Z, Wilusz J (1993) Cloning of a cDNA encoding an RNA binding protein by screening expression libraries using a Northwestern strategy. Anal Biochem 212(2):547–554

Raghow R (1987) Regulation of messenger RNA turnover in eukaryotes. TIBS 12:358–360

Rajagopalan LE, Malter JS (1994) Modulation of granulocyte-macrophage colony-stimulating factor mRNA stability in vitro by the adenosine-uridine binding factor. J Biol Chem 269 (39): 23882–23888

Rajagopalan LE, Burkholder JK, Turner J, Culp J, Yang NS, Malter JS (1995) Granulocyte-macrophage colony-stimulating factor mRNA stabilization enhances transgenic expression in normal cells and tissues. Blood 86(7):2551–2558

Ross HJ, Sato N, Ueyama Y, Koeffler HP (1991) Cytokine messenger RNA stability is enhanced in tumor cells. Blood 77(8):1787–1795

Rouault TA, Hentze MW, Haile DJ, Harford JB, Klausner RD (1989). The iron-responsive element binding protein: a method for the affinity purification of a regulatory RNA-binding protein. Proc Natl Acad Sci USA 86:5768–5772

Roy N, Laflamme G, Raymond V (1992) 5′ untranslated sequences modulate rapid mRNA degradation mediated by 3′ AU-rich element in *v-/c-fos* recombinants. Nucl Acids Res 20(21): 5753–5762

Ruby SW, Abelson J (1988) An early hierarchic role of U1 small nuclear ribonucleoprotein in spliceosome assembly. Science 242:1028–1035

Saini KS, Summerhayes IC (1991) Regulation of the turnover of mRNAs encoding cellular oncoproteins. Biochem Cell Biol 69:415–417

Salles FJ, Darrow AL, O'Connell ML, Strickland S (1992) Isolation of novel murine maternal mRNAs regulated by cytoplasmic polyadenylation. Genes Dev 6:1202–1211

Savant-Bhonsale S, Cleveland DW (1992) Evidence for instability of mRNAs containing AUUUA motifs mediated through translation dependent assembly of a >20S degradation complex. Genes Dev 6:1927–1939

Scheper W, Meinsma D, Holthuizen PE, Sussenbach JS (1995) Long-range RNA interaction of two sequence elements required for endonucleolytic cleavage of human insulin-like growth factor II mRNAs. Mol Cell Biol 15(1):235–245

Schiavi SC, Belasco JG, Greenberg ME (1992) Regulation of proto-oncogene mRNA stability. Biochem Biophys Acta 1114:95–106

Schiavi SC, Wellington CL, Shyu A, Chen CA, Greenberg ME, Belasco JG (1994) Multiple elements in the *c-fos* coding region facilitate mRNA deadenylation and decay by a mechanism coupled to translation. J Biol Chem 269(5):3441–3448

Shaw G, Kamen R (1986) A conserved AU sequence from the 3′ untranslated region of GM-CSF mRNA mediates selective mRNA degradation. Cell 46:659–667

Sheets MD, Fox CA, Hunt T, Vande Woude G, Wickens M (1994) The 3′- untranslated regions of *c-mos* and cyclin mRNAs stimulate translation by regulating cytoplasmic polyadenylation. Genes Dev 8:926–938

Shyu A, Greenberg ME, Belasco JG (1989) The *c-fos* transcript is targeted for rapid decay by tow distinct mRNA degradation pathways. Genes Dev 3:60–72

Singer RH, Penman S (1973) Messenger RNA in HeLa cells: Kinetics of formation and decay. J Mol Biol 78:321–334

Siomi H, Dreyfuss G (1995) A nuclear localization domain in the hnRNP A1 protein. J Cell Biol 129(3):551–560

St Johnston D (1995) The intracellular localization of messenger RNAs. Cell 81:161–170

Stephens J, Carter BZ, Pekala P, Malter JS (1992) Tumor necrosis factor α-induced glucose transporter (GLUT-1) mRNA stabilization: regulation by the adenosine uridine binding factor (AUBF). J Biol Chem 267(12):8336–8341

Stoeckle MY (1991) Post-transcriptional regulation of gro alpha, beta, gamma, and IL-8 mRNAs by IL-1 beta. Nucl Acids Res 19:917–920

Stoecklin G, Hahn S, Moroni C (1994) Functional hierarchy of AUUUA motifs in mediating rapid interleukin-3 mRNA decay. J Biol Chem 269:28591–28597

Szabo A, Dalmau J, Manley G, Rosenfeld M, Wong E, Henson J, Posner B, Furneaux HM (1991) HuD, a paraneoplastic encephalomyelitis antigen, contains RNA-binding domains and is homologous to *elav* and *sex-lethal*. Cell 67:629–639

Taniguchi T (1988) Regulation of cytokine gene expression. Annu Rev Immunol 6:439–464

Vakalopoulou E, Schaack J, Shenk T (1991) A 32-kilodalton protein binds to AU-rich domains in the 3′ untranslated regions of rapidly degraded mRNAs. Mol Cell Biol 11(6):3355–3364

Van Straaten F, Muller R, Curran T, Van Beveren C, Verma IM (1983) Complete nucleotide sequence of a human *c-onc* gene: deduced amino acid sequence of the human *c-fos* protein. Proc Natl Acad Sci USA 80:3183–3187

Veyrune JL, Carillo S, Vie A, Blanchard JM (1995) *c-fos* mRNA instability determinants present within both the coding and the 3′ noncoding region link the degradation of this mRNA to its translation. Oncogene 11:2127–2134

Voelkerding KV, Steffen DW, Zaidi SHE, Malter JS (1995) Posttranscriptional regulation of the p53 tumor suppressor gene during growth-induction of human peripheral blood mononuclear cells. Oncogene 10:515–521

Volloch V, Housman D (1981) Stability of globin mRNA in terminally differentiating murine erythroleukemia cells. Cell 23:509-514

Walden WE, Daniels-McQueen S, Brown PH, Gaffield L, Russell DA, Bielser D, Bailey LC, Thach RE (1988) Translational repression in eukaryotes: partial purification and characterization of a repressor of ferritin mRNA translation. Proc Natl Acad Sci USA 85:9503–9507

Wang X, Kiledjian M, Weiss IM, Liebhaber SA (1995) Detection and characterization of a 3′ untranslated region ribonucleoprotein complex associated with human α-globin mRNA stability. Mol Cell Biol 15(3):1769–1777

Weiss IM, Liebhaber SA (1994) Erythroid cell-specific determinants of α-globin mRNA stability. Mol Cell Biol 14(12):8123–8132

Wennborg A, Sohlberg B, Angerer D, Klein G, von Gabain A (1995) A human RNase E-like activity that cleaves RNA sequences involved in mRNA stability control. Proc Natl Acad Sci USA 92:7322–7326

Wharton RP, Struhl G (1991). RNA regulatory elements mediate control of *Drosophila* body pattern by the posterior morphogen *nanos*. Cell 67:955-967

Wilson T, Treisman R (1988) Removal of poly(A) and consequent degradation of *c-fos* mRNA facilitated by 3′ AU-rich sequences. Nature 336:396-399

Winstall E, Gamache M, Raymond V (1995) Rapid mRNA degradation mediated by the *c-fos* 3′ AU-rich element and that mediated by the granulocyte-macrophage colony-stimulating factor 3′ AU rich element occur through similar polysome-associated mechanisms. Mol Cell Biol 15(7):3796–3804

Wisdom R, Lee W (1991) The protein-coding region of *c-myc* mRNA contains a sequence that specifies rapid mRNA turnover and induction by protein synthesis inhibitors. Genes Dev 5:232–243

Wodnar-Filipowicz A, Moroni C (1990) Regulation of interleukin-3 mRNA expression in mast cells occurs at the post-transcriptional level and is mediated by calcium ions. Proc Natl Acad Sci USA 290:777–781

Wong GG, Witek JS, Temple RA, Wilkins KM, Leary AC, Luxenberg DP, Jones SS, Brown EL, Kay EM, Orr EC, Shoemaker C, Golde DW, Kaufman RJ, Hewick RM, Wang EA, Clark SC (1985) Human GM-CSF: molecular cloning of the complementary DNA and purification of the natural and recombinant proteins. Science 228:810–815

Wreschner DH, Rechavi G (1988) Differential mRNA stability to reticulocyte ribonucleases correlates with 3′ non-coding $(U)_nA$ sequences. Eur J Biochem 172:333–340

Zaidi SHE, Malter JS (1994) Amyloid precursor protein mRNA stability is controlled by a 29-base element in the 3′-untranslated region. J Biol Chem 269:24007–24013

Zaidi SHE, Malter JS (1995) Nucleolin and hnRNP C proteins specifically interact with the 3′ untranslated region of amyloid protein precursor mRNA. J Biol Chem 270(29):17292–17298
Zhang W, Wagner BJ, Ehrenman K, Schaefer AW, Demaria CT, Crater D, DeHaven K, Long L, Brewer G (1993) Purification, characterization, and cDNA cloning of an AU-rich element RNA-binding protein AUF-1. Mol Cell Biol 13(12):7652-7665
Zubiaga AM, Belasco JG, Greenberg ME (1995) The nonamer UUAUUUAUU is the key AU-rich sequence that mediates mRNA degradation. Mol Cell Biol 15(4):2219-2230

Translational Control by Polyadenylation During Early Development

H.B. OSBORNE[1] and J.D. RICHTER[2]

1 Introduction

The basic molecular mechanisms that control gene expression are the same in germ cells and somatic cells. One of the peculiarities of early development, however, is that the crossing of certain checkpoints is associated with changes in the principal mode of control of gene expression. Hence, during early development specific, modes of gene regulation acquire a particular importance whereas in somatic cells they appear more as part of an ensemble.

Studies using species as divergent as marine invertebrates, mammals, amphibians and flies have shown that oocyte growth is accompanied by an accumulation of maternal mRNAs that are destined for later use (review Davidson 1986). The oocytes of most species are arrested at the first prophase of meiosis. Completion of meiosis with the emission of the second polar body is either a hormonally triggered event, such as in the mouse or in *Xenopus laevis*, resulting in a mature, fertilisable egg, or it is triggered by fertilisation, as in the surf clam *Spisula* (for review see Browder 1984). In general, release from prophase I of meiosis is associated with an arrest of transcriptional activity in the oocyte. This blockage continues in the early embryo for a period that is species-dependent. In the mouse embryo transcription is silent until the two-cell stage, or about 24 h after fertilisation (reviewed in Schultz 1993); in the embryos of *Xenopus Laevis* (Newport and Kirschner 1982) and *Drosophila* (Edgar and Schubiger 1986), expression from the zygotic genome becomes effective only after the 12th mitotic division, which is 6–7 and 2–3 h after fertilisation, respectively. The developmental events that occur during this period of transcriptional silence, for instance the switch from the meiotic to the mitotic cell cycle and the definition of embryonic axes and polarity, are therefore controlled by maternal mRNAs or proteins. Some of these maternal mRNAs are under a temporally regulated translational control, mediated by changes in their polyadenylation status, which is the subject of this chapter.

[1]CNRS-UPR 41, Université de Rennes I, Campus de Beaulieu, 35042 Rennes Cédex, France
[2]Worcester Foundation for Experimental Biology, Shrewsbury, MA 01545, USA

Progress in Molecular and Subcellular Biology, Vol. 18
Ph. Jeanteur (Ed.)

2 Biological Importance of Translational Control in Development

Amongst the most striking examples of translational control in eukaryotes are the changes in protein synthesis that occur at meiotic activation and fertilisation of oocytes. In the sea urchin, the rate of protein synthesis increases 10- to 30-fold after fertilisation (Epel 1967; Regier and Kafatos 1977) without, however, an obvious change in the pattern of protein synthesis (Brandhorst 1976). In other species, despite only a small change in the rate of protein synthesis at these developmental checkpoints, there are important changes in the population of translated maternal mRNAs (Woodland 1974; Ballantine et al. 1979; Rosenthal et al. 1980; Cascio and Wassarman 1982). These changes in the rate of protein synthesis or the population of translated mRNA are the result of the recruitment of stored maternal mRNAs onto polysomes. Concomitant with this messenger recruitment are dynamic changes in poly(A) (Wilt 1973; Rosenthal et al. 1983; Dworkin et al. 1985; Huarte et al. 1987; Sallés et al. 1994). Evidence that polyadenylation could control mRNA recruitment in both mouse and *Xenopus* oocytes was obtained by Strickland et al. (1988) and McGrew et al. (1989). This causal relationship has been confirmed for other maternal mouse and *Xenopus* mRNAs (Paris et al. 1991b; Simon et al. 1992; Gebauer et al. 1994; Sheets et al. 1994; Stebbins-Boaz and Richter 1994).

For many mRNAs, length changes in the poly(A) tail occur at the two major developmental check points – oocyte maturation and fertilisation. These changes concern distinct populations of maternal messagers that, at least for *Xenopus*, can be grouped into four major classes (Richter 1995; see Fig. 1). Classes II and IV contain maternal mRNAs that undergo cytoplasmic polyadenylation during oocyte maturation. In most cases, prior to this active polyadenylation, the mRNA has only a short poly(A) tail, although in some cases, it can be as long as 50–100 nt (Osborne et al. 1991; McGrew et al. 1989). Nearly all polII transcripts are polyadenylated in the nucleus and exported to the cytoplasm with poly(A) tails of several hundred nucleotides, Hence, for those mRNAs with only a short poly(A) tail in the oocyte and that are polyadenylated at maturation (classes II and IV) or after fertilisation (class III), the initial poly(A) tail probably must first be shortened by a specific mechanism – one that is active only on these mRNAs. Shortening the poly(A) tail of these mRNAs ensures that their translation is repressed during oocyte growth. In *Xenopus*, several of the mRNAs that undergo poly(A) elongation at maturation code for cell cycle-regulatory proteins (Paris et al. 1991a; Sheets et al. 1994). It is thus easy to see the probable biological importance of their repressed translation in growing oocytes that are non-cycling cells.

This concept can be carried over to maternal *Xenopus* mRNAs that are deadenylated in the egg and embryo after fertilisation (class IV). At least some of the meiosis-specific gene products must be removed for the egg to undergo the first (and subsequent) mitotic divisions. For instance, expression of the c-*mos* protein in the two-cell *Xenopus* embryo causes an arrest of cell division (Sagata et al. 1989). Degradation and/or inactivation of these proteins is thus

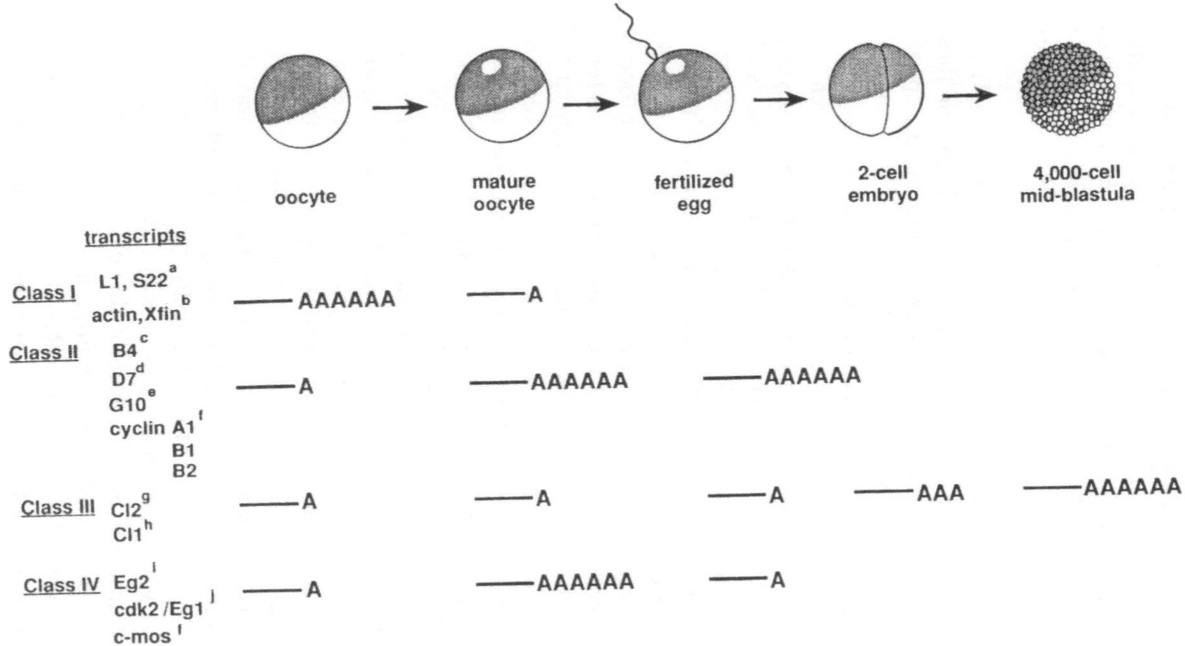

Fig. 1. Dynamics of poly(A) during *Xenopus* development [a]Varnum and Wormington (1990); [b]Fox and Wickens (1990); [c]Paris et al. (1991b); [d]Fox et al. (1989); [e]McGrew et al. (1989); [f]Sheets et al. (1994); [g]Simon et al. (1992); [h]Simon and Richter (1994); [i]Legagneux et al. (1992); [j]Stebbins-Boaz and Richter (1994)

evidently required (Watanabe et al. 1989), but deadenylation of the corresponding mRNAs (Sheets et al. 1994) ensures that their synthesis is also arrested. This situation is somewhat analogous to the immediate early genes that are expressed when cell proliferation is stimulated in quiescent cells. Prolonged or untimely expression of these proteins can lead to a transformed state (for review see H.R. Herschman 1991).

The remaining class of maternal *Xenopus* mRNAs (class I) are those that are deadenylated in the maturing oocyte. Deadenylation in these cells can be achieved by a default process that does not require any specific sequence information in the mRNA (see Sect. 4.3). Hence it seems unlikely that the translational arrest of the mRNAs deadenylated at this time is required for progession of the meiotic cell cycle. However, it has been known for many years that the translational capacity of the *Xenopus* oocyte is limited. The injection of increasing amounts of synthetic RNAs diverts the translational machinery from the maternal mRNAs to the exogeneous transcripts (Laskey et al. 1977; Richter and Smith 1981). The maturation of these RNA-loaded oocytes is retarded relative to that of mock injected oocytes (J.D. Richter, unpubl. results). Hence, the default deadenylation in maturing oocytes could be a means to liberate components of the translation apparatus for the synthesis of the meiosis-specific proteins. However, this would not appear to be an absolute requirement for oocyte maturation, as overloading *Xenopus* oocytes with poly(A) binding protein (PABP), which prevents the maturation-associated default deadenylation (see Sect. 4.3), did not affect the kinetics with which the oocytes matured (Wormington et al. 1996).

As mentioned above, in *Xenopus* and mouse, several of the identified mRNAs that are adenylated and recruited to the translational apparatus during oocyte maturation encode cell cycle-regulatory factors: c-*mos* (Sagata et al. 1989; Gebauer et al. 1994; Sheets et al. 1994), Eg1-cdk2 (Paris et al. 1991a) and the cyclins A1, B1 and B2 (Sheets et al. 1994). That polyadenylation of maternal mRNAs is required for oocyte maturation was demonstrated by Kuge and Inoue (1992), who incubated oocytes in cordycepin, which prevents ongoing polyadenylation. More specifically, the oocytes of both the mouse and *Xenopus*, in which the endogenous c-*mos* mRNA has been ablated by the injection of antisense oligonucleotides, do not complete meiosis (O'Keefe et al. 1989; Sagata et al. 1989). The maturation of these oocytes can be rescued if they are co-injected with c-*mos* mRNAs containing the wild-type 3′ UTR. Rescue is not achieved, however, if the mRNAs encoding c-*mos* carry mutations or deletions of the sequence elements required for cytoplasmic polyadenylation (Gebauer et al. 1994; Sheets et al. 1995).

The association between polyadenylation and efficient translation is not restricted to oocytes and early embryos. In several in vitro systems derived from somatic cells, the increase in translation for polyadenylated mRNAs relative to their non-adenylated counterparts is synergistic with the effect of RNA capping on translation (Gallie 1991). Particularly interesting, therefore, was the observation that the cap-dependent translation initiation factors eIF4B and eIF4F show a significant affinity for poly(A) (Gallie and Tanguay 1994). Also Munroe and Jacobson (1990a,b) have argued that the stimulation of translation by poly(A) may be due to its interaction with poly(A) binding protein (PABP) that would somehow facilitate ribosome entry at the 5′ end of the mRNA. PABP can clearly enhance protein synthesis in vitro (Grossi de Sa et al. 1988) and the data of Sachs and colleagues for yeast (Sachs and Davis 1989; Sachs and Deardorff 1992) demonstrate a role for PABP in the initiation of translation in this organism. Although these results allow the gap between the 5′ and 3′ ends of the mRNA to be conceptually closed, they do not explain the functional link between polyadenylation and increased translation. Furthermore, *Xenopus* oocytes and embryos contain only a limited amount of 70-kDa PABP (Zelus et al. 1989) that would appear to be insufficient to mediate the translational changes observed at maturation and after fertilisation (see also Sect. 4.3).

Recently, a more definitive step was made by Kuge and Richter (1995) towards understanding the functional link between polyadenylation and increased translation. These authors showed that an injected mRNA that undergoes polyadenylation during maturation of *Xenopus* oocytes also undergoes cap-ribose methylation. At this time, an increase in the cap-ribose methylation of maternal *Xenopus* mRNAs was also observed, similar to that previously reported for fertilised sea urchin eggs (Caldwell and Emerson 1985). It appears that active poly(A) elongation is important for this process because a pre-adenylated RNA does not induce cap-ribose methylation. Finally, using a reporter construct, Kuge and Richter (1995) showed that translational recruitment and cap methylation were temporally correlated. These authors propose that poly(A) polymerase may form a heteroduplex with a methyltransferase that is responsible for the cap-ribose methylation. Such a scenario appears to be the case in vaccina virus (Schnierle et al. 1992).

Hence, cap-ribose methylation may constitute a functional link between polyadenylation and the stimulation of protein synthesis if the ribose-methylated cap (^{7}mGpppNmpN or ^{7}mGpppNmpNmpN) had a competitive advantage over the monomethylated cap (^{7}mGpppN) for the recruitment to the translational apparatus. However, polyadenylation stimulated cap-ribose methylation may not be the only means by which poly(A) tails stimulate translation. For instance, RNAs injected into oocytes are translated with greater efficiency if they contain a poly(A) tail (Gallie et al. 1988; Vassalli et al. 1989). These are conditions that do not cause cap-ribose methylation (Kuge and Richter 1995).

3 Translational Repression in Oocytes

The purpose of presenting some elements here is to highlight the differences in the control of deadenylation between oocytes and embryos. As mentioned in

Section 2, the deadenylation of certain mRNAs in the growing oocyte is a means to ensure repression of the encoded protein until the resumption of meiosis triggers their polyadenylation. Work with mouse oocytes has led to a suggestion of how this sequence-specific deadenylation may be achieved.

During maturation of mouse oocytes, the mRNA encoding the tissue-type plasminogen activator (t-PA) protein becomes polyadenylated and is recruited onto polysomes (Vassalli et al. 1989). In growing oocytes and in full-grown primary oocytes, the t-PA transcripts undergo normal nuclear polyadenylation and have a poly(A) tail of 300–400 nt before being exported to the cytoplasm (Huarte et al. 1992). However, the t-PA mRNA present in the cytoplasm of primary oocytes has a short 3′ poly(A) tail of about 40 nt (Huarte et al. 1987). Therefore, this mRNA may be deadenylated in the cytoplasm of growing oocytes. By injecting various chimaeric polyadenylated RNAs, Huarte et al. (1992) showed that transcripts containing the 3′ UTR of t-PA mRNA were deadenylated in primary oocytes whereas those containing either the 3′ UTR of beta-actin or globin were not. When the injected oocytes were allowed to mature the RNA containing the t-PA 3′ UTR became readenylated.

Various deletions within the t-PA 3′ UTR showed that a region containing several motifs with the sequence AUUUUAAU and AUUUUA was required for both deadenylation in the primary oocyte and polyadenylation at maturation (see Sect. 4). These motifs resemble the UA-rich cytoplasmic polyadenylation elements (CPE) that are required for polyadenylation of maternal *Xenopus* mRNAs in maturing oocytes (see Sect. 4.1). From this work, Huarte et al. (1992) proposed that these UA-rich motifs play a dual role: they confer poly(A) shortening in the growing oocyte and poly(A) elongation during maturation. Hence, they suggested that the UA-rich element be called an adenylation control element (ACE) rather than a CPE.

In *Xenopus*, several of the maternal mRNAs that undergo CPE-driven polyadenylation during oocyte maturation have poly(A) tails of about only 50 to 100 nucleotides (Dworkin and Dworkin-Rastl 1985; Paris et al. 1991a; Sheets et al. 1994; Stebbins-Boaz and Richter 1994). Although this suggests that these mRNAs are partially deadenylated in the growing oocyte, it is not clear that the CPE is implicated in this poly(A) shortening. Indeed, when poly(A)$^+$ RNAs that contain the distal CPE-containing portion of the mRNA denoted as H4 were injected into *Xenopus* oocytes, they were not deadenylated (Fox and Wickens 1990).

Studies in the surf clam, *Drosophila*, and *Xenopus* have shown that the translational repression of mRNAs in oocytes, and their eventual translational activation upon the resumption of meosis or after fertilisation, can occur by a process independent of their adenylation status. In the clam *Spisula solidissima* the mRNAs encoding cyclin A and ribonucleotide reductase are translationally repressed in the oocyte (Standart et al. 1985; Swenson et al. 1986) and are present predominantly in the poly(A)$^-$ fraction (Rosenthal et al. 1983). Translational activation and polyadenylation of these mRNAs occurs at maturation which is triggered by fertilisation in this species (Standart et al. 1985; Swenson et al. 1986). In vitro derepression of the translation of these mRNAs can be achieved with antisense oligonucleotides directed to specific regions in their 3′ UTRs, called unmasking boxes. However, this translational derepression in vitro is not associated with polyadenylation of the RNA (Standart et al. 1990).

Translational repression of maternal mRNAs can also be linked to their localisation. This is the case for the mRNAs encoding the morphogens *nanos* (Gavis and Lehmann 1994) and *oskar* (Kim-Ha et al. 1995; Rongo et al. 1995) in *Drosophila* oocytes. The spatially restricted expression of these proteins is required for posterior body patterning in the *Drosophila* embryo (see St. Johnston and Nüsslein-Volhard 1992 for review). For *nanos* mRNA at least, translational activation that occurs at fertilisation is independent of polyadenylation (Sallés et al. 1994).

In *Xenopus*, translational repression has been considered to be a common fate for a large proportion (perhaps as much as 80%) of all maternal mRNAs in the growing oocyte (Davidson 1986). Masking of maternal mRNAs from the translational apparatus in the growing oocyte by packaging into mRNP particles has been well documented (for review see Standart 1992). Early work (Jenkins et al. 1978; Rosenthal et al. 1980; Richter and Smith 1984) showed that RNA-binding proteins were responsible for this masking rather than a modification of the mRNA, and several of these proteins have been identified (Darnbrough and Ford 1981; reviewed by Sommerville 1992). This form of translational repression by packaging into mRNP particles can occur without deadenylation of the masked mRNA (Crawford and Richter 1987; Kick et al. 1987; Cummings and Sommerville 1988). Also this masking of an mRNA from the translational apparatus is more efficient if the mRNA is transcribed in vivo, as opposed to being injected (Bouvet and Wolffe 1994). To date, no mRNA specificity for this mode of translational repression has been demonstrated. Indeed, the two most abundant proteins in these complexes, mRNP3 and mRNP4, can bind to many unrelated transcripts, including RNAs transcribed from the Bluscript vector (Murray et al. 1991). However, recently, Bouvet et al. (1995), using a SELEX assay, have identified a consensus hexanucleotide sequence to which the *Xenopus* Y-box protein FRGY2, which is identical to mRNP4, will bind specifically. This suggests that, although this protein and its closely related partner mRNP3 bind to any mRNA, the affinity with which these proteins bind may be influenced by the exact sequence of the RNA. Consequently, translational repression in oocytes, by packaging into mRNP particles, may not be a completely non-specific process.

Finally, the maternal mRNA encoding the *Xenopus* fibroblast growth factor receptor (XFGF-R) is also translationally repressed in immature oocytes (Musci et al. 1990). This inhibition of XFGF-R mRNA translation is specifically mediated by a *cis*-acting element(s) within a 180-nt portion of the 3′ UTR that is situated about 1 kb upstream of the AAUAAA hexanucleotide (Robbie et al. 1995). Synthetic RNA from which the 180-nt translation inhibitory element has been deleted is translated in immature oocytes, but this is not associated with a polyadenylation of the RNA.

Therefore in the growing oocyte, several specific mechanisms exist that mediate the translational repression of different, but perhaps overlapping, subpopulations of maternal mRNAs. One of these mechanisms involves partial deadenylation of the mRNA, possibly mediated by motifs that resemble the UA-rich CPEs.

4 Control of Polyadenylation During Oocyte Maturation

The resumption of meiosis in the fully grown oocyte, whether it be prior to or triggered by fertilisation, is a time of important changes in the population of maternal mRNAs that are polyadenylated and translated (Sagata et al. 1980; reviewed in Rosenthal and Wilt 1987). This is probably the stage of development where our knowledge of polyadenylation control at the molecular level is the most advanced.

4.1 Cytoplasmic Polyadenylation: Identification of Cis Elements

By the mid-1980s, it was clear that cytoplasmic polyadenylation was widespread in the animal kingdom, that it occurred only on specific mRNAs at certain times of development, and that it strongly correlated with translational activation. To take these observations to the next level, which was to determine whether cytoplasmic polyadenylation actually controlled translation, required much more knowledge about the polyadenylation process itself. To this end, it became necessary to identify the *cis* elements, and then the presumed interacting proteins that regulated polyadenylation or deadenylation. However, this required not only a full-length mRNA that would contain all the polyadenylation regulatory elements, but also a system in which this process could be readily examined. Although relatively large cDNA clones for mRNAs were available from sea urchin, clam and *Xenopus*, it was in the mouse oocyte that the first steps to map the *cis* elements for polyadenylation were taken.

In their analysis of tissue type plasminogen activator (t-PA), Huarte et al. (1985) discovered that although the mRNA encoding this protein was present in immature (primary), and mature (secondary) oocytes of the mouse, t-PA protein itself was detected only in the mature oocytes. This indicated that t-PA mRNA might be regulated at the translational level. This group subsequently showed that oocyte maturation was a time not only when t-PA mRNA was recruited for translation from an inactive pool, but that this recruitment was accompanied by poly(A) elongation (Huarte et al. 1987). These observations set the stage for the first attempt to disrupt cytoplasmic polyadenylation since the early studies of Mescher and Humphreys (1974), who prevented all polyadenylation by incubating sea urchin embryos with the chain-terminating molecule, cordycepin. In this case, however, Strickland et al. (1988) sought to ablate the cytoplasmic polyadenylation of t-PA mRNA specifically by injecting antisense oligodeoxynucleotides that were complementary to various portions of the message. Because of the activity of endogenous RNase H that destroys RNA-DNA hybrids, they were able to demonstrate that removal of the 3′103 nucleotides of t-PA mRNA prevented it from undergoing poly(A) elongation and translational activation. Two important questions arise from these experiments : were all the signals necessary for polyadenylation within these 103 nucleotides, and if so, what were the precise nucleotides that were responsible?

In a follow-up study by Vassalli et al. (1989), the 103-nucleotides region was attached to a reporter sequence that was then injected into mouse oocytes. After oocyte maturation, this chimaeric mRNA was not only polyadenylated to about the same extent as the endogenous t-PA mRNA, but was translationally activated as well. Thus, these investigators concluded that the 103 nucleotides contained all the information necessary to promote cytoplasmic polyadenylation and translation. Moreover, they also showed that one sequence, AAUAAA, was necessary, but not sufficient, for polyadenylation. Because this hexanucleotide is well known for its essential role in nuclear pre-mRNA cleavage and polyadenylation (Wahle and Keller 1992), it seemed likely that at least some of the molecular machinery that regulated nuclear polyadenylation would influence cytoplasmic polyadenylation as well.

In experiments parallel to those mentioned above, cytoplasmic polyadenylation was also being studied in *Xenopus* oocytes. Using cDNA clones for mRNAs that undergo polyadenylation and translational activation during *Xenopus* oocyte maturation (isolated by the Dworkin laboratory: Dworkin and Dworkin-Rastl 1985; Dworkin et al. 1985), McGrew et al. (1989) and Fox et al. (1989) injected in vitro-transcribed RNAs into oocytes that were then incubated with progesterone, the natural inducer of maturation in this species. Such mRNAs were not only polyadenylated but were also translationally activated. Based on these observations, it was comparatively straight-forward to inject mutated RNAs to determine the signals necessary for polyadenylation. As was the case with t-PA mRNA in mouse oocytes, the hexanucleotide AAUAAA was necessary, but not sufficient, for polyadenylation. Another sequence, residing fewer that 20 nucleotides 5′ of the hexanucleotide, was also required for both polyadenylation and translational activation. This sequence has the general structure of UUUUUUAU, and has subsequently been called the cytoplasmic polyadenylation element or CPE (McGrew and Richter 1990; Paris and Richter 1990). To distinguish this UUUUUUAU CPE from another CPE subsequently identified in mRNAs adenylated after fertilisation (see Sect. 5.1), the above sequence will be referred to here as a UA-rich CPE.

An examination of the 3′ UTRs of several mRNAs that are known to undergo cytoplasmic polyadenylation in maturing *Xenopus* oocytes shows that they are not identical, nor are their distances from the AAUAAA hexanucleotides constant (Fig. 2). For example, the CPE of c-*mos* mRNA consists

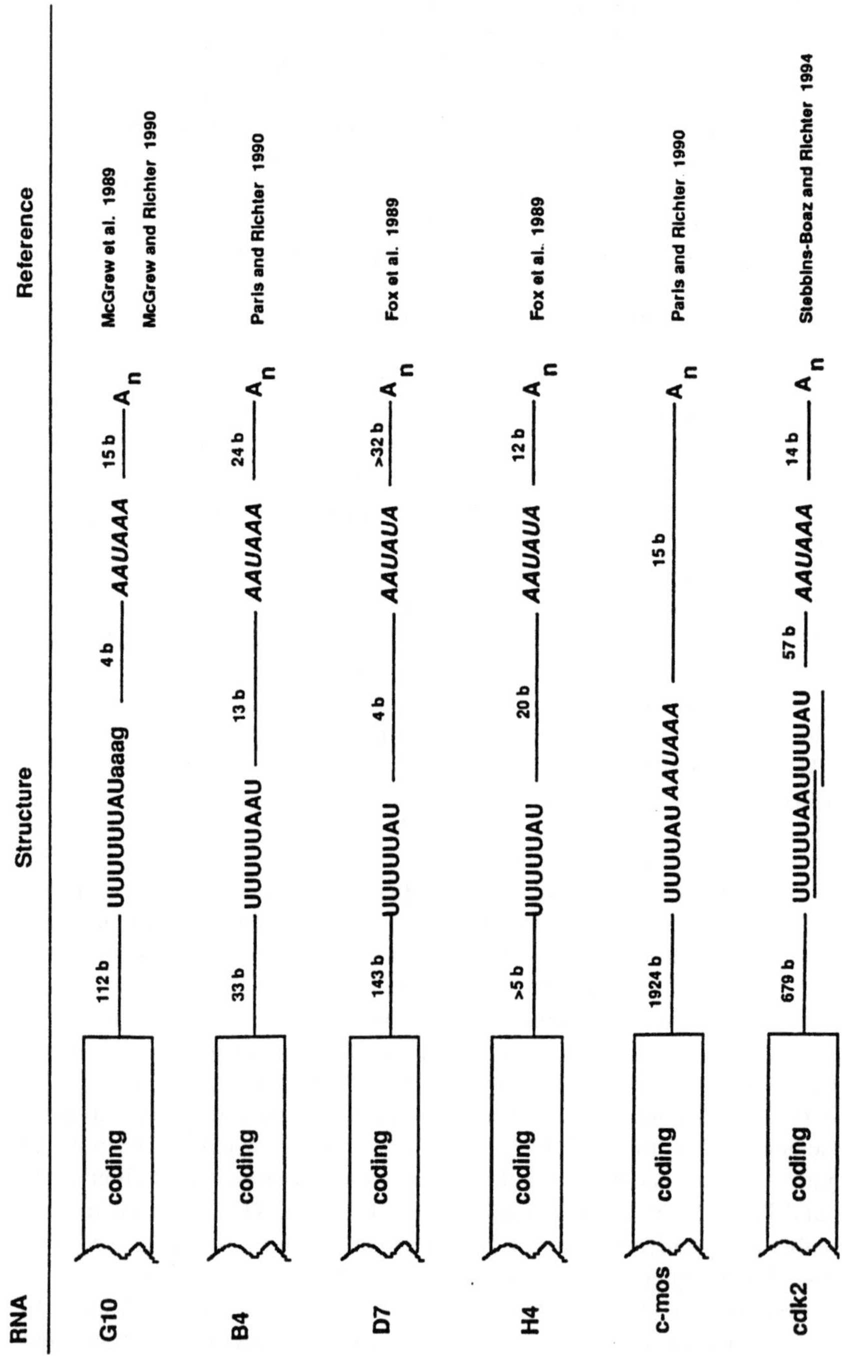

Fig. 2. Cytoplasmic polyadenylation signals of *Xenopus* RNAs

of UUUUAU and is immediately adjacent to the AAUAAA (Paris and Richter 1990). The CPE of B4 RNA, on the other hand, is UUUUUAAU, and is 13 bases from the hexanucleotide. Given this variation in the UA-rich CPE structure, what is the minimal sequence that is necessary to support polyadenylation? McGrew and Richter (1990) performed scanning mutagenesis of the G10 CPE and two important points emerged. First, with a single exception, any dinucleotide substitution of a part of the CPE (UUUUUUAU) with GG abolishes polyadenylation. The single exception is the 5′-most UU, so that the sequence GGUUUUAU still promotes polyadenylation, albeit somewhat weakly. This is the same sequence that promotes relatively weak polyadenylation of *Xenopus* c-*mos* mRNA (Paris and Richter 1990). Thus, UUUUAU may be considered the minimal UA-rich CPE. The second important point is that substitution of the AAAG following the AU dinucleotide with a *Kpn*I restriction site inhibits polyadenylation. Thus, at least in this case, the AAAG must be considered a part of the UA-rich CPE since its removal abrogates activity. In the case of the B4 CPE (UUUUUAU), nucleotide substitution experiments have also shown that at least four continuous Us are necessary for the polyadenylation of this RNA. The 3′-most AU dinucleotide, when replaced with GG, can still support very weak polyadenylation (Paris et al. 1991b). However, this "residual" polyadenylation activity may be due to another "cryptic" CPE that still is present in the B4 3′ UTR (Paris et al. 1991b).

Although the spacing between the CPE and AAUAAA varies among the RNAs, it may not have a substantial effect on polyadenylation during oocyte maturation. With G10 RNA, for example, McGrew and Richter (1990) increased the distance between the UA-rich CPE and the hexanucleotide with polylinker sequence by as much as 86 nucleotides with no obvious diminution of polyadenylation activity. In contrast, it should be borne in mind that the distance between the CPE and the AAUAAA hexanucleotide drastically alters both the timing and extent of polyadenylation in *Xenopus* embryos (see Sect. 5.1).

For all the *Xenopus* mRNAs so far examined, the residence of the CPE is always 5′ of the hexanucleotide (Fig. 2). Fox et al. (1989), however, asked whether this must be the case for polyadenylation to occur. They inserted a CPE 3′ of the hexanucleotide and showed that this promoted polyadenylation during maturation in mRNA-injected oocytes. Therefore, even though there are significant constraints on the sequence of the CPE to support polyadenylation, there is considerable flexibility when it comes to distance or polarity of the CPE relative to the AAUAAA.

Once the CPE was identified in *Xenopus*, similar sequences were found in mouse mRNAs that also undergo cytoplasmic polyadenylation. For example, hypoxanthine phosphoribosyl transferase (HPRT) mRNA (Paynton et al. 1988; Paris and Richter 1990; Paynton and Bachvarova 1994), t-PA mRNA (Huarte et al. 1992) and c-*mos* mRNA (Gebauer et al. 1994), all contain UA-rich CPEs in their 3′ UTRs. Indeed, the conservation of the UA-rich CPE has allowed Sallés et al. (1992) to construct a PCR-based cDNA library using oligo (dT) and a primer complementary to the CPE. Such a library could prove to be enormously useful in identifying additional mRNAs that are polyadenylated and translated during oocyte maturation.

4.2 Identification of Proteins Involved in Cytoplasmic Polyadenylation

To begin to analyse the proteins that regulate cytoplasmic polyadenylation, an in vitro system was required that could carry out this process in a manner similar to intact cells. Because *Xenopus* eggs retain several biochemical properties when made into cell extracts, such as cell cycle control (Murray and Kirschner 1989), and because they are readily available in abundance, they were the first to be used in this manner. Fortunately, *Xenopus* egg extracts not only polyadenylate exogenous RNA, they do so in a very faithful manner, that is, only RNAs with a CPE and AAUAAA are polyadenylated (McGrew and Richter 1990; Paris and Richter 1990). Based on this, then, the proteins that are important for polyadenylation could be examined in two ways : direct interaction with the *cis* elements, or enzymatic activity. Both of these approaches have proven to be fruitful.

CPE-binding activities have been detected by both RNA gel shift analysis (Paris and Richter 1990; Fox et al. 1992) and UV cross-linking (McGrew and Richter 1990; Paris et al. 1991b), the latter being particularly useful because it can distinguish between different CPE-binding proteins. For example, the CPE of G10 RNA, which is UUUUUUAUAAAG, is bound by a 82-kDa protein in egg extracts, but not in polyadenylation-deficient oocyte extracts (McGrew and Richter 1990). Thus, one might surmise that the 82-kDa protein could be synthesised de novo during the early stages of maturation to effect polyadenylation. Indeed, this is consistent with the observation that G10 polyadenylation is sensitive to cycloheximide treatment, but only soon after the oocytes are exposed to progesterone. This may be contrasted with the CPE of B4 RNA, UUUUUAAU, which is bound by a 62-kDa protein in both oocyte and egg extracts. This 62-kDa protein, however, is phosphorylated in egg extracts, which may be important for its activity (Paris et al. 1991b).

An understanding of the biochemistry of cytoplasmic polyadenylation would, of course, require purified components. To that end, the 62-kDa B4 CPE-binding (CPEB) protein was isolated by single-step RNA affinity chromatography. Amino acid sequencing of tryptic fragments of this protein led to the cloning of a cDNA which, when sequenced, revealed some interesting characteristics (Hake and Richter 1994). First, CPEB has striking homology to *Drosophila* Orb, which is important for egg chamber formation and the establishment of correct body axes in the embryo (Lantz et al. 1992; Christersen and McKearin 1994). This homology resides in the carboxy-terminal half of CPEB, where it is 62% identical to the fly protein. Within this region are two conserved domains that are likely to be important for function. The first of these are two RNA recognition motifs (RRMs), which, as the name implies, are probably responsible for recognising the RNA. The second is a cysteine/histidine repeat, whose conserved spacing is reminiscent of a zinc finger. This region might also be important for RNA binding, or perhaps even be involved in protein-protein interactions.

Although CPEB binds only to the UA-rich CPE (Hake and Richter 1994; Stebbins-Boaz et al. 1996) the major extant question is whether CPEB is involved in cytoplasmic polyadenylation. When this protein is immunodepleted from egg extracts, polyadenylation activity is completely destroyed, but can be at least partially restored when heterologous CPEB, synthesised in an in vitro translation system, is added to the depleted lysate (Hake and Richter 1994). Thus, CPEB is a specificity factor that mediates cytoplasmic polyadenylation. Perhaps not surprisingly, then, Orb has recently been shown to be critical for cytoplasmic polyadenylation in *Drosophila* (P. Schedl, pers. comm.).

Fox et al. (1992) have taken a different approach to assessing the factors involved in cytoplasmic polyadenylation. These investigators fractionated egg extracts into two compartments, one containing a UA-rich CPE-binding activity and the other a poly(A) polymerase activity. The RNA-binding activity had several properties that suggested it could contain CPSF (cleavage polyadenylation specificity factor), a heterotrimeric protein complex known to interact with the AAUAAA hexanucleotide and to be essential for nuclear polyadenylation (Wahle and Keller 1992). These properties included the observation that a gel shift with egg extract was similar to that with bovine CPSF when a UA-rich CPE-containing RNA was used as the substrate, and that bovine CPSF and bovine poly(A) polymerase (PAP) could catalyse CPE-dependent polyadenylation in vitro (Bilger et al. 1994).

The data of Bilger et al. (1994) are very strong in their implication of CPSF as a central component for cytoplasmic as well as nuclear polyadenylation. However, because many mRNAs are polyadenylated in a developmentally regulated manner, it would seem that CPSF and PAP would not be the only factors involved in this process. In addition, these results with CPSF could be construed as being at odds with the CPEB data summarised earlier.

One possible explanation for these seemingly disparate results is that CPEB and CPSF may each be a part of the apparatus that controls polyadenylation. That is, perhaps CPSF constitutes the "core" polyadenylation complex that recruits poly(A) polymerase. In vivo, however, CPSF might not be stably bound to the RNA, and might require CPEB to do so. This could explain differential mRNA polyadenylation during development. That is, unique CPEBs could recruit CPSF to different RNAs at specific times during oocyte maturation or embryogenesis. Indeed, at least two CPEB-binding proteins are thought to act at different times of development (Paris et al. 1991b; Hake and Richter 1994; Simon and Richter 1994).

Another factor that is important for cytoplasmic polyadenylation is, of course, poly(A) polymerase (PAP). Although there has been some dispute for a number of years as to whether there is a PAP specific for the cytoplasm of mammalian somatic cells, the results of Fox et al. (1992) suggest that at least the nuclear protein could substitute for a possibly unique cytoplasmic form in stimulating polyadenylation. These investigators showed that bovine nuclear PAP could complement an RNA-binding fraction from *Xenopus* eggs and restore full polyadenylation. Moreover, Western blot data by Ballantyne et al. (1995), using an antibody that recognises nuclear PAP, indicate that probably the same PAP is present in both the nucleus and cytoplasm of *Xenopus* oocytes.

Other evidence, however, shows that the cytoplasm of oocytes contains a unique PAP. Gebauer and Richter (1995) isolated a cDNA for a PAP that lacks any nuclear localisation signal and, when injected into oocytes, is localised exclusively in the cytoplasm. Most importantly, antibody raised against this unique cytoplasmic form of PAP inhibits polyadenylation when added to a *Xenopus* egg extract. Therefore, these results indicate that there are probably a number of forms of PAP, some of which are exclusively cytoplasmic. It remains to be determined whether the cytoplasmic PAP has unique enzymatic properties compared to the nuclear form.

4.3 Deadenylation in Maturing Oocytes

In *Xenopus*, mouse and several other species, polyadenylation during oocyte maturation is imparted by a functional CPE in conjunction with the AAUAAA nuclear polyadenylation signal (see previous section). During this period, other maternal mRNA, poly(A)$^+$ in the oocyte, are deadenylated and released from polysomes. In *Xenopus*, examples of such mRNAs are those encoding ribosomal proteins (rp-mRNAs) (Hyman and Wormington 1988); a zinc finger protein, Xfin (Ruiz i Altalba et al. 1987); and cytoskeletal actin (Hyman et al. 1984). In contrast to polyadenylation, deadenylation of these *Xenopus* maternal mRNAs during oocyte maturation can occur by a default process that does not require any specific sequence information.

One of the first indications of default deadenylation was that polyadenylated RNAs composed of transcribed plasmid sequences were deadenylated during oocyte maturation (Fox and Wickens 1990). Also, for the S22 rp-mRNA, two deletions, which together covered almost all of the 3′ UTR, did not affect the deadenylation of this RNA at maturation (Varnum and Wormington 1990). That a functional CPE was all that was required to protect an RNA from this default deadenylation was shown in two ways. Mutations either in the CPE or the AAUAAA motifs, which render the RNA refractory to polyadenylation during oocyte maturation, lead to its deadenylation in maturing oocytes (Fox and Wickens 1990). Also, the inclusion of the UA-rich CPE into the 3′ UTR of an mRNA, which is normally a substrate for default deadenylation, results in the maintenance or the elongation of the poly(A) tail (Varnum and Wormington 1990).

Although CPE-driven adenylation and default deadenylation both take place during oocyte maturation, they do not occur simultaneously. Careful time-course experiments showed that polyadenylation precedes deadenylation. In fact, this latter process is concomitant with germinal vesicle breakdown (GVBD) whereas the CPE-driven polyadenylation preceeds this event (Fox and Wickens 1990). Varnum et al. (1992) then showed that default deadenylation does not occur in enucleated oocytes. By making extracts of both isolated nuclei and cytoplasm, these authors showed that default deadenylation requires both nuclear and cytoplasmic components.

In time-course experiments, partially deadenylated intermediates are observed at points soon after GVBD, suggesting that the nuclease has the characteristics of a 3′ to 5′ exonuclease (Fox and Wickens 1990). The specificity of the complex is, however, unclear at the present. Poly(A) removal is specific in as much as the substrate molecules are not shortened beyond the 5′ end of the poly(A) tail. Somewhat surprisingly, 3′ terminal As are not required (Fox and Wickens 1990; Varnum and Wormington 1990) and as many as 15 non-A residues can terminate the transcript without affecting the efficiency with which the molecule is deadenylated although deadenylation efficiency is reduced for longer stretches of 3′ non-A residues (Fox and Wickens 1990). The length of the poly(A) tail is also important. RNAs that terminate with a poly(A) tail of 42 As are deadenylated, whereas those that only contain 20 3′ adenosines are not (Varnum and Wormington 1990). Lastly, *Xenopus* default deadenylation in maturing oocytes is abrogated by overexpression of PABP (Wormington et al. 1996). In these PABP-loaded oocytes, the decrease that would normally accompany oocyte maturation in the translation of both the maternally inherited actin mRNA, or a CPE-deficient mRNA encoding the luciferase protein, did not occur. Overexpression of mutant (non functional) PABP was ineffective in preventing the default deadenylation or release from polysomes. This is in contrast to the yeast poly(A) nuclease, which is activated by PABP (Sachs and Deardorff 1992). An explanation of these apparently opposing effects of PABP on deadenylation must probably await the purification and enzymatic characterisation of the *Xenopus* default poly(A) nuclease.

One model based on these studies in *Xenopus* is that RNAs containing a CPE are adenylated or protected from deadenylation during ooycte maturation, whereas those that do not contain a CPE are deadenylated by a default process. However, the situation is complicated by the identification of maternal *Xenopus* mRNAs that are adenylated only after fertilisation (Paris et al. 1988; Paris and Philippe 1990). As there is no evidence for a process of default polyadenylation in *Xenopus* embryos, these mRNAs presumably contain a CPE that is functional only after fertilisation. In conceptual terms, this means that there are CPEs that function at different stages of development. Work by Richter and colleagues, described in Section 5, showed that this temporal regulation is due to a CPE-silencing or masking element(s) in the 3′ UTR of these RNAs, which may make them a substrate for default deadenylation during oocyte maturation.

5 Control of Polyadenylation in Eggs and Embryos

The molecular mechanisms that control translation in the developing embryo have been most studied in *Xenopus*. The control of cytoplasmic adenylation in these embryos differs in certain respects to that previously described for oocyte maturation. The polyadenylation driven by the UA-rich CPE persists after fertilisation (Stebbins-Boaz and Richter 1994) and, not surprisingly, the 62-kDa CPEB factor is present in embryos, although at a reduced level (Hake and Richter 1994). Therefore, mRNAs containing a CPE (UA-rich) and an AAUAAA motif, inherited by the egg as poly(A)$^+$ transcripts, can potentially remain polyadenylated in the embryo. However, there are also mRNAs that are poly(A)$^-$ in the unfertilised egg and are adenylated after fertilisation (Paris and Philippe 1990). These mRNAs contain a CPE that become functional only after fertilisation (see below). In addition, other maternal mRNAs that contain a CPE and are adenylated in the mature oocyte are deadenylated after fertilisation (examples of such mRNAs are c-*mos* : Sheets et al. 1994, and a series of nine transcripts given the generic name of Eg mRNAs : Paris and Philippe 1990). Hence, these mRNAs, in addition to a CPE, must contain sequence information that directs them to be specifically deadenylated after fertilisation.

5.1 Polyadenylation in Embryos

Cl1 and Cl2 are two *Xenopus* mRNAs that are inherited by the egg as deadenylated transcripts and become polyadenylated after fertilisation (Paris et al. 1988; Paris and Philippe 1990). Cl1 mRNA encodes the *Xenopus* polypeptide chain release factor (eRF1) (Frolova et al. 1994). The function of Cl2 protein is unknown at present. By studying the adenylation behaviour of a series of mutant RNAs injected into *Xenopus* embryos, Simon et al. (1992) and Simon and Richter (1994) showed that a polyuridine track [Cl1–U18 (Tassan et al. 1993); Cl2–U27 (Simon et al. 1992)] situated in the 3′ UTR of these RNAs is required for this "embryonic" adenylation, in addition to the hexanucleotide AAUAAA. The minimum number of contiguous uridine residues that can function as a U-rich CPE is 12 (Simon et al. 1992). These two mRNAs differ slightly in the kinetics with which they are adenylated after fertilisation (Simon and Richter 1994), and this appears to be due to the distance separating the U-rich CPE from the AAUAAA hexanucleotide. Mutant RNAs in which this distance has been reduced are adenylated to a greater extent and more rapidly after fertilisation than the wild-type RNA (Simon and Richter 1994).

Surprisingly, when synthetic RNAs containing essentially only the Cl2 U-rich CPE and the AAUAAA hexanucleotide were injected into fully grown oocytes that were subsequently matured, the RNA became adenylated (Simon et al. 1992). Hence, the adenylation apparatus driven by this U-rich CPE is functional during maturation and the non-adenylation of the maternal mRNA is probably due to inhibition of the CPE by additional sequence elements. This was effectively shown to be the case for both Cl2 and Cl1 mRNAs. For the Cl2 mRNA CPE-silencing requires sequence information contained within a region of at least 400 nt that extends both 5′ and 3′ of the CPE (Simon et al. 1992). For the Cl1 mRNA, the silencing element is contained within a 180-nt region situated between the CPE and the AAUAAA hexanucleotide (Simon and Richter 1994). Comparison of these regions of Cl1 and Cl2 mRNA revealed no similarity of sequence or secondary structure (Simon and Richter 1994). Therefore, at present, it is not clear how these elements prevent the polyadenylation of Cl1 and Cl2 mRNA in the maturing oocyte. Also when the U-rich CPE in the Cl2 3′ UTR was replaced by the UA-rich CPE of B4 mRNA, the resulting chimaeric RNA was polyadenylated during oocyte maturation (Simon et al. 1992). There are two possible explanations of this result. The polyuridine track could itself be part of the silencing element. Alternatively, the CPE-silencing element in these RNAs may affect only the U-rich CPEs, possibly via the interaction of a repressor protein(s) that bind(s) to the silencing element and thereby inhibit(s) CPE function. In this context it should be noted that two proteins of 36 and 45 kDa interact specifically with the U-rich CPEs (Simon et al. 1994), whereas the protein CPEB (62 kDa) binds to the UA-rich B4 CPE (Hake and Richter 1994). Neither Cl1 nor Cl2 RNA binds the CPEB protein and B4 RNA does not associate with either the 36-kDa or the 45-kDa proteins.

The Cl1 protein is synthesised in growing oocytes (Tassan et al. 1993) and therefore this mRNA is presumably polyadenylated at this time. At maturation, the mRNA could be deadenylated by the default process due to the non-functional state of the CPE leading to the inheritance by the egg of poly(A)$^-$ Cl1 and Cl2 mRNAs. A number of maternal *Xenopus* mRNAs have been identified that contain potential U-rich CPEs in their 3′ UTR (referenced in Simon and Richter 1994; see also Bassez et al. 1990; Bouvet et al. 1994). Although many of these mRNAs are preferentially expressed in the embryo, others are polyadenylated and translated during oocyte maturation (Younglai et al. 1980; Paris and Philippe 1990; Osborne et al. 1991). Therefore, CPE silencing during oocyte maturation is not always associated with mRNAs whose 3′ UTR contains stretches of 12 or more uridine residues.

5.2 Embryo-Specific Deadenylation

The deadenylation of maternal mRNAs in *Xenopus* embryos has been studied so far using two of the Eg mRNAs: Eg2 and Eg1-cdk2. Legagneux et al. (1992) identified by UV-cross-linking two RNA-binding proteins (p53/p55) that specifically associate with the 3′ UTR of Eg1, Eg2 and Eg5 mRNA. These proteins are distinct from the RNP3 and RNP4 (FRGY2) "masking" proteins that have a similar electrophoretic mobility in SDS-polyacrylamide gels (Darnbrough and Ford 1981). They are also distinct from the CPEB protein (Hake and Richter 1994) that migrates as a 58-kDa protein in SDS-PAGE after

UV cross-linking (Paris et al. 1991b). The analysis of a series of deletion mutants identified a 17-nt segment of Eg2 3′ UTR that was required both for postfertilisation deadenylation of chimaeric RNAs and for the binding of the p53/p55 proteins (Bouvet et al. 1994). Deletion of this deadenylation element abrogated the deadenylation and allowed the chimaeric RNAs to become polyadenylated in the embryos. Therefore, all or part of this 17 nt segment constitutes a *cis*-acting element implicated in the developmental switch from adenylation to deadenylation of Eg2 mRNA. However, this 17-nt fragment alone is not sufficient to confer postfertilisation deadenylation on an RNA. This implies that either other unidentified sequence elements are required, or there are constraints on the position of the Eg2 deadenylation element, relative to other *cis*-acting elements (CPE or AAUAAA hexanucleotide).

A similar study has been performed by Stebbins-Boaz and Richter (1994) using Eg1-cdk2 mRNA. They showed that the last 188 nt of Eg1 3′ UTR contained all the sequence information required for polyadenylation during oocyte maturation and deadenylation after fertilisation. When injected as a poly(A)$^-$ transcript, this RNA was not adenylated in embryos. Reasoning that deadenylation and inhibition of polyadenylation could be two manifestations of the same activity they tested a series of deletion and substitution mutants for their ability to be polyadenylated in the embryo. In this way, they showed that sequence information both 5′ of the CPE and 3′ of the AAUAAA hexanucleotide participated in this inhibitory activity. Furthermore, these elements act synergistically. Individually, they partially inhibited the B4-CPE-dependent polyadenylation of a chimaeric RNA in embryos. Complete inhibition was achieved when both elements were present. The necessary sequence information in the 5′ element of 58 nt could not be further restricted to a single short segment. This 58-nt element has no sequence homology with the 17-nt Eg2 deadenylation element and it does not bind the associated p53/p55 proteins. The 3′ element was contained in a 14-nt A+U-rich fragment.

Both of these studies show that specific sequence information, so far only localised in the 3′ UTR, must be present in an mRNA for it to be deadenylated after fertilisation. In this respect, deadenylation in eggs and embryos is different from default deadenylation of CPE-deficient RNAs during oocyte maturation. However, neither of these studies allowed a clear distinction to be made between two possible models. (1) Silencing or masking of the CPE that would allow deadenylation by the same machinery as that which achieves default deadenylation. This is similar to the situation discussed earlier for the Cl1 and Cl2 mRNA during oocyte maturation. (2) Targeting of the mRNA to an embryonic deadenylation complex, distinct from the default deadenylation apparatus, that would override the CPE-driven polyadenylation. Recently, Legagneux et al. (1995) have obtained data in favour of this second model. Using an in vitro system, they showed that RNAs that contain the Eg2 deadenylation element are deadenylated much more rapidly than RNAs that are substrates for default deadenylation. If the sole action of the Eg2 deadenylation element was to silence the CPE, then the deadenylation kinetics of these two types of RNA substrate should have been the same. In addition, partially deadenylated intermediates were observed only with RNA substrates for default deadenylation. Hence, at least for Eg2 mRNA, the function of the deadenylation element is not simply to inactivate the CPE, but rather this sequence element and the associated factors appear to partake in an active process possibly involving an embryo-specific deadenylation complex.

The results obtained with the Eg2 and Eg1 chimaeric RNAs suggest that there is not one unique *cis*-acting element or set of elements that direct an mRNA for embryo-specific deadenylation. The corollary to this is that there may be more than one mechanism involved. Another alternative is that the different *cis*-acting elements work in synergy, as was observed for the two elements in Eg1-cdk2 mRNA. Hence, the number of different elements present would define the rapidity with which the poly(A) tail was removed and/or the rate of accumulation of the deadenylated product after fertilisation.

Complementary to the question of the mechanism of the embryo-specific deadenylation is that of its activation. Translation of one or more maternal mRNA, during the first hour after fertilisation, is required for the deadenylation of the endogenous Eg1-cdk2 mRNA and for the inhibition of polyadenylation of injected chimaeric RNAs containing the Eg1-cdk2 polyadenylation inhibitory sequence elements (Stebbins-Boaz and Richter 1994). Furthermore, when poly(A)$^+$ mRNAs, isolated from 1-h embryos, were injected into oocytes, the polyadenylation of RNAs containing the Eg1-cdk2 3′ UTR was repressed during oocyte maturation, whereas that of RNAs containing the B4 CPE and AAUAAA hexanucleotide was unaffected (Stebbins-Boaz and Richter 1994). The nature of the/these repressor protein(s) is as yet unknown. They are presumably not regulatory proteins (kinases or phosphatases) of the CPEB, which also binds to Eg1-cdk2 3′ UTR (Legagneux et al. 1992), as the UA-rich CPEs with which this protein is associated are functional in the embryo (Stebbins-Boaz and Richter 1994). Also they are probably not the p53/p55 Eg-RNA-binding proteins, as these are already present in oocytes and the unfertilised egg (Legagneux et al. 1992). Hence the emerging picture is one of mRNA deadenylation in the embryo being controlled by multiple factors, and some of these may themselves be encoded by translationally regulated mRNAs.

6 Conclusion

In the course of a lifetime in all animals, translation is probably never as major a regulator of gene expression as it is during gametogenesis and embryogenesis. Because of this, one might imagine that metazoans have evolved a conserved mechanism to control translation, both positively and negatively, during these stages of development. Clearly, cytoplasmic poly(A) elongation as a means to increase translational efficiency qualifies as such a mechanism because it occurs (or is thought to occur) in marine invertebrates (sea urchins, marine worms, and surf clams), insects (*Drosophila*), amphibians (frogs), and mammals (mice), as reviewed by Richter (1995). Interestingly, however, the signals that control cytoplasmic polyadenylation may be different among animals, or at least between vertebrates and invertebrates. We have discussed how the CPE mediates polyadenylation in frogs and mice, but in insects there apparently are at least two other sequences that control this process. In *Drosophila*, an undefined element in the 3′ UTR of bicoid mRNA controls polyadenylation in embryos (Sallés et al. 1994), while the sequence ACAUCAAAUUU, in the 5′ UTR of Mst87F mRNA, controls polyadenylation during *Drosophila* spermatogenesis (Kuhn et al. 1991). These sequences, plus the UA-rich and U-rich CPEs of *Xenopus*, indicate that at least four different elements are operative for cytoplasmic polyadenylation, and probably more (Rosenthal 1993). Why so many sequences when nuclear polyadenylation seems to require so few (Wahle and Keller 1992)? In a word, regulation. That is, nuclear polyadenylation is essentially a constitutive event, whereas cytoplasmic polyadenylation occurs only at certain times for certain mRNAs. By having a repertoire of sequences for cytoplasmic polyadenylation, the degree to which specific mRNAs can be adenylated, and hence recruited for translation, is increased. Conversely, mRNAs whose poly(A) tail is shortened are released from polysomes. This means of effecting a translational arrest of specific maternal mRNAs also appears to be conserved amongst many evolutionarily divergent species. We have reviewed data showing that deadenylation affects distinct subpopulations of mRNAs at different stages of development. For certain of these mRNA subpopulations, the translation is successively increased by polyadenylation and then arrested by deadenylation. Hence the rich repertoire of *cis* elements and complementary mechanisms that can control maternal mRNA translation via polyadenylation allows a fine tuning of this mode of gene regulation.

We shall conclude not by looking back, but by looking forward to the three major questions that remain. The first revolves around the biochemistry of cytoplasmic polyadenylation. Three factors, CPEB, CPSF and PAP, have been shown to be involved, but do they interact, and if so, how? Are there additional factors that are a part of the polyadenylation complex? The second question concerns translation. In a mechanistic sense, how does polyadenylation induce translation, and deadenylation cause message release from polysomes? We have discussed cap-ribose methylation as a possible enhancer of translation but certainly that cannot be all that is necessary because prepolyadenylated mRNAs injected into oocytes, which are not cap-ribose methylated, are translated more efficiently than their nonadenylated counterparts. In somatic cells, deadenylation is often the immediate forerunner of degradation, which evidently results in a decreased synthesis of the encoded protein. In embryos, this is not always the case, as some maternal *Xenopus* mRNAs are stable for several hours after their deadenylation (Duval et al. 1990). Hence, at present, the mechanistic connection between deadenylation and translational arrest is an open question. Finally, is there one or several *cis* element(s) required for deadenylation in embryos, and what are the factors involved? Also, for the mRNAs that are polyadenylated and then subsequently deadenylated, is there a dynamic competition between the polyadenylation and deadenylation machineries, or does one irreversibly displace the other? A review on this subject written 5 years hence will probably contain the answers to these questions.

Acknowledgements. We thank P. Bouvet, P. Schedl and M. Wormington for communicating results prior to publication, and to V. Legagneux for comments on the manuscript. Work in the laboratory of H.B.O. is supported by grants from the European Economic community Biotechnology Program and the Association pour la Recherche sur le Cancer and that in the laboratory of J.D.R. by grants from the National Institutes of Health, the National Science Foundation and the March of Dimes Birth Defects Foundation.

References

Ballantine JEM, Woodland HR, Sturgess EA (1979) Changes in protein synthesis during development of *Xenopus laevis*. J Embryol Exp Morphol 51:135–153

Ballantyne S, Bilger A, Astrom J, Virtanen A, Wickens M (1995) Poly(A) polymerases in the nucleus and cytoplasm of frog oocytes: dynamic changes during oocyte maturation and early development. RNA 1:64–78

Bassez T, Paris J, Omilli F, Dorel C, Osborne HB (1990) Post-Transcriptional regulation of ornithine decarboxylase in *Xenopus laevis* oocytes. Development 110:955–962

Bilger A, Fox CA, Wahle E, Wickens M (1994) Nuclear polyadenylation factors recognize cytoplasmic polyadenylation elements. Genes Dev 8:1106–1116

Bouvet P, Wolffe AP (1994) A role for transcription and FRGY2 in masking maternal mRNA within *Xenopus* oocytes. Cell 77:931–941

Bouvet P, Omilli F, Arlot-Bonnemains Y, Legagneux V, Roghi C, Bassez T, Osborne HB (1994) The deadenylation conferred by the 3′ untranslated region of a developmentally controlled mRNA in *Xenopus* embryos is switched to polyadenylation by deletion of a short sequence element. Mol Cell Biol 14:1893–1900

Bouvet P, Matsumoto K, Wolffe AP (1995) Sequence specific RNA recognition by the *Xenopus* Y-box proteins: an essential role for the cold shock domain. J Biol Chem 270:28297–28303

Brandhorst BP (1976) Two-dimensional gel patterns of protein synthesis before and after fertilisation of sea urchin eggs. Dev Biol 52:310–317

Browder LW (1984) Developmental biology. Saunders, Japan

Caldwell DC, Emerson CP (1985) The role of cap methylation in the translational activation of stored maternal histone mRNA in sea urchin embryos. Cell 42:691–700

Cascio SM, Wassarman PM (1982) Program of early development in the mammal: post-transcriptional control of a class of proteins synthesized by mouse oocytes and early embryos. Dev Biol 89:427–439

Christersen LB, McKearin DM (1994) orb is required for anteroposterior dorsoventral patterning during *Drosophila* oogenesis. Genes Dev 8:614–628

Crawford DR, Richter JD (1987) An RNA-binding protein from *Xenopus* oocytes is associated with specific message sequences. Development 101:741–749

Cummings A, Sommerville J (1988) Protein kinase activity associated with stored messenger ribonucleoprotein particles of *Xenopus* oocytes. J Cell Biol 107:45–56

Dale L, Matthews G, Tabe L, Colman A (1989) Developmental expression of the protein product of Vg1, a localized maternal mRNA in the frog *Xenopus laevis*. EMBO J 8:1057–1065

Darnbrough C, Ford P (1981) Identification in *Xenopus laevis* of a class of oocyte specific proteins bound to messenger RNA. Eur J Biochem 113:415–424

Davidson EH (1986) Gene activity in early development. Academic Press, New York

Duval C, Bouvet P, Omilli F, Roghi C, Dorel C, LeGuellec R, Paris J, Osborne HB (1990) Stability of maternal mRNA in *Xenopus* embryos: role of transcription and translation. Mol Cell Biol 10:4123–4129

Dworkin MB, Dworkin-Rastl E (1985) Changes in RNA titers and polyadenylation during oogenesis and oocyte maturation in *Xenopus laevis*. Dev Biol 112:451–457

Dworkin MB, Shrutkowski A, Dworkin-Rastl E (1985) Mobilization of specific maternal RNA species into polysomes after fertilization in *Xenopus laevis*. Proc Natl Acad Sci USA 82:7636–7640

Edgar BA, Schubiger G (1986) Parameters controlling transcriptional activation during early *Drosophila* development. Cell 44:871–877

Epel D (1967) Protein synthesis in sea urchin eggs: a "late" response to fertilization. Proc Natl Acad Sci USA 57:889–906

Fox CA, Wickens M (1990) Poly(A) removal during oocyte maturation: a default reaction selectively prevented by specific sequences in the 3′ UTR of certain maternal mRNAs. Genes Dev 4:2287–2298

Fox CA, Sheets MD, Wickens M (1989) Poly(A) addition during maturation of frog oocytes: distinct nuclear and cytoplasmic activities and regulation by the sequence UUUUUAU. Genes Dev 3:2151–2162

Fox CA, Sheets MD, Wahle E, Wickens M (1992) Polyadenylation of maternal mRNA during oocyte maturation: poly(A) addition in vitro requires a regulated RNA binding activity and a poly(A) polymerase. EMBO J 11:5021–5032

Frolova L, Le Goff X, Rasmussen HH, Cheperegin S, Drugeon G, Kress M, Arman I, Haenni AL, Celis JE, Philippe M, Justesen J, Kisselev L (1994) A highly conserved eukaryotic protein family possessing properties of polypeptide chain release factor. Nature 372:701–703

Gallie DR (1991) The cap and poly(A) tail function synergistically to regulate mRNA translation efficiency. Genes Dev 5:2108–2116

Gallie DR, Tanguay R (1994) Poly(A) binds to initiation factors and increases cap-dependent translation in vitro. J Biol Chem 269:17166–17173

Gallie G, Kawata E, Smith LD, Larkins BA (1988) Role of the 3′-poly(A) sequence in translational regulation of mRNAs in *Xenopus laevis* oocytes. J Biol Chem 263:5764–5770

Gavis ER, Lehmann R (1994) Translational regulation of *nanos* by RNA localization. Nature 369:315–318

Gebauer F, Richter JD (1995) Cloning and characterization of a *Xenopus* poly(A) polymerase. Mol Cell Biol 15:1422–1430

Gebauer F, Xu W, Cooper GM, Richter JD (1994) Translational control by cytoplasmic polyadenylation of c-*mos* mRNA is necessary for oocyte maturation in the mouse. EMBO J 13:5712–5720

Grossi de Sa MF, Standart N, Martins de Sa C, Akhayat O, Huesca M, Scherrer K (1988) The poly(A)-binding protein facilitates in vitro translation of poly(A)-rich mRNA. Eur J Biochem 176:521–526

Hake LE, Richter JD (1994) CPEB is a specificity factor that mediates cytoplasmic polyadenylation during *Xenopus* oocyte maturation. Cell 79:617–627

Herschman HR (1991) Primary response genes induced by growth factors and tumor promotors. Annu Rev Biochem 60:281–319

Huarte J, Belin D, Vassalli JD (1985) Plasminogen activator in mouse and rat oocytes: induction during meiotic maturation. Cell 43:551–558

Huarte J, Belin D, Vassalli A, Strickland S, Vassalli JD (1987) Meiotic maturation of mouse oocytes triggers the translation and polyadenylation of dormant tissue-type plasminogen activator mRNA. Genes Dev 1:1201–1211

Huarte J, Stutz A, O'Connel ML, Gubler P, Belin D, Darrow AL, Strickland S, Vassalli JD (1992) Transient translational silencing by reversible mRNA deadenylation. Cell 69:1021–1030

Hyman LE, Wormington WM (1988) Translational inactivation of ribosomal protein mRNAs during *Xenopus* oocyte maturation. Genes Dev 2:598–605

Hyman LE, Colot HV, Rosbash M (1984) Accumulation and behaviour of mRNA during oogenesis and early embryogenesis of *Xenopus laevis*. In: Malicinsk GM, Klein WH (ed) Molecular aspects of early development. Plenum Press, New York, pp 253–266

Jenkins NA, Kaumeyer JF, Young EM, Raff RA (1978) A test for masked messages: the template activity of messenger of ribonucleoprotein particles isolated from sea urchin eggs. Dev Biol 63:279–298

Kick D, Barett P, Cummings A, Sommerville J (1987) Phosphorylation of a 60-kDa polypeptide from *Xenopus* oocytes blocks messenger RNA translation. Nucl Acids Res 15:4099–4109

Kim-Ha J, Kerr K, Macdonald PM (1995) Translational regulation of *oskar* mRNA by Bruno, an ovarian RNA-binding protein, is essential. Cell 81:403–412

Kuge H, Inoue A (1992) Maturation of *Xenopus laevis* oocyte by progesterone requires poly(A) tail elongation of mRNA. Exp Cell Res 202:52–58

Kuge H, Richter JD (1995) Cytoplasmic 3′poly(A) addition induces 5′cap ribose methylation: implications for translational control of maternal mRNA, EMBO J 14:6301–6310

Kuhn R, Kuhn C, Borsch D, Glazer KH, Schafer U, Schafer M (1991) A cluster of four genes selectively expressed in the male germ line of *Drosophila melanogaster*. Mech Dev 35:143–151

Lantz V, Ambrosio L, Schedl P (1992) The *drosophila orb* gene is predicted to encode sex-specific germline RNA-binding proteins and has localized transcripts in ovaries and early embryos. Development 115:75–88

Laskey RA, Mills AD, Gurdon JB, Partington GA (1977) Protein synthesis in oocytes of *Xenopus laevis* is not regulated by the supply of messenger RNA. Cell 11:345–351

Legagneux V, Bouvet P, Omilli F, Chevalier S, Osborne HB (1992) Identification of RNA-binding proteins specific to *Xenopus* Eg maternal mRNAs: association with the portion of Eg2 mRNA that promotes deadenylation in embryos. Development 116:1193–1202

Legagneux V, Omilli F, Osborne HB (1995) Substrate-specific regulation of RNA deadenylation in *Xenopus* embryo and activated egg extracts. RNA 1:1001–1008

McGrew LL, Dworkin-Rastl E, Dworkin MB, Richter JD (1989) Poly(A) elongation during *Xenopus* oocyte maturation is required for translational recruitment and is mediated by a short sequence element. Genes Dev 3:803–815

McGrew LL, Richter JD (1990) Translational control by cytoplasmic polyadenylation during *Xenopus* oocyte maturation: characterization of cis and trans elements and regulation by cyclin/MPF. EMBO J 9:3743–3751

Mescher A, Humphreys T (1974) Activation of maternal mRNA in the absence of poly(A) formation in fertilised sea urchin eggs. Nature 249:138–139

Munroe D, Jacobson A (1990a) mRNA Poly(A) tail, a 3′ enhancer of translational initiation. Mol Cell Biol 10:3441–3455

Munroe D, Jacobson A (1990b) Tales of poly(A): a review. Gene 91:151–158

Murray AW, Kirschner MW (1989) Cyclin synthesis drives the early embryonic cell cycle. Nature 339:275–280

Murray MT, Krohne G, Franke WW (1991) Different forms of soluble cytoplasmic mRNA binding proteins and particles in *Xenopus laevis* oocytes and embryos. J Cell Biol 112:1–11

Musci TJ, Amaya E, Kirschner MW (1990) Regulation of fibroblast growth factor receptor in early *Xenopus* embryos. Proc Natl Acad Sci USA 87:8365–8369

Newport J, Kirschner M (1982) A major developmental transition in early *Xenopus* embryos: I. Characterization and timing of cellular changes at the midblastula stage. Cell 30:673–686

O'Keefe SJ, Wolfes H, Kiessling AA, Cooper GM (1989) Microinjection of antisense c-*mos* oligonucleotides prevents meiosis II in the maturing mouse egg. Proc Natl Acad Sci USA 86:7038–7042

Osborne HB, Duval C, Ghoda L, Omilli F, Bassez T, Coffino P (1991) Expression and post-transcriptional regulation of ornithine decarboxylase during early *Xenopus* development. Eur J Biochem 202:575–581

Paris J, Philippe M (1990) Poly(A) metabolism and polysomal recruitment of maternal mRNAs during early *Xenopus* development. Dev Biol 140:221–224

Paris J, Richter JD (1990) Maturation-specific polyadenylation and translational control: diversity of cytoplasmic polyadenylation elements, influence of poly(A) tail size, and formation of stable polyadenylation complexes. Mol Cell Biol 10:5634–5645

Paris J, Osborne HB, Couturier A, LeGuellec R, Philippe M (1988) Changes in the polyadenylation of specific stable RNAs during the early development of *Xenopus laevis*. Gene 72:169–176

Paris J, LeGuellec R, Couturier A, LeGuellec K, Omilli F, Camonis J, MacNeil S, Philippe M (1991a) Cloning by differential screening of a *Xenopus* cDNA coding for a protein highly homologous to cdc2. Proc Natl Acad Sci USA 88:1039–1043

Paris J, Swenson K, Piwnica-Worms H, Richter JD (1991b) Maturation-specific polyadenylation: in vitro activation by p34cdc2 and phosphorylation of a 58-kDa CPE-binding protein. Genes Dev 5:1697–1708

Paynton BV, Bachvarova R (1994) Polyadenylation and deadenylation of maternal mRNAs during oocyte growth and maturation in the mouse. Mol Reprod Dev 37:172–180

Paynton BV, Rempel R, Bachvarova R (1988) Changes in state of adenylation and time course of degradation of maternal mRNAs during oocyte maturation and early embryonic development in the mouse. Dev Biol 129:304–314

Regier JC, Kafatos FC (1977) Absolute rates of protein synthesis in sea urchins with specific activity measurements of radioactive leucine and leucyl-tRNA. Dev Biol 57:270–283

Richter JD (1995) Dynamics of poly(A) addition and removal during development. In: Mattews M, Sonenberg N, Hershey J (eds) Translational control. Cold Spring Harbour Press, New York, pp 481–503

Richter JD, Smith LD (1981) Differential capacity for translation and lack of competion between mRNAs which segregate to free and membrane-bound polysomes. Cell 27:183–192

Richter JD, Smith LD (1984) Reversible inhibition of translation by *Xenopus* oocyte specific proteins. Nature 309:378–380

Robbie EP, Peterson M, Amaya E, Musci TJ (1995) Temporal regulation of the *Xenopus* FGF-receptor in development: a translation inhibitory element in the 3′untranslated region. Development 121:1775–1785

Rongo C, Gavis ER, Lehmann R (1995) Localization of *oskar* RNA regulates *oskar* translation and requires *oskar* protein. Development 121:2737–2746

Rosenthal E (1993) Sequence analysis of translationally controlled maternal mRNAs from *Urechis caupo*. Dev Genet 14:485–491

Rosenthal E, Wilt F (1987) Selective messenger RNA translation in marine invertebrate oocytes, eggs and zygotes. In: Ilan J(ed) Translational regulation of gene expression. Plenum Press, New York, pp 87–110

Rosenthal ET, Hunt T, Ruderman JV (1980) Selective translation of mRNA controls the pattern of protein synthesis during early development of the surf clam *Spisula solidissima*. Cell 20:487–496

Rosenthal ET, Tansey T, Ruderman JV (1983) Sequence-specific adenylation and deadenylation accompany changes in the translation of maternal messenger RNA after fertilisation in *Spisula* oocytes. J Mol Biol 166:309–327

Ruiz i Altalba A, Perry-O'Keefe H, Melton DA (1987) Xfin: an embryonic gene encoding a multifingered protein in *Xenopus*. EMBO J 6:3065–3070

Sachs AB, Davis RW (1989) The poly(A) binding protein is required for poly(A) shortening and 60S ribosomal subunit-dependent translation initiation. Cell 58:857–867

Sachs AB, Deardorff JA (1992) Translation initiation requires the PAB-dependent poly(A) ribonuclease in Yeast. Cell 70:961–973

Sagata N, Shiokawa K, Yamana K (1980) A study on the steady-state population of poly(A)$^+$ RNA during early development of *Xenopus laevis*. Dev Biol 77:431–448

Sagata N, Watanabe N, Vande Woude GF, Ikawa Y (1989) The c-*mos* proto-oncogene product is a cytostatic factor responsible for meiotic arrest in vertebrate eggs. Nature 342:512–518

Sallés FJ, Darrow AL, O'Connell ML, Strickland S (1992) Isolation of novel murine maternal mRNAs regulated by cytoplasmic polyadenylation. Genes Dev 6:1202–1212
Sallés FJ, Lieberfarb ME, Wreden C, Gergen JP, Strickland S (1994) Coordinate initiation of *Drosophila* development by regulated polyadenylation of maternal messenger RNAs. Science 266:1996–1999
Schnierle BS, Kuhn R, Bosse F, Schafer U (1992) Cap-specific mRNA (nucleoside-$O^{2'}$-)-methyltransferase and poly(A) polymerase stimulatory activities of vaccinia virus mediated by a single protein. Proc Natl Acad Sci USA 89:2897–2901
Schultz RM (1993) Regulation of zygotic gene activation in the mouse. Bioessays 15:531–538
Sheets MD, Fox CA, Hunt T, Vande Woude G, Wickens M (1994) The 3′-untranslated regions of c-*mos* and cyclin mRNAs stimulate translation by regulating cytoplasmic polyadenylation. Genes Dev 8:926–938
Sheets MD, Wu M, Wickens M (1995) Polyadenylation of c-*mos* mRNA as a control point in *Xenopus* meiotic maturation. Nature 374:511–516
Simon R, Richter JD (1994) Further analysis of cytoplasmic polyadenylation in *Xenopus* embryos and identification of embryonic cytoplasmic polyadenylation element-binding proteins. Mol Cell Biol 14:7867–7875
Simon R, Tassan JP, Richter JD (1992) Translational control by poly(A) elongation during *Xenopus* development: differential repression and enhancement by a novel cytoplasmic polyadenylation element. Genes Dev 6:2580–2591
Sommerville J (1992) RNA-binding proteins: masking proteins revisited. Bioessays 14:337–339
St Johnston D, Nüsslein-Volhard C (1992) The origin of pattern and polarity in the *Drosophila* embryo. Cell 68:201–219
Standart N (1992) Masking and unmasking of maternal mRNA. Semin Dev Biol 3:367–379
Standart NM, Bray SJ, George EL, Hunt T, Ruderman JV (1985) The small subunit of ribonucleotide reductase is encoded by one of the most abundant translationally regulated maternal mRNAs in clam and sea urchin eggs. J Cell Biol 100: 1968–1976
Standart N, Dale M, Stewart E, Hunt T (1990) Maternal mRNA from clam oocytes can be specifically unmasked in vitro by antisense RNA complementary to the 3′untranslated region. Genes Dev 4:2157–2168
Stebbins-Boaz B, Richter JD (1994) Multiple sequence elements and a maternal mRNA product control cdk2 RNA polyadenylation and translation during early *Xenopus* development. Mol Cell Biol 14:5870–5880
Stebbins-Boaz B, Hake LE, Richter JD (1996) CPEB controls the cytoplasmic polyadenylation of cyclin, Cdkz and c-*mos* mRNAs and is necessary for oocyte maturation in *Xenopus*. EMBO J 15:2582–2592
Strickland S, Huarte J, Belin D, Vassalli A, Rickles RJ, Vassalli JD (1988) Antisense RNA directed against the 3′ noncoding region prevents dormant mRNA activation in mouse oocytes. Science 241:681–684
Swenson KI, Farrel KM, Ruderman JV (1986) The clam embryo protein cyclin A induces entry into M phase and the resumption of meiosis in *Xenopus* oocytes. Cell 47:861–870
Tannahill D, Melton DA (1989) Localized synthesis of the Vg1 protein during early *Xenopus* development. Development 106:775–785
Tassan JP, Le Goff X, Le Guellec R, Philippe M (1993) Cl1 the *Xenopus* homologue of *Saccharomyces cerevisae* SUP45, which is encoded by a maternal RNA, is not essential for translational fidelity in egg extracts. Biochem Soc Trans (London) 21:862–867
Varnum SM, Wormington WM (1990) Deadenylation of maternal mRNAs during *Xenopus* oocyte maturation does not require specific *cis*-sequences: a default mechanism for translational control. Genes Dev 4:2278–2286
Varnum SM, Hurney CA, Wormington WM (1992) Maturation-specific deadenylation in *Xenopus* oocytes requires nuclear and cytoplasmic factors. Dev Biol 153:283–290
Vassalli JD, Huarte J, Belin D, Gubler P, Vassalli A, O'Connell ML, Parton LA, Rickles RJ, Strickland S (1989) Regulated polyadenylation controls mRNA translation during meiotic maturation of mouse oocytes. Genes Dev 3:2163–2171

Wahle E, Keller W (1992) The Biochemistry of 3′-end cleavage and polyadenylation of messenger RNA precursors. Annu Rev Biochem 61:419–440

Watanabe N, Vande Woude GF, Ikawa Y, Sagata N (1989) Specific proteolysis of the c-*mos* proto-oncogene product by calpain on fertilization of *Xenopus* eggs. Nature 342:505–511

Wilt FH (1973) Polyadenylation of maternal RNA of sea urchin eggs after fertilization. Proc Natl Acad Sci USA 70:2345–2349

Woodland HR (1974) Changes in polysome content of developing *Xenopus* embryos. Dev Biol 40:90–101

Wormington M, Searfoss AM, Hurney CA (1996) Overexpression of poly(A) binding protein prevents maturation-specific deadenylation and translational inactivation. EMBO J 15:900–909

Yisraeli JK, Sokol S, Meltron DA (1990) A two-step model for the localization of maternal mRNA in *Xenopus* oocytes: involvement of microtubules and microfilaments in the translocation and anchoring of Vg1 mRNA. Development 108:289–298

Younglai EV, Godeau F, Mester J, Baulieu FE (1980) Increased ornithine decarboxylase activity during meiotic maturation in *Xenopus laevis* oocytes. Biochem Biophys Res Commun 96:1279–1281

Zelus BD, Giebelhaus DH, Eib DW, Kenner KA, Moon RT (1989) Expression of the poly(A) binding protein during development of *Xenopus laevis*. Mol Cell Biol 9:2756–2760

Function and Characterization of Poly(A)-Specific 3′ Exoribonucleases

A. VIRTANEN and J. ÅSTRÖM

1 Introduction

Poly(A) tails are commonly found at the 3′ end of various classes of RNA (reviewed in Brawerman 1981; Manley 1995b). They are evolutionarily widespread and appear on RNAs of several different organisms. In mammalian cells almost all mRNAs end with an approximately 200-adenosine-residue-long poly(A) tail. Poly(A) tails are also found on RNAs of other eukaryotes, including both metazoans and protozoa, as well as on RNAs of prokaryotes. However, the functional significance of the poly(A) tail is still under debate, and a single primary function for it is not evident. Instead, it seems likely that the poly(A) tail serves multiple biological functions linked to the processes of mRNA translation, turnover, and transport (Brawerman 1981; Jackson and Standart 1990; Munroe and Jacobson 1990b; Wickens 1990; Sachs and Wahle 1993; Beelman and Parker 1995; Cohen 1995; Curtis et al. 1995; Manley 1995b). These may all be manifestation of a singular molecular event requiring poly(A), such as its interaction with poly(A)-binding proteins. Understanding how, when, and where RNA poly(A) tails are synthesized, removed, and extended will form the basis for a comprehensive understanding of the functional significance of the RNA poly(A) tail. This chapter will summarize the current knowledge of poly(A)-specific exoribonucleases which may be involved in RNA poly(A) tail removal. However, before we go into poly(A)-degrading activities we will briefly summarize the biochemistry of polyadenylation and discuss the functional importance of poly(A) tail metabolism in various biological systems.

Department of Medical Genetics, Biomedical Center, Uppsala University, Box 589, 751 23 Uppsala, Sweden

Progress in Molecular and Subcellular Biology, Vol. 18
Ph. Jeanteur (Ed.)

2 Biochemistry of mRNA Polyadenylation

2.1 Polyadenylation in Mammalian Cells

2.1.1 Nuclear Polyadenylation

Mammalian mRNA poly(A) tails are added in the nucleus by a process defined as nuclear polyadenylation (Proudfoot 1991; Wahle and Keller 1992; Manley 1995a; Wahle 1995). The reaction proceeds in at least two steps, first the pre-mRNA is cleaved at a specific position [the poly(A) site] by an endoribonuclease activity (RNA cleavage reaction). Subsequently, an approximately 200- to 250-nucleotide-long poly(A) tail is added to the 3′ end of the upstream cleavage product [poly(A) addition; Moore and Sharp 1985]. Several sequence elements in the pre-mRNA are essential for the cleavage reaction (Proudfoot 1991): (1) the highly conserved hexanucleotide AAUAAA sequence element, located approximately 5 to 35 nucleotides upstream of the poly(A) site (Proudfoot and Brownlee 1976; Sheets et al. 1990), (2) the poly(A) site, usually following a CA dinucleotide; (3) a G/U- or U-rich sequence element located 5 to 50 nucleotides downstream of the poly(A) site (McLauchlan et al. 1985; Gil and Proudfoot 1984, 1987). Additional sequence elements located upstream of the hexanucleotide (USEs) are also present in some cases (Carswell and Alwine 1989; DeZazzo and Imperiale 1989). However, these USEs are not essential for the cleavage reaction. Rather, they contribute to the efficiency of the polyadenylation reaction, either positively or negatively (reviewed in Virtanen 1995). The poly(A) addition reaction requires only the AAUAAA hexanucleotide sequence element (Wigley et al. 1990).

Biochemical characterization of the polyadenylation apparatus has revealed a complex processing machinery including poly(A) polymerase (PAP), cleavage and polyadenylation specificity factor (CPSF), cleavage stimulatory factor (CstF), two cleavage factors (CF I, CF II) and a poly(A)-binding protein (PABII) (see Wahle and Keller 1992; Manley 1995a; Wahle 1995 and references therein). Five of these factors, PAP, CPSF, CstF, CF I and CF II are required during the RNA cleavage reaction. CPSF (Bienroth et al. 1991; Gilmartin and Nevins 1991; Murthy and Manley 1992; Jenny et al. 1994) and CstF (Takagaki et al. 1990; Gilmartin and Nevins 1991) are composed of multiple polypeptides, and both interact with the RNA during the cleavage reaction. CPSF interacts with the AAUAAA sequence element while CstF interacts with the G/U- or U-rich downstream element. The function of the three additional factors, PAP, CF I, and CF II, during the cleavage reaction is not yet completely established. Most likely, the CF I and CF II factors, either alone or in combination, possess the endoribonuclease activity. Poly(A) addition requires only three factors, PAP, CPSF, and PAB II (Wahle 1991a,b; Murthy and Manley 1992, 1995; Bienroth et al. 1993). During this reaction, CPSF remains associated with the AAUAAA sequence element and interacts with PAP. This interaction makes PAP specific for RNA substrates containing the AAUAAA sequence element.

However, it has been shown that poly(A) addition proceeds in two phases (Sheets and Wickens 1989). The first phase, which proceeds until 10 adenosines have been added, is dependent on the AAUAAA sequence element. During the second phase, the requirement for the hexanucleotide is weaker and, after addition of 10 to 30 residues, the requirement is released. PAB II associates with the growing poly(A) tail, transforming the initially slow phase to a more rapid elongation reaction (Wahle 1991a). PAB II is also thought to be involved in determining the length of the newly synthesized poly(A) tail (Wahle 1991a; Nemeth et al. 1995).

2.1.2 Cytoplasmic Polyadenylation

Poly(A) tail synthesis may also occur in the cytoplasm. In maturing oocytes of *Xenopus laevis,* the mechanism and requirements for cytoplasmic polyadenylation have been studied in more detail (see Wickens 1990; Wormington 1994 for recent reviews). McGrew et al. (McGrew et al. 1989; McGrew and Richter 1990) found that elongation of an existing poly(A) tail on the G10 mRNA (Dworkin and Dworkin-Rastl 1985) required a 50-nucleotide-long sequence element, containing the hexanucleotide AAUAAA element and a cytoplasmic polyadenylation UUUUUUAU sequence element (CPE) separated by five to ten nucleotides from each other (Simon et al. 1992). A similar element, AAUAAA, together with UUUUUAU, was identified by Fox et al. (1989).

The protein factors required for cytoplasmic polyadenylation in *X. laevis* have also been investigated. Wickens and coworkers (Fox et al. 1992; Bilger et al. 1994) identified two components, an RNA-binding activity and a poly(A) polymerase. These components are most likely cytoplasmic counterparts of the nuclear CPSF and PAP activities, respectively. Several factors interacting with the CPE element have been identified. Paris et al. (1991) identified a 58-kDa CPE-binding protein, Simon and Richter (1994) have characterized two polypeptides with molecular masses of 36 and 45 kDa (137) and Hake and Richter (1994) isolated and cloned a 62-kDa CPE-binding protein (CPEB). Their exact role in cytoplasmic polyadenylation remains to be established. It has been proposed that CPEB may function as a recruitment factor for CPSF to the RNA substrate (Hake and Richter 1994). In summary, it seems likely that the basic components of cytoplasmic and nuclear adenosine addition are similar, but that additional sequence elements (e.g. CPEs) and proteins (e.g. CPEB) are important for cytoplasmic polyadenylation.

2.2 Polyadenylation in Yeast

Nuclear polyadenylation in yeast shows both differences from and similarities with mammalian polyadenylation (reviewed in Wahle and Keller 1992; Wahle 1995). The main difference has been attributed to low conservation of essential poly(A) signal sequences in yeast. Three poorly conserved sequence elements have been identified (Russo et al. 1993). Two upstream elements, one distally located (UAG...UAUGUA, UAUAUA, or UUUUUAUA) and one proximally located (UUAAGAAC or AAAAAAAA). The poly(A) site is frequently located after a cytosine residue or at a PyAAA sequence. Downstream sequence elements have not yet been defined. Biochemical fractionation of polyadenylation-competent cell-free yeast extracts has revealed two cleavage factors, CF I and CF II, and three poly(A) addition factors, CF I, polyadenylation factor I (PF I), and PAP (Chen and Moore 1992). CF I is composed of subunits and is a functional analogue of the mammalian CPSF factor since CF I participates both during RNA cleavage and poly(A) addition. Two subunits of CFI have been identified, encoded by the genes *RNA14* and *RNA15* (Minvielle-Sebastia et al. 1994). Interestingly, these gene products are similar to two subunits of the mammalian CstF factor, which interacts with the downstream sequence elements.

3 The Functional Importance of RNA Poly(A) Tails

3.1 Poly(A) Tail Shortening and RNA Decay

The importance of poly(A) tail shortening during mRNA degradation has been extensively studied in both higher and lower eukaryotes (recently reviewed in Sachs 1993; Decker and Parker 1994; Beelman and Parker 1995; Ross 1995). In mammalian and yeast cells, differences in the half-life of mRNAs have been ascribed to their different rates of poly(A) removal. A particularly well-studied example is the mammalian c-fos mRNA (Fort et al. 1987; Rahmsdorf et al. 1987; Wilson and Treisman 1988; Shyu et al. 1989, 1991). In this case, an AU-rich sequence element (ARE), located in the 3′ UTR of the mRNA, has been identified as an important mRNA decay control element. The c-fos ARE determines the deadenylation rate (Fort et al. 1987; Shyu et al. 1989, 1991; Wilson and Treisman 1988) and thereby accelerates mRNA decay, apparently because degradation of the mRNA body is dependent on the removal of poly(A) (Chen and Shyu 1994). Lagnado et al. (1994) and Zubiaga et al. (1995) recently reported consensus sequences, UUAUUUA(U/A)(U/A) or UUAUUUA-UU, respectively, for the ARE sequence. Most studies of mammalian mRNA decay have been performed in vivo. Recently, Ross and colleagues (Ross and Kobs 1986; Brewer and Ross 1990) developed an in vitro RNA decay system. Using this cell-free system, they established that important characteristics of mRNA decay were reconstituted in vitro. For instance, Brewer and Ross (Brewer and Ross 1988) showed that c-myc mRNA degradation in vitro was preceded by poly(A) tail removal. This and similar systems (see Ross 1995 and references therein) will be extremely important in the future to elucidate the mechanisms of mRNA decay.

Several elegant experiments have revealed that the half-life of yeast mRNA is tightly linked to mRNA poly(A) tails. The pathways for RNA degradation have been carefully characterised and deadenylation-dependent and -independent pathways have been established (reviewed in Decker and Parker 1994; Beelman and Parker 1995). Deadenylation-dependent degradation is initiated by poly(A) shortening. Once a short poly(A) tail (5 to 15 adenosine residues) has been generated, degradation proceeds either by 3′ to 5′ degradation or by decapping at the mRNA 5′ end followed by 5′ to 3′ exonuclease degradation. In analogy with the mammalian case, sequence elements involved in controlling the initial deadenylation step have also been found in the yeast mRNA body (e.g., the 3′ UTR of the *MFA2* mRNA; Muhlrad and Parker 1992).

The role of polyadenylation has been neglected and not extensively studied in bacterial cells, although poly(A) polymerase was identified in *E. coli* as early as 1962 (August et al. 1962) and polyadenylated RNAs were known to be present (e.g., Nakazato et al. 1975; Srinivasan et al. 1975; Cao and Sarkar 1992b). Compared to the mammalian and yeast mechanisms, polyadenylation in bacteria does not seem to require cleavage of the pre-mRNA. Rather, polyadenylation commonly occurs at 3′ ends generated by transcriptional termination although polyadenylation at a site not known to be generated by termination has been observed (Cao and Sarkar 1992b). There is no evidence yet for a complex polyadenylation machinery, equivalent to that of eukaryotes, in bacteria (reviewed in Manley 1995b) and PAP is the only required component known so far. Recently, it was shown that PAP is encoded by the gene *pcnB* of *E. coli* (Cao and Sarkar 1992a). The gene *pcnB* was known to be involved in controlling the copy number of ColE1 and related plasmids, by controlling the decay of small antisense regulatory RNA (reviewed in Cohen 1995; Wagner and Simons 1994). It has now been demonstrated that these antisense RNA are destabilized by the addition of short poly(A) tails, which, in turn, results in an increased plasmid copy number since high amounts of antisense RNA inhibit replication (He et al. 1993; Xu et al. 1993; Xu and Cohen 1995). It has also recently been observed that several bacterial mRNAs are polyadenylated (O'Hara et al. 1995). In this case, deletion of the *pcnB* gene leads to an extended half-life of mRNAs (Hajnsdorf et al. 1995; O'Hara et al. 1995). Thus, in bacteria, the presence of a poly(A) tail at the RNA 3′ end destabilizes the RNA, probably by making the RNA accessable to 3′ exoribonucleases (e.g. PNPase and RNase II; Cohen 1995; Hajnsdorf et al. 1995; O'Hara et al. 1995; Xu and Cohen 1995).

In summary, it is evident that poly(A)-specific nucleases are important players during mRNA decay. It is also evident that these nucleases in mammalian cells communicate with signal sequence elements located in the mRNA body. These signal sequences (e.g., ARE motifs) are most likely recognized by RNA-binding factors that specifically interact with these elements and probably also with the degrading nucleases.

3.2 The Involvement of mRNA Poly(A) Tails During Translation

The details of poly(A) tail length control and its coupling to mRNA translational efficiency have been studied in several species during meiotic maturation and early development (reviewed in Richter 1991; Wickens 1990, 1992; Wormington 1994). One important characteristic of oocyte maturation is the absence of RNA transcription. Therefore, stage-specific gene expression during meiotic maturation must be achieved by a mechanism other than RNA transcription. Compelling evidence suggests that translational control is of particular importance, and it has been proposed that the mRNA poly(A) tail length plays an important role in regulating the translation efficiency of an mRNA. In short, two classes of mRNAs are present in maturing oocytes. One class is represented by mRNAs having short oligo(A) tails, 10–20 adenosine residues, while the other class is represented by long poly(A)-tailed mRNAs (Dworkin and Dworkin-Rastl 1990). Moreover, there seems to be a correlation between mRNA poly(A) tail length and translational competence. Support for this simplified view comes from the following observations (Huarte et al. 1987; Hyman and Wormington 1988; McGrew et al. 1989; Vassalli et al. 1989). Long poly(A)-tailed mRNAs are associated with polyribosomes while oligo(A)-tailed are not. During oocyte maturation, some oligo(A)-tailed mRNAs receive long poly(A) tails, and at the same time these mRNAs become associated with polyribosomes. Conversely, the poly(A) tail of some mRNAs is shortened to oligo(A) or removed completely. These mRNAs simultaneously dissociate from the polyribosomes.

Poly(A) tail elongation does not always result in translational activation (Dehlin et al. 1996). Translational repression of the human IFN-β transcript upon Sendai virus infection of Namalwa B cells (Burkitt lymphoma cell line) correlates with an extension of the IFN-β mRNA poly(A) tail. In this case, elongation of the poly(A) tail to about 400 adenosine residues results in the dissociation of the IFN-β mRNA from polyribosomes, indicating that the poly(A) tail may have to be within a certain length to efficiently stimulate translation.

In eukaryotic cells, the mRNA poly(A) tail is associated with poly(A)-binding protein (PABP). To investigate the importance of the complex between poly(A) and PABP, Sachs and Davis (1989) mutated the yeast gene encoding PABP. They found that cells lacking functional PABP were deficient in translational initiation, i.e., formation of polyribosomes. Thus, their study established a link between the mRNA poly(A) tail and the translational machinery. Interestingly, Sachs and Davis observed that the mRNA poly(A) tails were extended in PABP yeast mutants. This finding was unexpected, since PABP protects the mRNA poly(A) tail from degradation in in vitro decay systems (Bernstein et al. 1989; Bernstein and Ross 1989). Since lack of PABP caused longer poly(A) tails to accumulate, Sachs and Davis proposed that PABP was required for deadenylation, by a PABP-dependent nuclease (PAN). This nuclease activity was later identified in cell-free extracts of yeast. PAN has been extensively purified and shown to be a poly(A)-specific 3′ exonuclease (Lowell et al. 1992; Sachs and Deardorff 1992, 1995; Boeck et al. 1996 for further discussion see below). Recently, Caponigro and Parker (1995) reinvestigated the phenotypes of yeast PABP mutants and observed that mRNA poly(A) tails were extended in the absence of PABP. However, prolonged time-course experiments showed that the mRNA poly(A) tails were removed after a lag period, indicating that PABP-independent poly(A) specific nucleases are present in yeast.

Additional support for the involvement of the poly(A) tail during translation has been generated. Munroe and Jacobson (Munroe and Jacobson 1990a) investigated the translational effect by varying the mRNA poly(A) tail length. They found that the presence of a poly(A) tail stimulated in vitro translation in reticulocyte extract 2- to 2.5-fold. Gallie (1991) found by transient expression of reporter constructs in plant, mammalian, and yeast cells, that the cap and poly(A) tail interacted synergistically and significantly stimulated translation. More recently, two studies by Kuge and Richter (1995) and by Tarun and Sachs (1995) established functional interaction between the cap and the mRNA poly(A) tail. Kuge and Richter (1995) showed that the poly(A) tail influenced the methylated state of the cap. Tarun and Sachs (1995) used a poly(A)-and cap-dependent yeast in vitro translation system (Iizuka et al. 1994) and found that the PABP poly(A) tail complex functions to recruit the 40S ribosomal complex, suggesting that the cap and the poly(A) tail have overlapping functions.

Taken together, all these observations highlight the mRNA poly(A) tail as an important translational control element. They suggest that poly(A) tail length will influence translational efficiency and that the choice between adding or removing the poly(A) tails must be a finely regulated and balanced process, both in determining the exact length of the poly(A) tail as well as in selecting specific mRNAs to receive or lose their poly(A) tails. The key enzymatic activities of poly(A) tail synthesis and degradation will therefore be important factors involved in regulating mRNA translation efficiency.

4 Poly(A)-Specific 3′ Exoribonucleases

3′ exoribonuclease activities degrading poly(A) have been identified in cell-free extracts from several sources including mammalian, plant, trypanosomatid, yeast and bacterial cells. The key properties of such exoribonucleases should be high specificity for degrading poly(A) in the 3′ to 5′ direction and liberation of adenosine mononucleotides. Although several of these activities were described more than 10 years ago, only a few of them have been purified to homogeneity, molecularly cloned, and extensively characterized at the molecular level. These include PAN (Boeck et al. 1996) of yeast and several bacterial activities (reviewed in Deutscher 1993). During the past few years, poly(A) specific exoribonucleases have gained considerable attention, and it is likely that several activities will be molecularly characterized in the near future. The development of the techniques of in vitro transcription of radioactively labeled RNA substrates resembling natural mRNAs [i.e., a short sequence representing an mRNA body followed by a bona fide poly(A) tail] has simplified the identification and characterization of poly(A)-specific exoribonucleases. The early studies were hampered by the lack of such RNA substrates since the only available polyadenylated substrates were poly(A) and purified naturally occurring polyadenylated mRNAs. Table 1 summarizes the properties of several exoribonucleases degrading poly(A).

4.1 Mammalian Nucleases

A multitude of mammalian activities degrading poly(A) have been described. The first observations were made by Anderson and Heppel (1960) and by Lazarus and Sporn (Lazarus and Sporn 1967; Lazarus et al. 1968; Sporn et al. 1969) about three decades ago. Lazarus and Sporn (1967) identified a nuclear exoribonuclease in Ehrlich ascites tumor cells. Similar activities were also found in the nucleus of mouse liver, kidney, embryo, and mammary tumor (Sporn et al. 1969). The activity degraded single-stranded RNA to nucleoside 5′-monophosphates and was strongly inhibited by the presence of terminal 3′-phosphate or 2′,3′-cyclic phosphate groups. The exoribonuclease required Mg^{2+}, which could be replaced by Mn^{2+} but not by Ca^{2+} or Zn^{2+}. Poly(A) was preferentially degraded compared to poly(I), poly(U), poly(C) and microsomal RNA. Polynucleotides were degraded by a continuous, processive mechanism, rather than by dissociating from the substrate after each cleavage step (Lazarus et al. 1968; Sporn et al. 1969). Cleavage occurred by hydrolysis; most likely the enzyme catalyzed an attack on the 5′ phosphorous atom of the terminal nucleotide (Sporn et al. 1969).

During the 1970s and early 1980s, several poly(A)-specific nucleases, both endo- and exo-ribonucleases, were described by Müller, Schröder and colleagues (Müller 1976; Müller et al. 1976; Schröder et al. 1980a,b; reviewed in Bachmann et al. 1984). One calf thymus poly(A)-specific exoribonuclease, termed 2′3′-exoribonuclease since it hydrolyzed both 3′,5′- and 2′,5′-linked oligoriboadenylates, was extensively purified (Müller et al. 1980; Schröder et al. 1980b). The molecular weight was estimated by gel filtration to approximately 80 kDa and the nuclease most likely consists of two polypeptides, 58 and 31 kDa in molecular weight, respectively. The 2′,3′-exoribonuclease liberated 5′-AMP as the product, required divalent cations (Mg^{2+} or Mn^{2+}) for its activity, and degraded poly(A) more efficiently than poly(U), poly(C), poly(G), and 23 S rRNA. Poly(A) was degraded both in single-stranded form and when duplexed with poly(U).

In the late 1970s, three groups reported on mammalian nucleases degrading poly(A). Kwan (1977) reported on a cytoplasmic Mg^{2+}-dependent exoribonuclease from HeLa cells which, degraded, in addition to poly(A), also mRNA,

Table 1. Summary of 3′ exoribonuclease activities degrading poly(A)

Sources	Major characteristics	Reference
Mammalian nucleases		
Mouse tumor	Detected in various tissue and tumor cells; nuclear origin; liberates nucleoside 5′-monophoshate; requires Mg^{2+} or Mn^{2+}; processive mechanism of degradation; prefers poly(A) over poly(I), poly(U), poly(C), and microsomal RNA	Lazarus and Sporn (1967) Lazarus et al. (1968) Sporn et al. (1969)
Calf thymus	2′,3′-exoribonuclease; liberates nucleoside 5′-monophoshate; requires Mg^{2+} or Mn^{2+}; specificity for poly(A); degrades poly(A) complexed with poly(U); estimated MW 80 kDa; heterodimer of two polypeptide subunits, 58 and 31 kDa	Müller et al. (1980) Schröder et al. (1980b) reviewed in Bachmann et al. (1984)
HeLa cells	Cytoplasmic origin; liberates nucleoside 5′-monophoshate; requires Mg^{2+} and free 3′ OH group; processive mechanism of degradation; degrades poly(A), mRNA, rRNA, and poly(C)	Kwan (1977)
Rat liver	Present in rat liver microsomes; liberates nucleoside 5′-monophoshate; requires Mg^{2+} which can be partially replaced by Mn^{2+}; processive mechanism of degradation; inhibited by terminal 3′-phosphate group; degrades poly(A) and poly(U) efficiently, and poly(C) and yeast RNA poorly; estimated MW 90 kDa	Kumagai et al. (1979) Kumagai et al. (1985)
HeLa cells, calf thymus	Liberates 5′ AMP; requires Mg^{2+} and free terminal 3′ OH group; degrades only 3′-located poly(A); no sequence requirements beside poly(A); single strand-specific; leaves mRNA body unaffected; estimated MW 80 kDa	Åström et al. (1991) Åström et al. (1992) Åström et al. (1996 in prep.)
Rat liver, rat hepatoma cells	Liberates 5′ AMP and some ADP and ATP; requires Mg^{2+} or Mn^{2+}; substrate specificity unknown; copurifies with poly(A) polymerase; suggested to be the reverse reaction of poly(A) polymerization	Abraham and Jacob (1978)
Human K562 erythroleukimia cells	Polysome-associated; liberates nucleoside 5′-monophoshate; requires Mg^{2+}; requires terminal 3′ OH group; inhibited by terminal 3′-phosphate; poly(A) degradation inhibited by poly(A)-binding protein; not specific for poly(A); estimated MW 33 kDa	Caruccio and Ross (1994)
Plant nucleases		
Wheat germ	Present in in vitro translation extracts; requires Mg^{2+}; prefers poly(A) over poly(U) and mouse plasmacytoma mRNA	Abraham and Pihl (1978) Abraham et al. (1980)
Vigna ungiculata	Liberates primarily 5′-AMP; requires Mg^{2+} or Mn^{2+}; copurifies with poly(A) polymerase; degrades poly(A) efficiently and poly(C), poly(U), and tRNA poorly; estimated MW 63 kDa	Tarui and Minamikawa (1989)

Table 1 (*Contd.*)

Sources	Major characteristics	Reference
Trypanosomatid nucleases		
Crithidia luciliae *Leishmania donovani*	3′-nucleotidase / nuclease; localized in surface membrane; requires Zn^{2+}; degrades poly(A) efficiently and to some extent poly(U) and RNA; estimated MW 43 kDa (*Crithidia luciliae*); molecularly cloned and sequenced (*Leishmania donovani*)	Neubert and Gottlieb (1990) Campbell et al. (1991) Gbenle and Dwyer (1992) Debrabant et al. (1995)
Yeast nuclease		
Saccharomyces cerevisiae (PAN)	Requires poly(A)-binding protein associated with RNA as substrate; releases 5′-AMP; requires a free terminal 3′ OH group; requires Mg^{2+}; distributive mechanism of degradation; extensively purified; consists of two polypeptide subunits, 135 and 76 kDa; 135 kDa subunit gene molecularly cloned and sequenced	Boeck et al. (1996) Lowell et al. (1992) Sachs and Deardorff (1992)
Bacterial nuclease		Reviewed in Deutscher (1993)
RNase II	Identified in several bacteria; major poly(A)-degrading activity in *E. coli*; releases nucleoside 5′ monophosphates; requires Mg^{2+} or Mn^{2+}; cleavage by hydrolysis; no specificity for base composition; prefers poly(A) as substrate; estimated MW 70 kDa; gene molecularly cloned and sequenced	Reviewed in Shen and Schlessinger (1982) Zilhao et al. (1993)
PNPase	Identified in several bacteria; releases nucleoside 5′ diphosphates; requires inorganic orthophosphate and divalent cations, preferentially Mg^{2+}; cleavage by phosphorolysis; no specificity for base composition; estimated MW 260 kDa consisting of three subunits, 90 kDa each; gene molecularly cloned and sequenced	Reviewed in Littauer and Soreq (1982) Regnier et al. (1987)

rRNA, and poly(C). This cytoplasmic activity was processive in its action, required a 3′ OH for its activity, and released 5′-AMP as the product when poly(A) was used as the substrate. Kumagai et al. (1979, 1985) identified a 3′-exoribonuclease in rat liver microsomes, which degraded poly(A) and poly(U) efficiently. Poly(C) and yeast RNA were poor substrates, while DNA, double-stranded RNA, and poly(G) were not degraded by the microsome activity. The activity was purified to near homogeneity and the molecular weight was estimated to be approximately 90 kDa. Degradation occurred by a processive mechanism and released 5′-AMP as the product. The presence of a 3′-phosphate group inhibited the activity. Mg^{2+} was essential for hydrolysis and could partially be replaced by Mn^{2+}.

We have recently identified, using radioactively in vitro-transcribed RNA substrates with a 30-adenosine-residue-long poly(A) tail at the 3′ end, a HeLa cell poly(A)-specific 3′ exoribonuclease (Åström et al. 1991, 1992). The activity has the following properties: (1) efficient poly(A) degradation requires Mg^{2+} and a 3′ end-located poly(A) tail with a free 3′ hydroxyl group at the 3′ terminus; (2) degradation terminates when the RNA substrate is completely deadenylated; (3) degradation is independent of any specific sequences in the mRNA body; (4) partially deadenylated poly(A) RNA contains a 3′ located hydroxyl group and is substrate for further deadenylation; (5) poly(A) tail removal is mediated by a 3′ exonucleolytic activity since the 3′ end of the RNA substrate is degraded, mononucleotides are liberated, and oligo(A) cannot be detected; (6) the released nucleotide is 5′-AMP, and 5′-AMP is an efficient inhibitor of in vitro deadenylation. In summary, this activity has intrinsic properties predicted for a mammalian mRNA poly(A) tail removing 3′ exoribonuclease. Most importantly, only poly(A) is efficiently degraded and the mRNA body is unaffected by the nuclease. Based on these characteristics, we have proposed the following reaction pathway:

$$5'\text{-}(pX)_m(pA)_n pApA_{OH}\text{-}3' \longrightarrow 5'\text{-}(pX)_m(pA)_n pA_{OH}\text{-}3' + pA_{OH}.$$

We have recently identified a similar activity in calf thymus cell-free extract (Åström et al., prep.). This activity has been extensively purified and the molecular weight of the enzyme has been estimated to be approximately 80 kDa by gel filtration. The subcellular location of this activity has not yet been established.

In 1978, Abraham and Jacob reported that hydrolysis of poly(A) could be detected in highly purified preparations of poly(A) polymerase isolated from rat liver or rat hepatoma cells (Rose and Jacob 1976). Abraham and Jacob (1978) argued that this hydrolytic activity was an intrinsic activity of poly(A) polymerase, in essence the reverse reaction of polymerization, and they named the activity poly(A) hydrolase. The hydrolase required Mg^{2+} or Mn^{2+} and released 5′-AMP as the major product, although ADP and ATP were also formed. The poly(A) hydrolase was shown to be a 3′-exonuclease and degraded single-stranded poly(A). Hydrolysis of other substrates (e.g., other homopolymers or natural RNAs) was not investigated. Carruccio and Ross (1994) have recently purified to apparent homogeneity a human polyribosome-associated 3′ exoribonuclease. This exoribonuclease, which is a Mg^{2+}-dependent enzyme having a molecular weight of 33 kDa, efficiently degrades single-stranded RNA as well as poly(A). Thus, this activity is not specific for poly(A), although poly(A) can be hydrolyzed.

4.2 Plant Nucleases

Turnover of the mRNA poly(A) tail has been observed in wheat-germ extracts normally used for in vitro translation (Abraham and Pihl 1978; Abraham et al. 1980). The wheat-germ activity, termed poly(A) hydrolase, required Mg^{2+} and degraded poly(A) tails of purified in vivo-labelled mouse plasmacytoma cell mRNA. Recently, Tarui and Minamkawa (Tarui and Minamikawa 1989) described and purified a poly(A)-specific ribonuclease in plant seedling (*Vigna unguiculata*) cell-free extracts. The ribonuclease copurified with poly(A) polymerase. Only 3′ end-located poly(A) tails were degraded by the nuclease. Poly(C), poly(U), and tRNA were almost unaffected by the plant activity. Degradation required either Mg^{2+} or Mn^{2+} and released primarily 5′-AMP and minute amounts of 3′-AMP and 2′-AMP.

4.3 Trypanosomatid Nucleases

An enzyme called 3′-nucleotidase/nuclease has been identified in the surface membrane of several trypanosomatids (e.g., species of *Leishmania*, African trypanosomes, *Crithidia luciliae*). It has been proposed that 3′-nucleotidase/nuclease is required for providing the parasites with purine nucleosides, since they are incapable of de novo purine synthesis (Hammond and Gutteridge 1984; Gottlieb 1989). The *C. luciliae* 3′- nucleotidase/nuclease was recently purified to apparent homogenity (estimated molecular weight 43 kDa) and the properties were investigated (Neubert and Gottlieb 1990). The nuclease requires divalent cations (Zn^{2+}) for activity and most likely releases 5′-nucleotides as the product. The substrate specificity of the nuclease was investigated and it was found that poly(A) was the preferred substrate, poly(U) and RNA were intermediate substrates, while poly(G), poly(C), and single-stranded DNA were poor substrates. A similar 3′-nucleotidase/nuclease has been identified in *Leishmania donovani* (Campbell et al. 1991; Gbenle and Dwyer 1992). The gene for the *L. donovani* 3′-nucleotidase/nuclease has been molecularly cloned and sequenced (Debrabant et al. 1995). Nuclease activities with similar properties have been identified in plants and fungi (Fraser and Low 1993).

4.4 Yeast Nucleases

A PABP-dependent poly(A) ribonuclease (PAN) has been identified in yeast (Lowell et al. 1992; Sachs and Deardorff 1992; Boeck et al. 1996; see above). PAN is unique among poly(A)-specific exoribonucleases, since it requires a ribonucleoprotein complex, PABP complexed with poly(A), for its activity. Lowell et al. (1992) speculated that PABP could protect the phosphate backbone of poly(A), thereby providing a proper substrate for PAN. PABP may also act as a specificity factor for PAN [i.e., making PAN poly(A)-specific]. PAN is a Mg^{2+} dependent distributive nuclease; it requires a free 3′ OH group and releases 5′-AMP as the product. PAN has been extensively purified and a polypeptide of 135 kDa in molecular weight copurifies with the activity. A polypeptide of this size was microsequenced and a corresponding cDNA isolated (Sachs and Deardorff 1992). Recently, it has been reported (Sachs and Deardorff 1995; Boeck et al. 1996) that the 135-kDa polypeptide originally identified was heterogeneous in composition and that PAN activity was associated with two other polypeptides, 135 and 76 kDa in size. The new 135-kDa polypeptide has been named Pan2p, and the gene (PAN2) encoding it has been cloned and sequenced. Genetic analysis (deletion of the PAN2 gene) revealed that PAN2 was not essential for viability. However, mRNA poly(A) tails were extended in PAN2 mutants, suggesting that PAN activity was reduced in these mutants.

4.5 Bacterial Nucleases

Several bacterial 3′ exoribonucleases have been identified in bacteria, polynucleotide phosphorylase (PNPase), RNase II, D, BN, T, PH and R, and oligoribonuclease (reviewed in Deutscher 1993). None of these nucleases exclusively degrades poly(A). However, two of them, RNase II and PNPase, will be discussed here, since they degrade poly(A) efficiently and have been suggested to be involved in degradation of polyadenylated RNAs (as discussed above). The detailed studies over the years of RNase II and PNPase are excellent examples of careful biochemical characterization of exoribonucleases.

RNase II (see Shen and Schlessinger 1982 and references therein) is the major poly(A)-degrading activity in *E. coli*, where, it has been suggested that RNase II accounts for more than 95% of poly(A) degradation (Zaniewski et al. 1984; Kelly and Deutscher 1992). RNase II hydrolyzes RNA and releases nucleoside 5′ monophosphates. It requires Mg^{2+} or Mn^{2+} for degradation. The molecular weight has been determined to be 70 kDa, and the gene encoding RNase II has been cloned and sequenced (Zilhao et al. 1993). RNase II degrades single-stranded RNA and without high specificity for a particular base composition. However, the efficiency of degradation follows the order poly(A), poly(U), poly(C), and tRNA. RNA with substantial secondary structure is not significantly hydrolyzed.

PNPase (see Littauer and Soreq 1982 and references therein) degrades poly(A) by phosphorolysis in the presence of inorganic orthophosphate. It releases NDPs [ADP in the case of poly(A) degradation] and degrades single-stranded substrates efficiently and, to a lesser extent, multistranded or highly structured RNA. Thus, PNPase is not specific for poly(A). RNA with 3′-terminal phosphate group is resistant to degradation. PNPase is a multifunctional enzyme and catalyzes at least three additional reactions, polymerization of ribonucleoside diphosphates to polyribonucleotide releasing inorganic orthophosphate (i.e. the reverse reaction of phosphorolysis), nucleoside diphosphate-Pi exchange, and transfer of nucleoside phosphate moieties. PNPase has been purified to homogeneity and the gene encoding PNPase has been cloned and sequenced (Regnier et al. 1987). In *E. coli*, the molecular weight of PNPase is approximately 260 kDa. PNPase is composed of three identical subunits, each 90 kDa in molecular weight. Mg^{2+} is required for phosphorolysis and can partially be replaced by Mn^{2+}. To some extent, other divalent cations can replace Mg^{2+}.

5 Regulation of Poly(A) Tail Shortening

An important aspect of the biogenesis of the poly(A) tail is how a specific mRNA is targeted for deadenylation under different conditions. During meiotic maturation of *X. laevis* oocytes, cytoplasmic poly(A) tail elongation requires specific sequence elements in the 3′ UTR of the mRNA while deadenylation does not (Wickens 1992). Consequently, deadenylation has been described as a default reaction (Fox and Wickens 1990; Varnum and Wormington 1990). The net result, whether to polyadenylate or deadenylate, will therefore be dependent on elements that control the rate of poly(A) addition (i.e., CPE together with AAUAAA). During oocyte maturation, mRNAs lacking a CPE lose their poly(A) tails (Wickens 1990). However, these mRNAs are not degraded, but remain intact, and can even be polyadenylated at a later time (Wickens 1990).

Timing of polyadenylation and deadenylation can be regulated. Sheets et al. (1994) demonstrated that c-mos, cyclin, and actin mRNA gain and lose poly(A) at different times during meiotic maturation of *X. laevis* oocytes. The poly(A) tail of c-mos mRNA, initially two discrete lengths of 40 and 75 nucleotides, was extended to a heterogenous length averaging 120 residues by the time of germinal vesicle breakdown (GVBD). Thereafter, the c-mos mRNA poly(A) tail was shortened to 50 adenosine residues again. Cyclin B1 and B2 mRNAs, on the other hand, were both polyadenylated after GVBD. Finally, actin mRNA was not polyadenylated during oocyte maturation; instead this mRNA was completely deadenylated after GVBD. This is in agreement with the findings that deadenylation during oocyte maturation requires both nuclear and cytoplasmic components, in that both actin and c-mos mRNA poly(A) tails shorten only after nuclear breakdown (Varnum et al. 1992). Isolated nuclear or cytoplasmic extracts are not capable of in vitro deadenylation, although the mixed extracts are active (Varnum et al. 1992). Moreover, enucleated oocytes are incapable of deadenylation during meiotic maturation. Taken together, these findings indicate that finely tuned regulation mechanisms control poly(A) tail length during meiotic maturation of *X. laevis* oocytes.

Deadenylation before meiotic maturation has been studied in mouse oocytes, and may parallel the progressive shortening of poly(A) tails in mammalian cells. Vassalli and coworkers (Vassalli et al. 1989; Huarte et al. 1992) have found that the mRNA for tissue type plasminogen activator (tPA), is first polyadenylated in the nucleus and then becomes deadenylated when it enters the cytoplasm of the primary oocytes, creating a translationally inactive mRNA with a short poly(A) tail. The sequence found to control this deadenylation reaction overlapped and may coincide with the sequence required for subsequent polyadenylation of tPA mRNA in the cytoplasm, during meiotic maturation. The investigators therefore named this sequence adenylation control element (ACE).

Deadenylation has also been observed in the early *X. laevis* embryo. After fertilization of the *X. laevis* egg, the pattern of polyadenylation and deadenylation changes. c-mos mRNA (Sheets et al. 1994) for example, loses its 50-residue poly(A) tail after fertilization. Bouvet and Wolffe (1994), studying deadenylation of Eg2 mRNA, identified a 17-nucleotide sequence in the 3′ UTR of the Eg2 mRNA that is required for rapid deadenylation after fertilization. Eg2 mRNA, like c-mos, is polyadenylated during oocyte maturation and then rapidly deadenylated after fertilization (Paris and Philippe 1990). This implies that the CPEs that function during oocyte maturation either are inactive at fertilization or are overhelmed by a sequence-specific deadenylation activity. After deadenylation, the EG2 mRNA is stable and remains present in the embryo until mid blastula transition when zygotic transcription initiates (Duval et al. 1990; Bouvet et al. 1991). Legagneux et al. (1992) have identified, by UV-cross linking, a protein doublet p53–55 which interacts with this putative deadenylation signal. Although, mutational analysis showed that deadenylation of Eg2 correlated with the p53–55 cross link, the role of these proteins in deadenylation is not known.

The mechanism by which deadenylation is controlled has not been elucidated. If cytoplasmic polyadenylation competes with the deadenylation reaction, net length could be modulated by regulating the rate of either addition or removal. The finding that *cis* elements required for cytoplasmic polyadenylation and deadenylation may overlap suggests a direct biochemical link between these competing reactions.

6 Summary and Perspectives

In this chapter we have discussed several aspects of RNA poly(A) tail metabolism. These include the biochemistry of poly(A) tail synthesis, the functional importance of poly(A) tail metabolism during RNA decay, and translation and the characterization of poly(A) specific exoribonucleases. We have tried to emphasize the functional significance of poly(A)-specific exoribonucleases in various systems, as already indicated in the title, *Function and Characterization of Poly(A)-Specific 3′ Exoribonucleases*. However, we want to point out here at the end that the knowledge about the functional significance of the poly(A) specific exoribonucleases is still very vague, although the importance of deadenylation has been very well documented during RNA decay and oocyte maturation/early development. In particular, the studies of small antisense RNA in bacterial cells have been enlightening. These studies elegantly link polyadenylation, poly(A) tail removal, RNA decay, and biological significance together.

What can we expect in future? One route of research, the unambiguous identification of poly(A)-specific 3′ exoribonuclease, will certainly be fruitful to explore, since the majority of the nucleases we have discussed remain to be identified at the molecular level. The recent progress in recombinant DNA technology and nucleic acid enzymology will have (and already has) a major impact on a successful tour along this route. Molecular cloning and expression of recombinant nucleases will allow detailed mechanistic studies of the enzymes and the generation of specific molecular probes (nucleic acids and antibodies) directed against the nucleases. These probes will be important reagents to study the tissue and subcellular location of the enzymes. Antibodies will be of particular importance, since they will make it possible to identify proteins that interact with the nucleases within the cells, i.e., to test whether complex "nuclease machineries" similar to the polyadenylation apparatus exist. One would predict that such complexes may exist, due to the simple fact that sequence elements located far away from the poly(A) tail affect deadenylation during mRNA decay in eukaryotic cells. Another reason to expect the existence of a "nuclease machinery" comes from the studies of polyadenylation. In this case, a complex apparatus is required to identify a subset of mRNA terminally located 3′ OH groups by recognizing signal sequences within the pre-mRNA. Evidence for specific components interacting with signal sequences important for deadenylation has already started to accumulate.

Another important subject to investigate will be the elucidation of how the processes of polyadenylation and deadenylation compete with each other during oocyte maturation and early development. The development of in vitro systems for both poly(A) addition and removal will make it possible to investigate if such competition mechanisms exist, and how different reaction conditions would affect the outcome. Critical questions to raise will be: When will competition occur? Will competition occur at every single adenosine residue? Important parameters to investigate will therefore be whether the reactions are processive or distributive in their nature.

Finally, we want to conclude at the end of this chapter that: "It might in fact be so, that what is most important for gene expression after the pre-mRNA is formed, is the end of the message" (Åström 1995).

Acknowledgements. We would like to thank Drs. E. Bridge, L. Kirsebom, J. Martinez, H. Öhman, C. Phillips, F. Söderbom, A-C. Thuresson and M. Wickens for helpful comments and suggestions during the preparation of this chapter. We acknowledge financial support from the Swedish Cancer Society, the Swedish Medical Research Council, the Swedish Natural Science Research Council, the Wellcome Trust and grants from Uppsala University.

References

Abraham AK, Jacob ST (1978) Hydrolysis of poly(A) to adenine nucleotides by purified poly(A) polymerase. Proc Natl Acad Sci USA 75:2085–2087

Abraham AK, Pihl A (1978) Formation of ATP from the poly-adenylated region of eukaryotic messenger RNAs. FEBS Lett 87:121–124

Abraham AK, Pihl A, Jacob ST (1980) Turnover of the poly(A) moiety of mRNA in wheat-germ extract. Eur J Biochem 110:1–5

Anderson EP, Heppel LA (1960) Purification and properties of a leukemic cell phosphodiesterase. Biochim Biophys Acta 43:79–89

Åström J (1995) Polyadenylation and deadenylation of messenger RNA in eukaryotic cells. Thesis at Uppsala Univ Fac Sci Technol

Åström J, Åström A, Virtanen A (1991) In vitro deadenylation of mammalian mRNA by a HeLa cell 3′ exonuclease. EMBO J 10:3067–3071

Åström J, Åström A, Virtanen A (1992) Properties of a HeLa cell 3′ exonuclease specific for degrading poly(A) tails of mammalian mRNA. J Biol Chem 267:18154–18159

August T, Oritz PJ, Hurwitz J (1962) Ribonucleic acid-dependent ribonucleotide incorporation. J Biol Chem 237:3786–3793

Bachmann M, Schröder HC, Messer R, Müller WEG (1984) Basespecific ribonucleases potentially involved in heterogeneous nuclear RNA processing and poly(A) metabolism. FEBS Lett 171:25–30

Beelman CA, Parker R (1995) Degradation of mRNA in eukaryotes. Cell 81:179–185

Bernstein P, Ross J (1989) Poly(A), poly(A) binding protein and the regulation of mRNA stability. Trends Biochem Sci 14:373–377

Bernstein P, Peltz SW, Ross J (1989) The poly(A)-poly(A)-binding protein complex is a major determinant of mRNA stability in vitro. Mol Cell Biol 9:659–670

Bienroth S, Wahle E, Suteer-Crazzolara C, Keller W (1991) Purification of the cleavage and polyadenylation factor involved in the 3′-processing of messenger RNA precursors. J Biol Chem 266:19768–19776

Bienroth S, Keller W, Wahle E (1993) Assembly of a processive messenger RNA polyadenylation complex. EMBO J 12:585–594

Bilger A, Fox CA, Wahle E, Wickens M (1994) Nuclear polyadenylation factors recognize cytoplasmic polyadenylation elements. Genes Dev 8:1106–1116

Boeck R, Tarun S, Rieger M, Deardorff JA, Müller-Auer S, Sachs AB (1996) The yeast Pan2 protein is required for poly(A)-binding proteinstimulated poly(A)-nuclease activity. J Biol Chem 271:432–438

Bouvet P, Wolffe AP (1994) A role for transcription and FRGY2 in masking maternal mRNA within *Xenopus* oocytes. Cell 77:931–941

Bouvet P, Paris J, Philippe M, Osborne HB (1991) Degradation of a developmentally regulated mRNA in *Xenopus* embryos is controlled by the 3′ region and requires the translation of another maternal mRNA. Mol Cell Biol 11:3115-3124

Brawerman G (1981) The role of the poly(A) sequence in mamalian messenger RNA. Crit Rev Biochem 10:1–38

Brewer G, Ross J (1988) Poly(A) shortening and degradation of the 3′ A + U-rich sequences of human c-myc mRNA in a cell-free system. Mol Cell Biol 8:1697–1708

Brewer G, Ross J (1990) Messenger RNA turnover in cell-free extracts. In: Dahlberg JE, Abelson JN (eds) Methods in enzymology, vol 181. Academic Press, San Diego, pp 202–203

Campbell TA, Zlotnick GW, Neubert TA, Sacci JB, Gottlieb M (1991) Purification and characterization of the 3′-nucleotidase / nuclease from promastigotes of *Leishmania donovani*. Mol Biochem Parasitol 47:109–118

Cao G-j, Sarkar N (1992a) Identification of the gene for an *Escherichia coli* poly(A) polymerase. Proc Natl Acad Sci USA 89:10380–10384

Cao G-j, Sarkar N (1992b) Poly(A) RNA in *Escherichia coli*: nucleotide sequence at the junction of the lpp transcript and the polyadenylate moeity. Proc Natl Acad Sci USA 89:7546–7550

Caponigro G, Parker R (1995) Multiple functions for the poly(A)-binding protein in mRNA decapping and deadenylation in yeast. Genes Dev 9:2421–2432

Carswell S, Alwine JC (1989) Efficiency of utilization of the simian virus 40 late polyadenylation site: Effects of upstream sequences. Mol Cell Biol 9:4248–4258

Caruccio N, Ross J (1994) Purification of a human polyribosome-associated 3′ to 5′ exoribonuclease. J Biol Chem 269:31814–31821

Chen C-Y A, Shyu A-B (1994) Selective degradation of early-response-gene mRNAs: functional analyses of sequence features of the AU-rich elements. Mol Cell Biol 14:8471–8482

Chen J, Moore C (1992) Separation of factors required for cleavage and polyadenylation of yeast pre-messenger RNA. Mol Cell Biol 12:3470–3481

Cohen SN (1995) Surprises at the 3′ end of prokaryotic RNA. Cell 80:829–832

Curtis D, Lehmann R, Zamore PD (1995) Translational regulation in development. Cell 81:171–178

Debrabant A, Gottlieb M, Dwyer DM (1995) Isolation and charachterization of the gene encoding the surface membrane 3′.nucleotidase / nuclease of *Leishmania donovani*. Mol Biochem Parasitol 71:51–63

Decker CJ, Parker R (1994) Mechanisms of mRNA degradation in eukaryotes. Trends Biochem Sci 19:336–340

Dehlin E, Gabain A v, Alm G, Dingelmayer R, Resnekov O (1996) Repression of beta interferon gene expression in virus-induced cells is correlated with a poly(A) tail elongation. Mol Cell Biol 16:468–474

Deutscher MP (1993) Ribonuclease multiplicity, diversity, and complexity, J Biol Chem 268:13011-13014

DeZazzo JD, Imperiale MJ (1989) Sequences upstream of AAUAAA influence poly(A) site selection in a complex transcription unit. Mol Cell Biol 9:4951–4961

Duval C, Bouvet P, Omilli F, Roghi C, Dorel C, et al. (1990) Stability of maternal mRNA in *Xenopus* embryos: role of transcription and translation. Mol Cell Biol 10:4123–4129

Dworkin MB, Dworkin-Rastl E (1985) Changes in RNA titers and polyadenylation during oogenesis and oocyte maturation in *Xenopus laevis*. Dev Biol 112:451–457

Dworkin MB, Dworkin-Rastl E (1990) Functions of maternal mRNA in early development. Mol Reprod Dev 26:261–297

Fort P, Rech J, Piechaczyk M, Bonnieu A, Jeanteur P, Blanchard JM (1987) Regulation of c-fos gene expression in hamster fibroblasts: initiation and elongation of transcription and mRNA degradation. Nucl Acid Res 15:5657–5667

Fox CA, Wickens M (1990) Poly(A) removal during oocyte maturation: a default reaction selectively prevented by specific sequences in the 3′ UTR of certain maternal mRNAs. Genes Dev 4:2287–2298

Fox CA, Sheets MD, Wickens MP (1989) Poly(A) addition during maturation of frog oocytes: distinct nuclear and cytoplasmic activities and regulation by the sequence UUUUUAU. Genes Dev 3:2151–2162

Fox CA, Sheets MD, Wahle E, Wickens M (1992) Polyadenylation of maternal mRNA during oocyte maturation: poly(A) addition requires a regulated RNA binding activity and a poly(A) polymerase. EMBO J 11:5021–5032

Fraser MJ, Low RL (1993) Fungal and mitochondrial nucleases. In: Linn SM, Lloyd RS, Roberts

RJ (eds) Nucleases. Cold Spring Harbor Lab Press, Cold Spring Harbor, New York, pp 171–207

Gallie DR (1991) The cap and poly(A) tail function synergistically to regulate mRNA translational efficiency. Genes Dev 5:2108–2116

Gbenle GO, Dwyer DM (1992) Purification and properties of 3′nucleotidase of *Leishmania donovani.* Biochem J 285:41–46

Gil A, Proudfoot NJ (1984) A sequence downstream of AAUAAA is required for rabbit β-globin 3′ end formation. Nature 312:473–474

Gil A, Proudfoot NJ (1987) Position-dependent sequence elements downstream of AAUAAA are required for efficient rabbit β-globin mRNA 3′ end formation. Cell 49:399–406

Gilmartin GM, Nevins JR (1991) Molecular analysis of two poly(A) site-processing factors that determine the recognition and efficiency of cleavage of the pre-mRNA. Mol Cell Biol 11:2432–2438

Gottlieb M (1989) The surface membrane 3′-nucleotidase/nuclease of trypanosomatid protozoa. Parasitol Today 5:257–260

Hajnsdorf E, Braun F, Haugel-Nielsen J, Régnier P (1995) Polyadenylylation destabilizes the rpsO mRNA of *Escherichia coli.* Proc Natl Acad Sci USA 92:3973–3977

Hake LE, Richter JD (1994) CPEB is a specificity factor that mediates cytoplasmic polyadenylation during *Xenopus* oocyte maturation. Cell 79:617–627

Hammond DJ, Gutteridge WE (1984) Purine and pyrimidine metabolism in the trypanosomatidae. Mol Biochem Parasitol 13:243–261

He L, Söderbom F, Wagner EG, Binnie U, Binns N, Masters M (1993) PcnB is required for the rapid degradation of RNAI, the antisense RNA that controls the copy number of ColE1-related plasmids. Mol Microbiol 9:1131–1142

Huarte J, Belin D, Vassalli A, Strickland S, Vassalli J-D (1987) Meiotic maturation of mouse oocytes triggers the translation and polyadenylation of dormant tissue-type plasminogen activator mRNA. Genes Dev 1:1201–1211

Huarte J, Stutz A, Oconnell ML, Gubler P, Belin D, et al. (1992) Transient translational silencing by reversible messenger RNA deadenylation. Cell 69:1021–1030

Hyman LE, Wormington WM (1988) Translational inactivation of ribosomal protein mRNAs during *Xenopus* oocyte maturation. Genes Dev 2:598–605

Iizuka N, Najita L, Franzusoff A, Sarnow P (1994) Cap-dependent and cap-independent translation by internal initiation of mRNAs in cell extracts prepared from *Saccharomyces cerevisiae.* Mol Cell Biol 14:7322–7330

Jackson RJ, Standart N (1990) Do the poly(A) tail and 3′ untranslated region control mRNA translation? Cell 62:15–24

Jenny A, Hauri H-P, Keller W (1994) Characterization of cleavage and polyadenylation specificity factor and cloning of its 100-kilodalton subunit. Mol Cell Biol 14:8183–8190

Kelly KO, Deutscher MP (1992) The presence of only one of five exoribonucleases is sufficient to support the growth of *Escherichia coli.* J Bacteriol 174:6682–6684

Kuge H, Richter JD (1995) Cytoplasmic 3′ poly(A) addition induces 5′ cap ribose methylation: implications for translational control of maternal mRNA. EMBO J 14:6301–6310

Kumagai H, Igarashi K, Tanaka K, Nakao H, Hirose S (1979) A microsomal exoribonuclease from rat liver. Biochim Biophys Acta 566:192–199

Kumagai H, Abiko T, Ono C, Marumo Y, Enomoto S, et al. (1985) Purification and mode of action of a microsomal endoribonuclease from rat liver. Biochim Biophys Acta 827:424–430

Kwan CN (1977) A cytoplasmic exoribonuclease from HeLa cells. Biochim Biophys Acta 479:322–331

Lagnado CA, Brown CY, Goodall GJ (1994) AUUUA is not sufficient to promote poly(A) shortening and degradation of an mRNA: the functional sequence within AU-rich elements may be UUAUUUA(U/A)(U/A). Mol Cell Biol 14:7984–7995

Lazarus HM, Sporn MB (1967) Purification and properties of a nuclear exoribonuclease from Ehrlich ascites tumor cells. Proc Natl Acad Sci USA 57:1386–1393

Lazarus HM, Sporn MB, Bradley DF (1968) A new kinetic model for polynucleotide metabolism. Proc Natl Acad Sci USA 60:1503–1510

Legagneux V, Bouvet P, Omilli F, Chevalier S, Osborne HB (1992) Identification of RNA-binding proteins specific to *Xenopus* Eg maternal mRNAs: association with the portion of Eg2 mRNA that promotes deadenylation in embryos. Development 116:1193–1202

Littauer UZ, Soreq H (1982) Polynucleotide phosphorylase. In: Boyer PD (ed) The enzymes, vol XV Academic Press, New York, pp 517–553

Lowell JE, Rudner DZ, Sachs AB (1992) 3′-UTR-dependent deadenylation by the yeast poly(A) nuclease. Genes Dev 6:2088–2099

Manley JL (1995a) A complex protein assembly catalyzes polyadenylation of mRNA precursors. Curr Biol 5:222–228

Manley JL (1995b) Messenger RNA polyadenylylation: a universal modification. Proc Natl Acad Sci USA 92:1800–1801

McGrew LL, Richter JD (1990) Translational contol by cytoplasmatic polyadenylation during *Xenopus* oocyte maturation: characterization of *cis* and *trans* elements and regulation by cyclin/MPF.EMBO J 9:3743–3751

McGrew LL, Dworkin-Rastl E, Dworkin MB, Richter JD (1989) Poly(A) elongation during *Xenopus* oocyte maturation is required for translational recruitment and is mediated by a short sequence element. Genes Dev 3:803–815

McLauchlan J, Gaffney D, Whitton JL, Clements JB (1985) The consensus sequence YGTGTTYY located downstream from the AATAAA signal is required for efficient formation of mRNA 3′ termini. Nucl Acids Res 13:1347–1368

Minvielle-Sebastia L, Preker PJ, Keller W (1994) RNA14 and RNA15 proteins as components of a yeast pre-mRNA 3′-end processing factor. Science 266:1702–1705

Moore CL, Sharp PA (1985) Accurate cleavage and polyadenylation of exogenous RNA substrate. Cell 41:845–855

Muhlrad D, Parker R (1992) Mutations affecting stability and deadenylation of the yeast MFA2 transcript. Genes Dev 6:2100–2111

Munroe D, Jacobson A (1990a) mRNA poly(A) tail, a 3′ enhancer of translational initiation. Mol Cell Biol 10:3441-3455

Munroe D, Jacobson A (1990b) Tales of poly(A): a review. Gene 91:151–158

Murthy KGK, Manley JL (1992) Characterization of the multisubunit cleavage-polyadenylation specificity factor from calf thymus. J Biol Chem 267:14804–14811

Murthy KGK, Manley JL (1995) The 160-kD subunit of human cleavage-polyadenylation specificity factor coordinates pre-mRNA 3′-end formation. Genes Dev 9:2672–2683

Müller WEG (1976) Endoribonuclease IV A poly(A)-specific ribonuclease from chick oviduct 1. Purification of the enzyme. Eur J Biochem 70:241–248

Müller WEG, Seibert G, Steffen R, Zahn RK (1976) Endoribonuclease IV. 2. Further investigation on the specificity. Eur J Biochem 70:249–258

Müller WEG, Schröder HC, Zahn RK, Dose K (1980) Degradation of 2′-5′-linked oligoriboadenylates by 3′-exoribonuclease and 5′-nucleotidase from calf thymus. Hoppe-Seyler's Z Physiol Chem 361:469–472

Nakazato H, Venkatesan S, Edmonds M (1975) Polyadenylic acid sequences in *E. coli* messenger RNA. Nature 256:144–146

Nemeth A, Krause S, Blank D, Jenny A, Jenö P, et al. (1995) Isolation of genomic and cDNA clones encoding bovine poly(A) binding protein II. Nucl Acids Res 23:4034–4041

Neubert TA, Gottlieb M (1990) An inducible 3′-nucleotidase/nuclease from the trypanosomatid *Crithidia luciliae*. Purification and characterization. J Biol Chem 265:7236–7242

O'Hara EB, Chekanova JA, Ingle CA, Kushner ZR, Peters E, Kushner SR (1995) Polyadenylylation helps regulate mRNA decay in *Escherichia coli*. Proc Natl Acad Sci USA 92:1807–1811

Paris J, Philippe M (1990) Poly(A) metabolism and polysomal recruitment of maternal mRNAs during early *Xenopus* development. Dev Biol 140:221–224

Paris J, Swenson K, Piwnica-worms H, Richter JD (1991) Maturation-specific polyadenylation: in vitro activation by $p34^{cdc2}$ and phosphorylation of a 58-kD CPE-binding protein. Genes Dev 5:1697–1708

Proudfoot N (1991) Poly(A) signals. Cell 64:671–674

Proudfoot NJ, Brownlee GG (1976) 3′ non-coding region sequences in eukaryotic messenger RNA. Nature 263:211–214

Rahmsdorf HH, Schönthal A, Angel P, Litfin M, Ruther U, Herrlich P (1987) Post-transcriptional regulation of c-fos mRNA expression. Nucl Acid Res 15:1643–1659

Regnier P, Grunberg-Manago M, Portier C (1987) Nucleotide sequence of the pnp gene of *Escherichia coli* encoding polynucleotide phosphorylase. J Biol Chem 262:63–68

Richter JD (1991) Translational control during early development. BioEssays 13:179–183

Rose KM, Jacob ST, (1976) Nuclear poly(A) polymerase from rat liver and a hepatoma. Eur J Biochem 67:11–21

Ross J (1995) mRNA stability in mammalian cells. Microbiol Rev 59:423–450

Ross J, Kobs G (1986) H4 histone mRNA decay in cell-free extracts initiates at or near the 3′ terminus and proceeds 3′ to 5′. J Mol Biol 188:579–593

Russo P, Li WZ, Guo Z, Sherman F (1993) Signals that produce 3′ termini in CYC1 mRNA of the yeast *Saccharomyces cerevisiae*. Mol Cell Biol 13:7836–7849

Sachs A (1993) Messenger RNA degradation in eukaryotes. Cell 74:413–421

Sachs A, Wahle E (1993) Poly(A) tail metabolism and function in eucaryotes. J Biol Chem 268:22955–22958

Sachs AB, Davis RW (1989) The poly(A) binding protein is required for poly(A) shortening and 60S ribosomal subunit-dependent translation initiation. Cell 58:857–867

Sachs AB, Deardorff JA (1992) Translation initiation requires the PAB-dependent poly(A) ribonuclease in yeast. Cell 70:961–973

Sachs AB, Deardorff JA (1995) Translation initiation requires the PAB-dependent poly(A) ribonuclease in yeast (Erratum). Cell 83/6

Schröder HC, Dose K, Zahn RK, Müller WEG (1980a) Isolation and characterization of the novel polyadenylate- and polyuridylate- degrading acid endoribonuclease V from calf thymus. J Biol Chem 255:5108–5112

Schröder HC, Zahn RK, Dose K, Müller WEG (1980b) Purification and characterization of a poly(A)-specific exoribonuclease from calf thymus. J Biol Chem 255:4535–4538

Sheets MD, Wickens M (1989) Two phases in the addition of a poly(A) tail. Genes Dev 3:1401–1412

Sheets MD, Ogg SC, Wickens MP (1990) Point mutations in AAUAAA and the poly(A) addition site: effects on the accuracy and efficiency of cleavage and polyadenylation in vitro. Nucl Acids Res 18:5799–5805

Sheets MD, Fox CA, Hunt T, Woude GV, Wickens M (1994) The 3′-untranslated regions of c-mos and cyclin mRNAs stimulate translation by regulating cytoplasmic polyadenylation. Genes Dev 8:926–938

Shen V, Schlessinger D (1982) RNases, I, II, and IV of Escgerichia coli. In: Boyer PD(ed) The enzymes, vol XV Academic Press, New York, pp 501–515

Shyu A-B, Greenberg ME, Belasco JG (1989) The c-fos transcript is targeted for rapid decay by two distinct mRNA degradation pathways. Genes Dev 3:60–72

Shyu A-B, Belasco JG, Greenberg ME (1991) Two distinct destabilizing elements in the c-fos message trigger deadenylation as a first step in rapid mRNA decay. Genes Dev 5:221–231

Simon R, Richter JD (1994) Further analysis of cytoplasmic polyadenylation in *Xenopus* embryos and identification of embryonic cytoplasmic polyadenylation element-binding proteins. Mol Cell Biol 14:7867–7875

Simon R, Tassan J-P, Richter JD (1992) Translational control by poly(A) elongation during *Xenopus* development: differentail represion and enhancement by a novel cytoplasmic polyadenylation element. Genes Dev 6:2580–2591

Sporn MB, Lazarus HM, Smith JM, Henderson WR (1969) Studies on nuclear exoribonucleases. III. Isolation and properties of the enzyme form normal and malignant tissues of the mouse. Biochemistry 8:1698–1705

Srinivasan PR, Rmanarayanan M, Rabbani E (1975) Presence of polyriboadenylate sequences in pulse-labeled RNA of *Escherichia coli.* Proc Natl Acad Sci USA 72:2910–2914

Takagaki Y, Manley JL, MacDonald CC, Wilusz J, Shenk T (1990) A multisubunit factor, CstF, is required for polyadenylation of mammalian pre-mRNAs. Genes Dev 4:2112–2120

Tarui Y, Minamikawa T (1989) Poly(A) polymerase from Vigna unguiculata seedlings. A bifunctional enzyme responsible for both poly(A)-polymerizing and poly(A)-hydrolyzing activities. Eur J Biochem 186:591–596

Tarun SZ, Sachs AB (1995) A common function for mRNA 5′ and 3′ ends in translation initiation in yeast. Genes Dev 9:2997–3007

Varnum SM, Wormington WM (1990) Deadenylation of maternal mRNAs during Xenopus oocyte maturation does not require specific *cis*-sequences: a default mechanism for translational control. Genes Dev 4:2278–2286

Varnum SM, Hurney CA, Wormington WM (1992) Maturation-specific deadenylation in *Xenopus* oocytes requires nuclear and cytoplasmic factors. Dev Biol 153:283–290

Vassalli J-D, Huarte J, Belin D, Gubler P, Vassalli A, et al. (1989) Regulated polyadenylation controls mRNA translation during meiotic maturation of mouse oocytes. Genes Dev 3:2163–2171

Virtanen A (1995) Strategies for regulating nuclear pre-mRNA polyadenylation. In: Lammond AI (ed) pre-mRNA processing. Landes, Austin, pp 135–149

Wagner EG, Simons RW (1994) Antisense RNA control in bacteria, phages, and plasmids. Annu Rev Microbiol 48:713–742

Wahle E (1991a) A novel poly(A)-binding protein acts as a specificity factor in the second phase of messenger RNA polyadenylation. Cell 66:759–768

Wahle E (1991b) Purification and characterization of a mamalian polyadenylate polymerase involved in the 3′ end processing of messenger RNA precursors. J Biol Chem 266:3131–3139

Wahle E (1995) 3′-end cleavage and polyadenylation of mRNA precursors. Biochim Biophys Acta 1261:183–194

Wahle E, Keller W (1992) The biochemistry of 3′-end cleavage and polyadenylation of messenger RNA precursors. Annu Rev Biochem 61:419–440

Wickens M (1990) In the beginning is the end: regulation of poly(A) addition and removal during early development. Trends Biochem Sci 15:320–324

Wickens M (1992) Forward, backward, how much, when: mechanisms of poly(A) addition and removal and their role in early development. Semin Dev Biol 3:399–412

Wigley PL, Sheets MD, Zarkower DA, Whitmer ME, Wickens M (1990) Polyadenylation of mRNA: minimal substrates and a requirement for the 2′ hydroxyl of the U in AAUAAA. Mol Cell Biol 10:1705–1713

Wilson T, Treisman R (1988) Removal of poly(A) and consequent degradation of c-fos mRNA facilitated by 3′ AU-rich sequences. Nature 336:396–399

Wormington M (1994) Unmasking the role of the 3′ UTR in the cytoplasmic polyadenylation and translational regulation of maternal mRNAs. BioEssays 16:533–535

Xu F, Cohen SN (1995) RNA degradation in *Escherichia coli* regulated by 3′ adenylylation and 5′ phosphorylation. Nature 374:180–183

Xu F, Lin-Chao S, Cohen SN (1993) The Escerichia coli pcnB gene promotes adenylylation of antisense RNAI of ColE1-type plasmids in vivo and degradation of RNAI decay intermediates. Proc Natl Acad Sci USA 90:6756–6760

Zaniewski R, Petkaitis E, Deutscher MP (1984) A multiple mutant of *Escherichia coli* lacking the exoribonucleases RNase II, RNase D and RNase BN. J Biol Chem 259:11651–11653

Zilhao R, Camelo L, Arraiano CM (1993) DNA sequencing and expression of the gene rnb encoding *Escherichia coli* ribonuclease II. Mol Microbiol 8:43–51

Zubiaga AM, Belasco JG, Greenberg ME (1995) The nonamer UUAUUUAUU is the key AU-rich sequence motif that mediates mRNA degradation. Mol Cell Biol 15:2219–2230

Subject Index

MIX
Papier aus verantwortungsvollen Quellen
Paper from responsible sources
FSC® C105338

If you have any concerns about our products,
you can contact us on
ProductSafety@springernature.com

In case Publisher is established outside the EU,
the EU authorized representative is:
Springer Nature Customer Service Center GmbH
Europaplatz 3, 69115 Heidelberg, Germany

Printed by Libri Plureos GmbH
in Hamburg, Germany